The Genesis of the Abstract Group Concept

The Genesis of the Abstract Group Concept

A Contribution to the History of the Origin of
Abstract Group Theory

Hans Wussing

Translated by Abe Shenitzer
with the editorial assistance of Hardy Grant

The MIT Press
Cambridge, Massachusetts
London, England

This book was printed in the German Democratic Republic.

Library of Congress Cataloging in Publication Data

Wussing, Hans.

The genesis of the abstract group concept.

Translation of: Die Genesis des abstrakten Gruppenbegriffes.
Bibliography: p. 295
Includes indexes.
1. Groups, Theory of—History. I. Title.
QA171. W8713 1984 512'.22'09 83-18765
ISBN 0-262-23109-3

Contents

Preface

The development of mathematics in the last 150 years presents a large field for historical research. We cannot even claim to have accurately staked out the historical terrain presented by the nineteenth century alone.

The mathematical development of earlier periods has been investigated, in part, down to the subtlest details, and examined and summarized in a great many publications. On the other hand, in spite of Klein's *Vorlesungen über die Entwicklung der Mathematik im 19. Jahrhundert* and N. Bourbaki's *Eléments d'histoire des mathématiques* of 1960, it is fair to say that the historiography of modern mathematical development has barely begun.

The lopsided relation between the explosive development of mathematics in the nineteenth and twentieth centuries and the modest historiography of that period provides a strong stimulus to consider the newer mathematics from a historical viewpoint. If we define the aim of all historiography as the discovery of the laws of scientific development and the clarification of developmental tendencies of present and future promise rather than the preparation of historical inventories, then the modern period must necessarily become the main subject of historical investigations. It is generally conceded that the radically different view of the nature, aims, and methods of mathematics underlying the substance and methodology of modern mathematics began to develop in the nineteenth century. If historiography is to fulfill its role, then this consensus must be described with all possible precision rather than in the summary manner of past presentations. This calls for numerous and varied investigations concerned with the history of problems, biography, the social role of mathematics, the philosophy of mathematics, the forms of scientific organization, and so on.

The conviction which guided the author's choice of theme and its treatment is that the history of the origin and development of fundamental mathematical concepts has to be an integrating component of any presentation of the development of mathematics in the nineteenth and twentieth centuries and that this way of looking at the historical material is ideally suited to the discovery of the inner connections of mathematical development.

I wish to express my sincere thanks to a number of people whose knowledge aided me in the design and execution of this work with its emphasis on the historical evolution of problems. Special thanks are due to Professor G. Harig (director of the Karl Sudhoff Institute for the History of Medicine and the Natural Sciences of the Karl Marx University in Leipzig), Professor H. Salié (the Mathematical Institute of the Karl Marx University in Leipzig), Professor K. Schröter and

H. Reichardt (the Mathematical Institute of the Humboldt University in Berlin), Professor A. P. Juškevič (Institute for the History of the Natural Sciences and Technology of the Soviet Academy of Sciences in Moscow), and Professor V. Kořinek (director of the Mathematical Institute of the Czechoslovak Academy of Sciences in Prague).

I owe my very deepest thanks to my wife.

Leipzig, April 1965

Discussions of the manuscript of this book by a small number of colleagues in my country and abroad have shown that there is a great deal of interest in its subject and have strengthened my hope that this study, already completed in 1965, will contribute to a broad discussion of the aims, the methods, and the very subject of mathematical historiography and of its relation of the natural sciences. In this connection I wish to pay tribute to the memory of my prematurely deceased teacher Gerhard Harig for many varied and fruitful suggestions and for the great and constant interest that he took in this work.

The printing of the book was made possible by the unfailing cooperation of the VEB Deutscher Verlag der Wissenschaften and especially its editor-in-chief Mr. L. Boll. Thanks are also due to VEB lectors Mrs. B. Mai, Dipl.-Math., and Miss E. Arndt, Dipl.-Math., to Mrs. T. Gedlich and Mrs. J. Greschuchna of the Karl Sudhoff Institute for the History of Medicine and the Natural Sciences of the Karl Marx University in Leipzig, and to Mrs. I. Letzel, librarian of the Mathematical Institute of the Karl Marx University in Leipzig.

Leipzig, fall 1968

Preface to the American Edition

A welcome recent trend has been the rapid growth of the historiography of mathematics. In terms of the pace of that development, it has been a relatively long time since this book was written (1964–1966). Nevertheless, MIT Press and I have undertaken to publish a translation based on the printed version issued in 1969 by VEB Deutscher Verlag der Wissenschaften, Berlin. What justifies this in that, since that time, there have been only relatively minor modifications and revisions of my conclusions on the history of the group concept. In particular, my account of Galois's progress toward that concept is borne out by the writings of Galois (E. Galois, *Ecrits et mémoires mathématiques.* Édition critique intégrale de ses manuscrits et publications par R. Bourgne et J.-P. Azra. Préface de J. Dieudonné. Paris, 1962), published only in 1962 and unavailable to me when I wrote my own book. In 1977, W. Purkert was able to describe with great precision Dedekind's position regarding the application of the abstract group concept in Galois theory (W. Purkert, „Ein Manuskript Dedekinds über Galois-Theorie," in NTM, 13 (2), 1977).

Beyond the history of group theory, the historiography of algebra in the nineteenth century has made significant gains. It is my duty and pleasure to mention a number of works of my colleagues and friends on the history of algebra that followed the *Genesis des abstrakten Gruppenbegriffes* and were inspired by my work to a certain extent. Soon after *Genesis* there appeared in the ČSSR a valuable monograph, completed in 1970, on the origins of modern algebra—L. Nový, *Origins of Modern Algebra* (Prague, 1973). Already at that time Dr. Nový and I were in close contact. In the meantime my doctoral student W. Purkert of Leipzig embarked on a profound study of the history of the field concept along lines analogous to my history of the group concept that provided the substance for a doctoral dissertation (W. Purkert, „Zur Genesis des abstrakten Körperbegriffes", Autoreferat der Dissertation in NTM 10 (1), 10 (2), 1973). Of the many significant works that have appeared in connection with the Gauss-Jubiläum in the GDR, I wish to mention the studies of K.-R. Biermann and H. Reichardt, both of Berlin, and the publication, on my initiative, of the Gauss Tagebuch. The Jubiläum has also provided the impulse for the extremely thorough papers of O. Neumann on the history of the interaction between algebra and number theory in the nineteenth century (O. Neumann, „Zur Genesis der algebraischen Zahlentheorie", in NTM, 16 (2), 17 (1), 17 (2), 1979/80).

I would like to call the attention of the reader of this book on the history of the group concept to a few comprehensive accounts of the history of mathematics that include treatments of group theory, and to a number of valuable monographs on its history. These include *History of Mathematics*, written by a group of authors under the editorship of A. P. Juškevič (in Russian); M. Kline, *Mathematical Thought from Ancient to Modern Times* (New York, 1972); P. Dugac, *Richard Dedekind et les fondements des mathématiques* (Paris, 1979); J. Dieudonné, (*Abrégé d'histoire des mathématiques*, 1700–1900 (Paris, 1978); H. Mehrtens, *Die Entstehung der Verbandstheorie* (Hildesheim, 1979); H. Wussing, *Vorlesungen zur Geschichte der Mathematik* (Berlin, 1979); E. Knobloch, *Der Beginn der Determinantentheorie* (Hildesheim, 1980); B. Chandler, W. Magnus, *History of Combinatorial Group Theory—A Case Study of the History of Ideas* (Berlin, Heidelberg, New York, Tokyo 1982). A comprehensive history of ideal theory will soon be published in the United States: H. M. Edwards, *Genesis of Ideal Theory*.

My sincere thanks go to the translator, professor Abe Shenitzer, who maintained friendly contact with me and completed a far from simple task with admirable conscientiousness.

Translator's Note

I wish to thank Hardy Grant for the saintly patience with which he has read and reread the translation, weeding out linguistic infelicities and shoring up faulty logic; Wilhelm Magnus, for serving as a court of last resort on mathematical technicalities and subtleties of the German language; Walter Kaufmann-Bühler, for reading a draft of the translation and eliminating many mistranslations; Trueman MacHenry, for improving my versions of some particularly difficult passages; Miriam Shenitzer for translating most of the French quotations; and last, but not least, the author, for his technical assistance and his warm and stimulating appreciation of our efforts to produce a faithful translation of his pioneering work.

Introduction

There can be no doubt that in recent times the term "structure" has come to occupy a central position in mathematics. This involves not merely its increased use in the modern literature but, more important, the recognition of the study of structures as a fundamental tool for the promotion of a unified development of mathematics. The conscious tendency to think in terms of structures has even produced its own characterization of mathematics. An extreme characterization of this kind, advanced by the Nicolas Bourbaki group of mathematicians, primarily in France, sees mathematics as a hierarchy of structures.

The investigation of structures—as both aim and method—is not restricted to mathematics; it has become the leading idea in the assessment and prognosis of modern natural science. For example, the report *The Modern Tendencies of Scientific Research* prepared in 1960 for UNESCO by the French atomic physicist P.-V. Auger (born in 1899) and many leading scientists and institutions, is based on the view that the study of structures is the central task of the modern physical sciences as well as of mathematics.

In algebra, "structure" has been a household term for a long time. An algebraic structure is defined as a set with one or more operations. In all modern textbooks groups are used to clarify and examplify the idea of an algebraic structure. It is a common conviction that the new conception of the nature of algebra, which first gained currency in the twenties and is associated with the names of E. Noether, E. Artin, H. Hasse, W. Krull, B. L. van der Waerden, and others, was based on the identification of fundamental structures such as groups, rings, modules, fields and ideals and that "modern algebra"—the programmatic name of the new trend—has given rise to a fundamentally new and extremely fruitful development of algebra.

At first, "modern algebra" met with considerable resistance. It is interesting that, in countering this opposition, van der Waerden and Hasse, the foremost proponents of "modern algebra," pointed to group theory, which by then had already attained the state toward which all of algebra was supposed to move. In fact the contemporary state of group theory bore witness to the far-reaching and fruitful results which might be confidently expected upon general acceptance of the new method and outlook. The validity of this viewpoint has long since become a truism in algebra as well as elsewhere. In this light, the history of the rise of group theory is the history of a mathematical discipline that has aided the advance of algebra in two distinct ways: as the earliest example of the evolution of an abstract algebraic structure and, insofar as it has aimed at the study of algebraic structures, as "midwife" of modern algebra.

While it was (and is) correct to invoke the development of group theory as a methodological example, the actual historical process gives only partial support to the above of the evolution of abstract group theory. It is usually asserted, or at least implied, that the abstract group concept arose at the end of the nineteenth century by pure abstraction from the concept of a permutation group and that the latter derived from the theory of algebraic equations. This view is found in the historical notes in modern monographs on group theory and in textbooks of algebra, and is even the basis of the article "Gruppe, II. Geschichte" in the encyclopedia *Mathematisches Wörterbuch* (Berlin and Leipzig, 1961). Insofar as they deal with group theory, virtually all the histories of mathematics follow this view. In particular, it is reflected in E. T. Bell's *The Development of Mathematics* (2nd ed., New York and London, 1945), in D. J. Struik's *A Concise History of Mathematics* (New York, 1948), and even in N. Bourbaki's *Eléments d'histoire des mathématiques* (Paris, 1960).

It is obviously true that the concept of a permutation group derived from the development of the theory of algebraic equations and from Galois theory. But this development, associated with the names of A. Vandermonde, J.-L. Lagrange, P. Ruffini, N. H. Abel, E. Galois, J.-A. Serret, and C. Jordan, is just *one* of the historical roots of group theory. The mathematical literature of the nineteenth century, and especially the work of decisive importance for the evolution of the abstract group concept written at the century's end (see III.4.2.), make it abundantly clear that that development had *three* equally important historical roots, namely, the theory of algebraic equations, number theory, and geometry. Abstract group theory was the result of a gradual process of abstraction from implicit and explicit group-theoretic methods and concepts involving the interaction of its three historical roots. I stress that my inclusion of number theory and geometry among the sources of causal tendencies for the development of abstract group theory is grounded in the historical record and is not the result of a backward projection of modern group-theoretic thought.

Thus one function of the present work is to provide, from a comprehensive study of the sources, a relatively detailed account of the origin of abstract group theory, while another aim is to offer a reconstruction of the actual historical record that will help set aside a prejudiced historical interpretation.

To the best of my knowledge, there exists no comprehensive account of the history of group theory that gives proper weight to all of its historical roots, or even one based on the deficient conception of the supposedly exclusive role of the theory of permutation groups. The discovery of two other genuine sources of this theory has significantly influenced the plan of the present work, my mode of research, and the adopted manner of presentation.

The existence of two additional roots of abstract group theory has been obscured mainly by the fact that the group-theoretic modes of thought in number theory and geometry remained implicit until the end of the middle third of the nineteenth century; they made no use of the term "group" and, in the beginning, had virtually no link to the contemporary development of the theory of permutation groups. Thus the historian is forced to search large areas of the development of mathematics for patterns of reasoning, methods and concepts equivalent to modern group-theoretic methods and concepts, and to trace the evolution of such implicit group-theoretic thought to the stage of explicit group theory. This task also extends to the theory of algebraic equations, which is why the account of the history of permutation groups starts with J.-L. Lagrange and A. Vandermonde rather than with E. Galois, who first used the word "group" in its modern technical sense.

In spite of the difficulties involved in the search for implicit group-theoretic modes of thought, we have one fairly reliable point of departure, albeit of a mathematical rather than historical nature. Since the automorphisms of structures form groups, we must begin our search wherever we come across investigations of automorphisms of number-theoretic, geometric or algebraic systems. While this is only a necessary condition, it is strictly that; in no case has a study of sets with operations led to group theory unless mediated by the study of their automorphisms. Examples are the explicitly manipulated composition of forces based on the parallelogram law, an operation adopted in the beginning of the nineteenth century, and the traditional operations on the positive integers.

While mathematical considerations help to identify instances of implicit group-theoretic thought, obviously the true criterion must be historical. Historical research must not reduce to the injection of advanced group-theoretic thinking into earlier mathematical results; it must, rather, involve spotting those paths of development of implicit group theory that have made a causal contribution to the rise of explicit group theory. This distinction, in my opinion, is of great importance for all mathematical historiography. To put it in the form of a play on words: We are not concerned with the historical manifestation of some logical development but with the logical manifestation of the historical development.

I shall give two examples, relating to older studies on the history of group theory, to clarify this. In his paper "On the Foundations of Geometry" [576] of 1898, H. Poincaré stated that the actual basis of the ancient euclidean proofs was the concept and properties of a group, and that "the group concept is as old as mathematics" [576, p. 39], an assertion that in my opinion has no historical basis. Again,

G. A. Miller, a leading American group theorist of the beginning of the twentieth century, published in 1921 a short sketch [522] of the history of group theory according to which its beginning coincides with the beginning of mathematics. (See, however, his other relevant work [521].) To be sure, the positive rational numbers are closed under multiplication. But this, and the fact that the author of the Rhind papyrus (about 1700 B.C.) computed with fractions, hardly supports Miller's contention that the Egyptian calculations with fractions are an instance of implicit group theory. Miller goes so far as to claim that the difficulties associated with the intellectual mastery of the concepts of zero and of negative numbers in the early stages of mathematics were due the traditional adherence to the multiplicative group of numbers, and to the great difficulty of transition to the additive group [522, p. 76].

The identification and description of implicit group-theoretic modes of thought which have had a "historical-causal" effect on the evolution of abstract group theory afford, in my opinion, the only opportunity of objectifying the historical process under investigation. This is especially true if we use our understanding of the dynamics of the mathematical development that has made implicit and explicit group theory an additional tool of investigation. In this way, without embarking on a special study, we are also bound to touch on the problem of the inner dynamic of the mathematical development, as well as the problem of the extension of concepts as the subject matter of historical research bearing on the dialectic of concepts.

The degree of evolution and autonomization of the group concept must, in each instance, be determined on the basic of the historical record. These determinations form the basis of the proposed exposition of the true roots of abstract group theory and fix its starting point.

As a result of the well-known investigations of A. Speiser (an account of which is also included in his textbook of group theory), we know of modes of thought associated with geometric ornaments that are thousands of years old and admit of group-theoretic interpretations. Again, in connection with a study of magic squares, Manuel Moschopulos (c. 1300), a Byzantine scholar, employed operations resembling permutations, and the Jewish scholar Levi ben Gerson, in his "Practice of the Calculator" of 1321, made explicit use of mathematical induction to prove that (in modern terms) the number of permutations of n objects is $n!$. But these and other instances of group-theoretic thinking, brought to light by historians of mathematics, have not been the starting points of coherent lines of development leading to group theory. They belong to the mathematical results which do not stand in a historical-causal relation to the origin of abstract group theory. While such instances may eventually be classified as prefigurations or early anchor points of implicit group-theoretic thinking, they remain outside the scope of the present investigation.

Thus our exposition begins with the rise of coherent developments of substantial importance for the history of the rise of group theory. For the theory of algebraic equations and for number theory this is the end of the eighteenth, for geometry the beginning of the nineteenth century. The events in question took place mainly in Western and Central Europe. This points emphatically to the close connection between the Industrial Revolution and a period of sudden growth in mathematics and in the natural sciences—a historical interaction that I do not propose to investigate in detail. I shall merely note that the Industrial Revolution brought about new and rapidly increasing demands for applicable mathematical methods, and thus generated a social attitude that kept alive the interest in abstract questions, methods and constructs and provided for the partial but significant start of a number of mathematical disciplines—including group theory—before their findings could be applied to problems of natural science as well as other practical problems.

The desire to reconstruct the actual historical course of events has made it necessary for me to consult the original sources in all cases. For the same reason, I occasionally quote lengthy passages from these sources. This helps reproduce the actual intellectual attitudes at critical junctures in the process of emancipation of the group concept. It should not surprise anyone that many historical judgments, many repeatedly cited "first" or "decisive" quotations and even references to the literature have turned out to be deficient, inadequate or downright false. In view of my intent to provide a positive exposition of the history of the rise of group theory, I have relegated critical comments on sources to the notes. Space considerations have prevented me from providing explicit corrections of faulty references to the literature.

A complete history of group theory concerned with the reconstruction of the historical process and its inner problem-historical dynamic must pay a measure of attention to the biographical element and to relevant group-theoretic papers and manuscripts that have failed to influence the course of the historical development. In the present work I have paid little attention to either of these areas. I have noted biographical elements only to the extent to which they are directly relevant to my exposition; one instance of such biographical material is the entrance of a scientist into an intellectual environment conducive to the pursuit of group theory. Also, little attention has been paid to ineffective group-theoretic works and manuscripts, regardless of whether their ineffectiveness was due to publication in a then obscure journal, or to insertion in a *Nachlass* whose evaluation came "too late," or to other causes. Obviously, this does not constitute or imply an evaluation of the intrinsic merits of the work in question.

My last remark pertains to conventions of composition. A reference to material in this book is given by part, chapter, and section numbers and is enclosed in parantheses. A direct or indirect quotation or a reference to a relevant paper in the bibliography is given by a bibliographical-entry number, possibly followed by page number, enclosed in square brackets. For greater readability, all notes have been relegated to the back of the book. They are referred to in the text by numerical superscripts.

The Genesis of the Abstract Group Concept

I Implicit Group-Theoretic Ways of Thinking in Geometry and Number Theory

1 Divergence of the Different Tendencies Inherent in the Evolution of Geometry during the First Half of the Nineteenth Century: The New View of the Nature of Geometry

1 Features of the Development of Geometry in the Beginning of the Nineteenth Century

The rise of analytic geometry in the seventeenth century completed a methodological break with the geometry of ancient times. But the first fundamental steps aimed at setting aside the older view of the *nature* of geometry were taken only at the end of the eighteenth century. It was algebra and analysis, not geometry, that took the lead in overcoming classical views. The coss-algebra and the letter algebra of the sixteenth and seventeenth centuries were radical departures; and if we ignore a few of the contributions of Eudoxos and Archimedes, then we may say that the seventeenth- and eighteenth-century work that roughed out the mode of thought based on function concept, and that perfected infinitesimal methods, made analysis a completely new discipline.

While fundamental changes in geometry lagged behind such changes in algebra and analysis, geometry developed explosively[1] both in scope and in depth until the end of the nineteenth century. There is a remarkable and striking contrast between the development of algebra and analysis on the one hand and of geometry on the other. In algebra and analysis there were periods of vaulting growth in scope—during the sixteenth and seventeenth centuries—followed considerably later—at the beginning of the nineteenth century—by periods of evolution in depth. In geometry there was continuous interaction between growth in scope and growth in depth, with virtually simultaneous development in both respects.

Since the present book aims to describe the historical evolution of mathematical problems, we cannot do more than make a few factual remarks about its causes.

Seen as a whole, the extraordinary growth of geometry in the nineteenth century stemmed directly from the Industrial Revolution's soaring demand for mathematically trained engineers. Since the scholastic educational patterns associated with feudalism and absolutism could not possibly meet the new engineering requirements, it was natural for descriptive geometry to become the core of an educational program designed for their quick satisfaction. This "language of the engineer"—a term coined by G. Monge (1746–1818), creator of descriptive geometry and guiding spirit of the Paris *Ecole Polytechnique* founded at the height of the Great French Revolution—became dominant in the polytechnic schools (predecessors of technical colleges and universities) founded throughout Europe and modeled on the original Paris school. Predictably, descriptive geometry exerted a strong influence on the direction of mathematical training in *Gymnasiums* and universities.[2] In this sense, the social usefulness of geometry, which became apparent, in particular, in the case of descriptive geometry, prepared the ground for the development of all geometry.

The work of J.-V. Poncelet (1788–1867) and the development of projective geometry were immediate realizations of the powerful thrust imparted by Monge and the *Ecole Polytechnique.* At the same time, during the first half of the nineteenth century, geometry developed along other lines that, although not directly linked to the progress of projective geometry, nevertheless originated in the revolution in geometry begun at the end of the eighteenth century.

Actually, what we witness in geometry during that period is not the realization of one particular trend but the evolution—in seemingly divergent directions—of several implicit tendencies about to break free. Fundamental modes of thought in and about geometry were no longer dictated by habit. The geometry whose content, method, and aims were conceived in terms of the millenial euclidean tradition began to change. Once the idea had been abandoned that geometry is something unique (its content, methods and goals interpretable only in the ancient tradition stemming from Euclid), concepts such as coordinate, length, parallelism and distance; the habit of regarding the point as the fundamental element of all geometry; and the very view of geometry as the art of measurement, all turned out to be capable of generalization or in need of critical review. The dissolution of a seemingly unavoidable way of thinking had the following effect: Through a dialectic evolution the contradictions inherent in the concept of the unity of geometry produced a wealth of "geometries" determined only by their attributes—"noneuclidean," "*n*-dimensional," "line," "projective," and so on.

Certain aspects of this evolution are noteworthy in that they were starting points of an implicit group-theoretic mode of thought in geometry. These starting points were in turn further developed through application of group theory itself. The four most important of these aspects are

i. the elimination of the seemingly indissoluble link between geometry and metric, and the related rise of the question of the connection between projective geometry and metric geometry;
ii. the extension of the coordinate concept beyond traditional, parallel (cartesian) coordinates;
iii. the development of noneuclidean geometries;

iv. the turn toward abstraction due to the introduction of an arbitrarily large (finite) number of dimensions.

2 The Rise of Projective Geometry

One consequence of the rapid spread of descriptive geometry (of the kind developed by Monge) was the fact that as early as the beginning of the nineteenth century it was common to distinguish between metric and "descriptive" (that is, incidence) properties of figures invariant under (orthogonal) projections. Ideas inspired by *De la corrélation des figures de géométrie* (1801) and *Géométrie de position* (1803) —two works of Monge's student L. N. M. Carnot (1753–1823)—and, above all, the publication of Poncelet's pioneering *Traité des propriétés projectives des figures* (1822) made into a central problem the investigation of the differences between metric and incidence properties. This was particularly pursued by the French school grouped around the "Annales" of J.-D. Gergonne (1771–1859) who, in 1826, proposed the development of a geometry based exclusively on incidence properties [240, p. 209].

Whereas Monge, in his *Leçons de géométrie descriptive* (1794–95), relied on the general use of orthogonal projections, the chief tool of Poncelet's *Traité* was the more general concept of a central projection. It was Poncelet who introduced the fundamental distinction between "projective" and "nonprojective" properties of figures, that is, between properties that are always preserved by central projections and properties that are usually destroyed by such projections. All incidence properties, but only special metric properties, are projective. Poncelet was able to give (*Traité*, 1st ed., §20) the general metric function of a projective nature, which implied, among other things, the projective invariance of the cross ratio of four collinear points. In this way he deduced from a basic conceptual framework a result first found by Pappus and again, in 1817, by Ch.-L. Brianchon.[3]

By considering properties of figures invariant under central projections, as well as properties invariant under other projections, Poncelet anticipated the main idea of a subsequent development. This approach, and the *analytic* treatment of geometric figures—that is, the shift from synthetic projection to the analytic study of coordinate transformations with a view to finding their invariants—made it possible to apply invariant theory, rooted in number theory, to the classification of geometric objects.

It is against this historical background that we shall look for signs of group-theoretic thinking in the systematic study of "geometric relations" (geometrische Verwandtschaften) begun in the first half of the nineteenth century, and in the concurrent consolidation of invariant theory. The authoritative influence of A. F.

Möbius, J. Steiner, J. Plücker, M. Chasles, and others centered on the first of these issues, and that of G. Boole, A. Cayley, J. Sylvester, L. Cremona, and others on the second (see III.1). Of course, in developing projective geometry, Poncelet, and later Möbius, Steiner, and others, used metric considerations, and the cross ratio, crucially important in the definition of projective coordinates, itself continued to depend on a metric definition. The gap was not closed until the appearance of *Geometrie der Lage* (1847) and *Beiträge zur Geometrie der Lage* (three parts: 1856, 1857, 1860) of Chr. v. Staudt (1798–1867), and the last inconsistency was removed only in 1871 by F. Klein (1849–1925), who gave a nonmetric definition of projective coordinates based on the (nonmetric) cross ratio. This met the demand, dictated by methodological considerations, for an exclusively projective development of projective geometry.

But all this failed to clarify the inner connection between metric geometry and projective geometry. Such clarification, involving the explicit use of group theory, was achieved only by Klein in his Erlangen Program of 1872. One of the consequences of this work was that the metrics later named for Cayley could now be assigned a logical position in the overall structure of geometry (see III.1–2).

3 Extension of the Coordinate Concept

The development of projective geometry in depth was closely linked to the discarding of the traditional view that limited the coordinate concept to parallel (cartesian) coordinates. Especially significant for the rise of group-theoretic views was the extension of this concept beyond *point* coordinates, for this overcame yet another, seemingly unbreakable Euclidean tradition, which viewed the point as *the* fundamental element of all geometry. This change alone made possible an understanding of the principle of duality, an idea that at first had been surrounded by a kind of mystical fog and, at the same time, had been the subject of a vehement conflict over priority.

Until the nineteenth century the only coordinates used were point coordinates; and apart from the occasional use of curvilinear coordinates—for example, by G. W. Leibniz in 1692—point coordinates were primarily parallel. Moreover it was not until the beginning of the nineteenth century—largely as a result of the forceful stand taken by A. M. Legendre (*Eléments de Géométrie*, 1794)—that mathematicians began to view coordinates as numbers rather than as intervals. (In G. S. Klügel's *Mathematisches Wörterbuch* of 1803, for example, coordinates are defined as "straight lines" (gerade Linien) that is, as intervals.[4]

Without going into details (and, in particular, into questions of priority), I must stress here the fundamental contribution of J. Plücker (1801–68) in overcoming the traditional idea of coordinates. As early as 1829 Plücker took a very general view of point coordinates,[5] and his *System der analytischen Geometrie* of 1835 introducted the designation "parallel point coordinate" (Parallelkoordinate) as against other types of point coordinates. At the same time he used homogeneous coordinates without restricting himself to point coordinates. (In this connection we must mention the earlier use of homogeneous point coordinates by Möbius and their later, consistent use by Cayley and Hesse.)

Plücker's *System der analytischen Geometrie* of 1835 contains a complete analysis of all possible systems of linear point coordinates in 3-space defined by the fact that in such coordinate systems planes are given by linear equations. About 1830 we find many instances of the use of nonrectangular linear point coordinates—for example, in Möbius's *Der barycentrische Calcul* and in the works of K. W. Feuerbach and E. Bobillier—but Plücker was the first to single out the *class* of linear point coordinates, and to prove that any two such coordinate systems are linked by a linear fractional transformation. This idea of a class of coordinate systems connected by a closed system of transformations is an instance of group-theoretic thinking. Plücker's analysis also marks his transition to line and plane coordinates. This development of line geometry was a direct step—undertaken jointly by Plücker and his assistant F. Klein—toward the elaboration of the Erlangen Program (see III.2).

In sum, what was general about the coordinate concept as it evolved in the nineteenth century was the view that a coordinate system of a geometric manifold consists of independent parameters of widely varying geometric meaning. Finally, in another move toward abstraction, a "space" (Raum) became a "number manifold" (Zahlenmannigfaltigkeit), of dimension determined by the number of independent parameters. This formulation was stressed by S. Lie in connection with his concept, developed in the eighties, of a transformation group. In fact, even his doctoral dissertation of 1871 begins with a direct reference to the change, inspired by Plücker's example, in the prevailing view of the nature of geometry. We quote:

* It is well known that there are close connections between the rapid development of geometry in our century and the philosophical ideas about the nature of analytic geometry spelled out in the most general form in Plücker's earliest papers.

For those who have entered into the spirit of Plücker's work there is nothing essentially new in the idea that one can use as an element of the geometry of space any curve which depends on three parameters. [470, p. 105]

---

* Die rasche Entwicklung der Geometrie in unserem Jahrhundert steht, wie bekannt, in einem intimen Abhängigkeitsverhältnisse zu philosophischen Betrachtungen über das Wesen der Carte-
(* continued on p. 30)

The developing revision of the coordinate concept inspired many inquiries into the conditions to be satisfied by a number manifold that is to resemble a coordinatized space. We find here important preliminaries for the development of topology in the middle of the nineteenth century and of topological groups at the century's end.

In epistemological terms, the view of space as a number manifold separated the study of objective physical space from the study of mathematical spaces, of physics from geometry—a fact of decisive importance for group theory in the last third of the nineteenth century. The development of noneuclidean geometry in the nineteenth century had a similar effect.

4 Noneuclidean Geometry and the Epistemological Problem of Space

This issue, though highly significant for group theory, can be discussed without detailed consideration of the history of the discovery of noneuclidean geometry and of the relevant contributions of Lambert, Taurinus, Schweikart, Gauss, Lobachevski, Bolyai, and others.

Discussion of the "proofs" that derived the parallel postulate from the other euclidean axioms led at first to the discovery of a number of its equivalent alternatives. But later, through the development of a noneuclidean (hyperbolic) geometry independent of the parallel postulate, this discussion ran into difficult philosophical and epistemological problems. In spite of the fact that the mathematical core of the matter was reasonably well understood, the gravity of these epistemological dilemmas can be inferred from the very names bestowed on the noneuclidean geometry by its founders: "astral science of magnitude" (astralische Größenlehre) (Schweikart), "logarithmic-spherical geometry" (logarithmisch-sphärische Geometrie) (Taurinus) and "imaginary geometry" (воображаемая геометрия) or "pangeometry" (пангеометрия) (Lobachevski). What had to be clarified was that the development of various geometries was a mathematical problem essentially independent of the structure of physical space and that the question of which geometry best fits this space mathematically can be decided only by experiment. It is well known that Gauss and Lobachevski adopted this essentially materialist viewpoint. Gauss embarked on terrestrial measurements of the angle sum of a triangle, and Lobachevski planned corresponding planetary measurements.

(* continued from p. 29)
sischen Geometrie – Betrachtungen, die in ihrer allgemeinsten Form von Plücker in seinen ältesten Arbeiten ausgesprochen worden sind.

Für den, der in den Geist der Plückerschen Werke eingedrungen ist, liegt nichts wesentlich Neues in der Idee, daß man als Element für die Geometrie des Raumes eine beliebige Kurve benutzen kann, die von drei Parametern abhängt.

The elaboration of the distinction between geometric spaces and physical space was hampered by Kantian philosophy. The *Kritik der reinen Vernunft* implied that euclidean geometry (as well as 3-dimensional geometry) was a necessary mode of thought.[6] Such was the predominance of Kantian philosophy in Germany that, as is well known, Gauss's fear of the "clamor of the Boeotians" (Geschrei der Böotier) forced him into silence. That clamor did in fact materialize as soon as N. I. Lobachevski (1792–1856)[7] started to publish papers on noneuclidean geometry. At about that time Kantian ideas penetrated into Russia; but meanwhile the czarist government, with the creation (1825) of a "Ministry for Religious and National Education," launched a campaign aimed at excluding all of the extremely diverse philosophies of Western Europe as infected with "materialism."

Thus the contemporary Russian scene provides an ideal background for studying the frontline of the epistemological debate over noneuclidean geometry [243, 302, 708]. Fundamental to the debate was Kant's philosophy of space and time; and in the discussion of these ideas the University of Charkov, and, secondarily, the University of Kazan (where Lobachevski taught), took leading roles. The objections to Kant came from two viewpoints. The first regarded the space built mathematically on Euclid's axioms as an objective category intrinsic to nature; but though opposed to Kant's ideas, this notion also tended to bar the recognition of noneuclidean geometry. The second viewpoint was strongly influenced by eighteenth century English empiricists and French sensualists, who had regarded space as a subjective category derived from experience. While Lobachevski also proceeded from experience, he thought of space as neither objective nor subjective but rather as an a posteriori concept of which we are aware as a result of the process of motion of physical bodies. He also thought that physics or astronomy may be able to decide experimentally, through the study of motion, which of the two geometries is valid.

Not later than 1829 Lobachevski realized that, essentially, his investigations led him to study logical and epistemological questions of mathematics [708, p. 475]. In his own thought, he did not consummate the break between the concepts of mathematical and physical space; yet it was the recognition of just this distinction that displaced the Kantian view of space and thus paved the way for the universal acceptance of noneuclidean geometry. Lobachevski was not recognized in Russia; in 1866, ten years after his death, Bunyakovski characterized his geometry as nonsense. Lobachevski tried to gain recognition in Germany, but apart from Gauss's appreciation he failed to achieve it before 1842—this in spite of the publication (1837) of his "Géométrie imaginaire" in *Crelles Journal* and of his "Geometrische Untersuchungen zur Theorie der Parallellinien".

The epistemological question posed by (hyperbolic) noneuclidean geometry was clarified in Germany only after Kantian philosophy lost its preeminence. This clarification was accomplished by mathematicians and scientists whose natural leanings were materialist. Here B. Riemann (1826–66) played a leading role. In his differential-geometric studies of the intrinsic geometry of surfaces (in n-dimensional space) Riemann introduced the line element as a quadratic differential form. This formulation included the ordinary euclidean metric and pointed to the existence of the two noneuclidean geometries known today, in Klein's terminology, as hyperbolic and elliptic, respectively. In order to separate geometry from physics Riemann used the term "space" (Raum) to denote objective physical space and the term "manifold" (Mannigfaltigkeit) to denote mathematical space. Thus in Riemann's terminology the epistemological problem was (and is) to determine the nature of space as a manifold.

The high points of this clarification were Riemann's habilitation lecture "Über die Hypothesen, welche der Geometrie zugrunde liegen" (1854) and the relevant lectures and papers of Helmholtz, dating from the period between 1866 and 1870. One of the best known is "Über die Thatsachen, welche der Geometrie zugrunde liegen," a paper that Helmholtz submitted to the Göttinger Gesellschaft der Wissenschaften (the title is a conscious parallel to Riemann's). It was Helmholtz who devised the thought experiment, frequently invoked to this day, involving the world picture of thinking 2-dimensional beings who move about in a spherical world and naively use geometric axioms impressed on them by their particular environment.

The separation of geometry and physics, and the elimination of obstacles due to constraining philosophical systems, provided the epistemological basis for the definitive classification of "geometries" undertaken in the Erlangen Program of 1872.

5 *The n-Dimensional Manifold as Space*

The development of n-dimensional geometry was hampered by inadequate philosophical views, above all again by the views of the Kantians. But a concurrent obstacle to the universal acceptance of multidimensional geometries was, of course, the habit of 3-dimensional geometric thinking. It is not surprising to find this in the work of Möbius, around 1827;[8] it is surprising to find it in the case of E. E. Kummer (1810–93), who, in the sixties, heaped scorn and ridicule on this movement at a time when it was being systematically advanced by such mathematicians as A. Cayley (1834), H. Grassmann (1844), A.-L. Cauchy (1847), and B. Riemann (1854). Earlier, n-dimensional manifolds had been spontaneously considered by, among others, Lagrange, Jacobi, and Gauss ([703, pp. 684–692; 697, pp. 469–481]; and see III.2).

The development of n-dimensional geometry, which also imparted an essential impulse to the effort to classify all of geometry by group-theoretic means, was fed by two sources. One source, in the work of Lagrange, Jacobi, Gauss, and others, was the close connection between mathematics and physics. Specifically, physics (especially mechanics and electromagnetism) increasingly found itself forced to consider simultaneously more than two physical magnitudes; this trend, beginning at the end of the eighteenth century, accelerated early in the nineteenth. Moreover, the necessity of dealing mathematically with directed physical quantities gave rise to the vector and quaternion calculus of W. R. Hamilton (1805–65). The other source of n-dimensional geometry—in the work of Grassmann and Cayley—was pure intellectual daring, though in the abstract form that it here assumed, especially in Grassmann's formulation ("Lineale Ausdehnungslehre", 1844), n-dimensional geometry had at first little influence. Grassmann's ideas stemmed from Leibniz's concept of a "geometric analysis" (geometrische Analyse) while Cayley's were generalizations of quadratic and ternary forms.

6 *The New Conceptual Content of "Geometry"*

In sum, geometry presented to the mathematicians of the first half of the nineteenth century a picture of explosive development both in scope and in depth. Gone was the apparent monolithic unity still present at the end of the eighteenth century. A number of intrinsic tendencies acquired, as it were, a life of their own, and developed in divergent directions as independent "geometries" or "geometric methods."

Among the geometers and mathematicians of the midnineteenth century the problem of the inner connections among the different geometries and geometric methods produced much perplexity. But the very effort to resolve the existing chaos and overcome its increasingly disturbing effect provided a significant incentive for the development of future group-theoretic methods. At first, the mathematicians' conscious pursuit of a synthesis was marked by indecision in choosing from among the basic mathematical concepts that were newly available and suitable at least in outline. Then they may be said to have settled for the Cayley-Sylvester invariant theory, which we recognize today as implicitly group-theoretic. Finally, as a result of an explicit link with the theory of permutation groups, their pursuit reached fulfillment in the Erlangen Program of 1872 (see III.1–2).

The geometers of that time fully realized that all aspects of geometry were in a state of revolutionary change, whose nature they could not initially assess, in mathematical terms, with any degree of precision. The introductions to contemporary mathematical monographs contain a wealth of reflections on the subject matter of geometry, and many others are found in the actual texts. The clearest expression of the effort to define geometry in the light of particular new developments appears in the "Aperçu historique" (1837) of M. Chasles (1793–1880), who was at once a productive geometer and one of the most acute observers of contemporary trends. The timeliness of Chasles's summary is reflected in its fuller title: "Aperçu historique sur l'origine et le développement des méthodes en géométrie, particulièrement de celles que se rapportent à la géométrie moderne" [153].

In view of the newest developments—especially the development of descriptive geometry and of synthetic projective geometry—Chasles rejected the traditional definition of geometry as "*the science whose object is the measurement of extension*" ("*la science qui a pour objet la mesure de l'étendue*") [153, p. 288]. Chasles found that, far from being tenable, this definition was "*incomplete and insufficient*" ("*incomplète et insuffisante*") [153, p. 288], and he replaced it by a description of geometry as the science of measurement and order of extension. We quote:

* One must do more, and replace, as well, the simple idea of *measurement* with the complex idea of *measurement* and *order* which alone gives the word *geometry* a true and complete meaning. [153, p. 290]

Though rather vague, Chasles's formulation is a serious effort to capture the essence of a development converging toward the study of "geometric structures." It was group theory that, thirty-five years later, turned out to provide the decisive tool; but the path to that outcome proved long indeed.

* Il fallait faire plus, et remplacer aussi l'idée simple de *mesure*, par l'idée complexe de *mesure* et *d'ordre*, qui est indispensable pour donner au mot *Géométrie* un sens vrai et complet.

2 The Search for Ordering Principles in Geometry through the Study of Geometric Relations (Geometrische Verwandtschaften)

1 The Study of Geometric Relations

Descriptive geometry and projective geometry emphasized those relations among geometric figures that were associated with particular formations. This point of view was formulated[9] by Carnot in his *Géométrie de position* and is reflected in his *Méthode des relations contingentes*[10] developed in that work. The main principle of Carnot's method was that two geometric figures connected by a projection share a body of properties. But interest increasingly shifted from the shared properties to the transformations themselves, that is, to the links that imply particular relations among figures. This trend first took clearly recognizable form in Poncelet's "principle of continuity" (principe de continuité) elaborated in that author's *Traité*, which assigned equivalent status to two figures connected by a "continuous" transformation. This principle had been tacitly applied by Monge. But its use by Poncelet focused attention on the question of "continuous" transformations including, in any case, central projections. In this way, the study of the geometric relations of figures became the study of the associated transformations. Between 1830 and 1870 transformations became the object of many prolific, specialized (and largely independent) investigations, which gave rise to the theories of circular transformations, spherical transformations, inversions, affinities, collineations, and so on. (Of course, some of these transformations were not entirely new; in fact, their sporadic use dates back to the sixteenth century.)

While this development was still in progress, the study of geometric relations gradually entered a third phase, which investigated the logical connections among transformations. This led to the problem of classifying transformations, and to the group-theoretic synthesis of geometry.

2 Möbius's Classification Efforts in Geometry

The high point of these tentative beginnings, unaffected by the simultaneous but seemingly unrelated development of the theory of permutations, was the geometric work of August Ferdinand Möbius (1790–1868). Möbius was in a sense outside the mathematical community, and his contacts with it were relatively loose. Even so, his investigations in geometry encompass all contemporary developments in the field. This explains why Möbius, at first largely ignored, subsequently won the highest esteem when it was recognized that his ideas, developed in quiet isolation, anticipated the later evolution of geometry and even the Erlangen Program. Möbius had studied under Gauss and later maintained contact with him. Nevertheless, while he was professor of astronomy at Leipzig, and (from 1 May, 1816) director of the observatory there, he was cut off from the centers of contemporary mathematical thought—and even neglected to make full use of his admittedly limited opportunities to study the mathematical literature.[11] These facts can only enhance our appreciation of his contributions. But it was not until the second half of the nineteenth century, after an incubation period necessitated by the strangeness of their form and content, that Möbius's relatively unconventional writings began to exert a significant influence.

In view of the theme of this book, we wish to stress two elements of Möbius's geometric thought that attest to the inner logic and inevitability of mathematical development:

i. He made a significant contribution to the shaking up of the traditional coordinate concept.
ii. Although completely unaware of the group concept, he managed, as if guided by instinct, to delineate the group-theoretic organization of geometry later set forth in the Erlangen Program of 1872.

Both of these features are already present in his first major work, *Der barycentrische Calcul, ein neues Hülfsmittel zur analytischen Behandlung der Geometrie dargestellt und insbesondere auf die Bildung neuer Classen von Aufgaben und die Entwicklung mehrerer Eigenschaften der Kegelschnitte angewendet* (Leipzig, 1827). In the preface Möbius motivates the barycentric calculus by the observation that "the mechanical discipline of centers of gravity has frequently been used as an aid in the discovery of purely geometric truths" ("die der Mechanik zugehörige Lehre vom Schwerpuncte schon oftmals als Hülfsmittel zur Erfindung rein geometrischer Wahrheiten benutzt worden ist") [523, p. 5], and refers, quite rightly, to Archimedes, Pappus and Guldin, as well as to L. N. M. Carnot and l'Huilier. He starts with the concept of the mechanical center of gravity and the theorem that asserts that every system of mass points (he writes: gewichtige Punkte) has exactly one center of gravity. He then compares physical relations with the corresponding mathematical operations—for example, mechanical equilibrium with the mathematical equation. Next he defines by means of his "coefficients" (Coefficiente) what we now call barycentric coordinates.[12]

* The observation that it is possible to assign weights to three arbitrary points in the plane in such a way that a given fourth point of the plane can be regarded as their center of gravity, and that the ratios of the three weights can be determined from the disposition of the four points in just one way, led me to a new method for the determination of the positions of points in the plane.

I called the three points which thus serve to determine all others, fundamental points, the lines connecting them fundamental lines, and the triangle formed by them the fundamental triangle; but the ratios of the weights of the fundamental points—or their coefficients, as I shall call these weights from now on—which must hold for the point to be determined are its coordinates. I proceeded similarly to determine a point in space, in which case one required four fundamental points and therefore six fundamental lines and one fundamental pyramid. [523, p. 6]

Möbius was well aware of his break with the conventional view of coordinates. On a number of occasions, both in the introduction to and in the development of his analytic geometry,[13] Möbius paid special attention to the comparison between his own method and "the usual method of parallel coordinates" ("die gewöhnliche Methode der parallelen Coordinaten") [523, pp. 6–7] and their mutual computational interchangeability. Thus in the introduction he states that

* The aforementioned is only a brief sketch of an analytic geometry in which instead of the usual method of rectangular coordinates we use the barycentric determination by means of fundamental points. [523, p. 8]

In the text itself ("Drittes Kapitel: Neue Methode die Lage von Puncten zu bestimmen, " §28) he states, among other things:

** The determination of the position of points on a line, in a plane or in space requires two types of elements. Some elements are the same for all points—for example, the axes of an ordinary coordinate system. Others, called coordinates in the most general sense, depend on the various positions of the points relative to the elements of the former type, and so vary from point to point.

The present method of determination of points is essentially the following. Certain points, called fundamental, are taken as fixed elements, and the point to be determined is viewed as their center of gravity. Thus the variable elements, or coordinates, are the ratios which must hold for the coefficients of the fundamental points in order for the point to be determined to be their center of gravity. [523, p. 50]

(* see p. 36)
Die Bemerkung, dass irgend dreien Puncten einer Ebene immer solche Gewichte beigelegt werden können, dass ein gegebener vierter Punct der Ebene als Schwerpunct derselben betrachtet werden kann, und dass diese drei Gewichte in Verhältnissen zu einander stehen, die aus der gegenseitigen Lage der vier Puncte nur auf eine Weise bestimmbar sind, führte mich weiter zu einer neuen Methode, die Lage von Puncten in einer Ebene zu bestimmen. Die drei Puncte, welche somit zur Bestimmung aller übrigen dienten, nannte ich Fundamentalpuncte, die sie verbindenden Geraden Fundamentallinien, und das von ihnen gebildete Dreieck das Fundamentaldreieck; die Verhältnisse aber, die für den zu bestimmenden Punct zwischen den Gewichten der Fundamentalpuncte oder ihren Coefficienten, wie ich die Gewichte von jetzt an hiess, stattfinden müssen, waren die Coordinaten dieses Punctes. Auf ähnliche Art verfuhr ich zur Bestimmung eines Punctes im Raume, wo vier Fundamentalpuncte nöthig waren, und es daher sechs Fundamentallinien und eine Fundamentalpyramide gab.

* Das Bisherige ist nur ein kurzer Abriss einer analytischen Geometrie, wo statt der Methode der parallelen Coordinaten die barycentrische Bestimmung durch Fundamentalpuncte angewendet ist.

(** see p. 38)

It is clear that the coordinates introduced by Möbius are homogeneous. Thus Möbius, together with J. Plücker, paved the way for the general use of homogeneous coordinates. (Their universal acceptance was due to Hesse. An important reason for their acceptance was the elegance of the formal apparatus associated with their use.[14])

Already in the introduction to *Der barycentrische Calcul* we encounter a second element of Möbius's geometric thought. Möbius explains that the generality of barycentric coordinates turned him from the consideration of similarity of figures to the investigation of the concept of affinity, introduced but then abandoned[15] by Euler, and finally stimulated his interest in the general study of geometric relations.

This is what is involved. Relative to a system of fundamental points, each point of a figure can be expressed in terms of barycentric coordinates. By referring these coordinates to another system of fundamental points one obtains a new figure connected to the old figure by a "geometric relation." In Möbius's words,

* It is clear that two such figures are not similar but are connected by a more general relation. It occurred to me that this relation was identical with the relation referred to by Euler, in his *Introductio*, as an "affinity." [523, p. 9]

(Euler, and with him Möbius, use the term "affinity" in the modern sense.)

Möbius goes on to explain that this coordinate transformation moved him

** to find many more such relations among figures. This is how the second half of my book came into being. It deals with the study of geometric relations—a discipline that, in the sense used here, contains the foundation of all geometry and, when presented in full generality and detail, may be extremely difficult. [523, p. 9]

** see p. 37)

Zu der Bestimmung der Lage von Puncten, sei es in einer gegebenen Geraden, oder in einer gegebenen Ebene oder im Raume überhaupt, werden Stücke von zweierlei Art erfordert. Die einen bleiben für alle Puncte dieselben, wie z. B. die Axen eines gewöhnlichen Coordinatensystems. Die andern, welche Coordinaten im allgemeinsten Sinne heissen, hängen von der verschiedenen Lage der Puncte gegen die ersten ab, und sind daher von einem Puncte zum andern veränderlich.

Die jetzt zu erörternde Methode, Puncte zu bestimmen, besteht nun im Wesentlichen darin, dass, als unveränderliche Stücke, gewisse Puncte genommen werden, die ich Fundamentalpuncte nennen will, und der zu bestimmende Punct als Schwerpunct derselben gedacht wird. Die veränderlichen Stücke oder die Coordinaten sind demnach hier die Verhältnisse, welche zwischen den Coeffizienten der Fundamentalpuncte stattfinden müssen, damit der zu bestimmende Punct dieser Puncte Schwerpunct sei.

* Man sieht augenblicklich, dass zwei solche Figuren einander nicht ähnlich sind, sondern in einer allgemeineren Beziehung zu einander stehen. Es ergab sich mir, dass diese Beziehung einerlei sei mit derjenigen, welche von Euler in seiner *Introductio* Affinität genannt wird.

** noch mehrere dergleichen Beziehungen zwischen Figuren auszumitteln, und somit entstand der zweite Abschnitt meines Buchs, welcher von den geometrischen Verwandtschaften handelt, einer Lehre, welche in dem hier gebrauchten Sinne die Grundlage der ganzen Geometrie in sich fasst, die aber auch eine der schwierigsten sein möchte, wenn sie in völliger Allgemeinheit und erschöpfend vorgetragen werden soll.

In view of the Erlangen Program, these words of Möbius have a prophetic ring. While he knew nothing whatever of the tools of group theory, which then existed only in fragmentary form,[16] Möbius spelled out with utmost clarity the goal of classifying geometric relations:

* Thus far only the simplest types of relations have been investigated. They are the relations which occur in elementary geometry. Their principal value is that each of them leads to a particular class of problems in which, given sufficiently many elements of a figure, one is required to find one or more additional elements. [523, pp. 9–10]

While it is true that Möbius left no doubt concerning the newness of the problem of "geometric relations," he formulated it throughout in the language of the traditional geometric mode of thought prevailing at the time: Given certain geometric elements, to determine other such elements. The invariant-theoretic viewpoint is not yet clearly present in his writings.

Already in the introduction to *Der barycentrische Calcul* Möbius lists the "geometric relations" of congruence, similarity, affinity, and collineation and clarifies their mutual connections. He states that there is little difference between congruence and similarity—an assertion corresponding to the properties of the principal group of the Erlangen Program. Affinities are more general and include similarities and congruences as special cases—an assertion corresponding to the relation between the affine and equiform (or principal) group. Finally, in saying that collineations are the most general relations, Möbius asserts that (to use modern terms) affine geometry is a subgeometry of projective geometry.

This idea of the inclusion of certain geometries in other geometries was presented by Möbius a second time, in a beautiful form, in the introduction to a minor work called "Über eine allgemeinere Art der Affinität geometrischer Figuren" [529]. He says:

** In my "barycentrischer Calcul" I tried to show that in addition to the mutual relations of figures encountered in the most elementary parts of geometry, according to which two figures are congruent and similar or just similar, there are other such connections or relations that, while more general than the familiar ones, nevertheless belong to elementary geometry in the sense that, under these more general relations, it is still true that collinear points correspond to collinear points. I called these more general relations affinities and collineations. Congruence, which I treated as a particular relation, is essentially a special case of affinity.

---

* Im Gegenwärtigen sind nur die einfachsten Arten der Verwandtschaften betrachtet worden, diejenigen nämlich, welche auch in der niederen Geometrie in Anwendung kommen können. Der Hauptnutzen dieser der niederen Geometrie angehörigen Verwandtschaften besteht aber darin, dass jede derselben zu einer besonderen Classe von Aufgaben hinführt, wo aus gewissen in hinreichender Anzahl gegebenen Stücken einer Figur ein oder mehrere andere Stücke gefunden werden sollen.

(** see p. 40)

In the case of collineations, the name itself suggests that their sole characteristic is the correspondence between straight lines. On the other hand, in the case of affinities we have an additional characteristic, namely, the fact that, depending on whether we are dealing with figures in the plane or in space, the ratios of the areas or the volumes, respectively, of two parts of one figure are the same as the ratios of the corresponding parts of the other figure.

Just as the retention of the first of the two characteristics of an affinity yields the more general relation of collineation, so too an affinity yields a more general relation if we disregard the first but retain the second of the two characteristics.

This program, formulated in the introduction to *Der barycentrische Calcul*, is realized in the second of its three parts, "Von den Verwandtschaften der Figuren und den daraus entspringenden Classen geometrischer Aufgaben" ("On the Relations of Figures and on the Resulting Classes of Geometric Problems"). Starting with the definitions of congruence, similarity, affinity, and collineation, and using the formal apparatus of his special calculus, Möbius studied in detail the mutual dependence of these geometric relations. But this study contains no fundamentally new ideas significant for the history of group theory.

In subsequent works Möbius enlarged the content of the program formulated in the introduction to *Der barycentrische Calcul*. In a smaller work, "Von den metrischen Relationen im Gebiete der Lineal-Geometrie" [528] of 1829, Möbius determined the logical position, in the total system of his geometric relations, of geometric problems solvable by straightedge alone—rather than by straightedge and compass; the corresponding geometry had been called "Lineal-Geometrie" by J. H. Lambert ("Die freye Perspektive," Zürich, 1759). After stating that what matters here is only incidence of points and lines, Möbius concluded that this

---

(** see p. 39)

In meinem ‚barycentrischen Calcul' habe ich zu zeigen gesucht, dass ausser den schon in den ersten Elementen der Geometrie in Betrachtung kommenden Beziehungen, in welchen Figuren zu einander stehen können, und wonach zwei Figuren gleich und ähnlich, oder ähnlich allein, heissen, es noch einige andere dergleichen Beziehungen oder Verwandtschaften gebe, die, obschon von allgemeinerer Beschaffenheit, als jene schon bekannten, doch in das Gebiet der Elementargeometrie noch gehören, indem auch bei ihnen Puncten der einen Figur, welche in einer Geraden liegen, in einer Geraden liegende Puncte der anderen Figur entsprechen. Ich nannte diese allgemeineren Verwandtschaften Affinität und Collineationsverwandtschaft; denn die Gleichheit, die ich an jenem Orte zwar ebenfalls als eine besondere Verwandtschaft behandelt habe, ist im Grunde nur als eine specielle Art der Affinität zu betrachten.

Bei der Collineationsverwandtschaft ist, wie schon ihr Name ausdrücken soll, gedachtes Entsprechen gerader Linien das alleinige sie charakterisirende Merkmal. Bei der Affinität hingegen kommt als zweites Merkmal noch hinzu, dass je zwei Theile der einen Figur – Flächentheile oder Theile des Raumes, nachdem die Figur in einer Ebene oder im Raume überhaupt enthalten ist, – sich ihrem Inhalte nach eben so zu einander verhalten, wie die entsprechenden Theile der anderen Figur.

So wie nun aus der Affinität die allgemeinere Verwandtschaft der Collineation entspringt, wenn wir von den oben erwähnten zwei Merkmalen nur das erste beibehalten; so wird die Affinität gleichfalls in eine allgemeinere Verwandtschaft übergehen, wenn wir das erste Kennzeichen nicht mehr berücksichtigen und nur das zweite noch festhalten.

study is identical with the study of "collineation relations" (Collineationsverwandtschaften). Since his barycentric concepts and symbolism had not been generally accepted, Möbius had to rely, in the technical part of the work, on traditional formalism, that is, on ordinary coordinate geometry with the usual parallel coordinates. Also, he restricted himself to the case of the plane.[17]

The second phase of Möbius's creative activity was devoted chiefly to applied mathematics. He worked on problems involving systems of lenses, equilibrium of forces (the basis of his textbook on statics of 1837), celestial mechanics, and systems of crystals. It is of great interest that these practical problems forced Möbius to continue his study of geometric relations and that specific questions relating to these problems required the investigation of further geometric relations.

An additional important development from this period was Möbius's generalization of the traditional concept of addition. Through the parallelogram rule, composition of forces yields a force and composition of motions yields a motion. Thus composition of successive "operations" of a given kind involves the use of a composition rule, and the difficult mathematical task, implicit in the above physical examples, is to express the composition rule in a suitable calculus. (The difficulty of this task is illustrated by the extreme efforts of H. Grassmann to grasp the essence of "addition of segments," efforts that eventually led to the concept of vector and to vector calculus.) Möbius's interests in applied mathematics led him also to various rules of composition. In 1838 he published "Ueber die Zusammensetzung unendlich kleiner Drehungen"; in 1844, "Ueber die Zusammensetzung gerader Linien und eine daraus entspringende neue Begründungsweise des barycentrischen Calculs"; and in 1850, "Ueber einen Beweis des Satzes vom Parallelogramm der Kräfte". While these works greatly furthered the "composition concept" in the sense that they helped create a whole spectrum of composition laws for a variety of operations, they did very little directly to prepare the later general group concept, and even less to bring forth group-theoretic methods. At the time, "composition" of operations could not of itself lead to the group concept; for the outcome of any composition on a set necessarily belonged, by definition or on physical grounds, to the original set. What was missing at this level of development of geometry was recognition of the fact that a composition on a set singles out a subset closed relative to the composition. It was this phenomenon that later proved decisive in the study of permutations.

The question of composition rules gave rise to Möbius's third creative period. Beginning in 1853 he published works on special geometric transformations: two of these were "Ueber eine neue Verwandtschaft zwischen ebenen Figuren" (1853) and "Die Theorie der Kreisverwandtschaft in rein geometrischer Darstellung" (1855). These were followed by numerous papers dealing with involution of points. In addition to dealing with the various special transformations, Möbius attempted in these works to assign to each of the "geometries" associated with particular transformations its proper place in the scheme of congruence, similarity, affinity, and collineation. For lack of technical resources these attempts could not possibly come to full fruition; nevertheless, they provided a further impulse to the conceptual synthesis of the edifice of geometry. Beginning in 1858, at an advanced age, Möbius embarked on a study of so-called "elementary relations" (Elementarverwandtschaften) [530][18] more general than collineations; of the modern disciplines, these transformations are closest (more or less) to topology.

> * We shall say of two figures that they are connected by an elementary relation if to every element of one figure which is infinitesimal in all dimensions there corresponds just such an element of the second figure, and if the elements of the second figure corresponding to contiguous elements in the first figure are likewise contiguous; or, what amounts to the same thing, when to each point of one figure there corresponds a point of the other figure such that to every two infinitely close points of one figure there correspond two infinitely close points of the other figure.
>
> Thus an elementary relation can link a curve only to a curve, a surface only to a surface and a solid only to a solid. [524, p. 435]

We are quoting from a work published in 1863 and called "Theorie der elementaren Verwandtschaft." From Möbius's *Nachlass* we infer that this was part of a larger work that was Möbius's entry in a mathematical contest of the Paris Academy.

Möbius's work on geometric relations met a poor response. This was due not to a lack of interest in his stated objective but rather to the imperfect realization of so profound a program. In spite of the fact that Möbius was Gauss's student and the two kept in touch, Gauss said of *Der barycentrische Calcul* that the calculus by itself was noteworthy without being praiseworthy.[19] It was Felix Klein who, retrospectively, paid homage to Möbius's strenuous attempts at conceptual mastery of "geometry." Between 1885 and 1887 Klein helped prepare the collected works of Möbius for publication. He "grasped their inner connection" (dessen Werke "nach ihrem inneren Zusammenhang erfaßte") [346, p. 497] and found in them the model of his own Erlangen Program except for the explicit use of the required group-theoretic tools.

---

* Zwei geometrische Figuren sollen einander elementar verwandt heissen, wenn jedem nach allen Dimensionen unendlich kleinen Elemente der einen Figur ein dergleichen Element in der anderen dergestalt entspricht, dass von je zwei an einander grenzenden Elementen der einen Figur die zwei ihnen entsprechenden Elemente der anderen ebenfalls zusammenstossen; oder, was dasselbe ausdrückt: wenn je einem Puncte der einen Figur ein Punct der anderen also entspricht, dass von je zwei einander unendlich nahen Puncten der einen auch die ihnen entsprechenden der anderen einander unendlich nahe sind.

Einer Linie kann hiernach nur eine Linie, einer Fläche nur eine Fläche, und einem körperlichen Raume nur ein körperlicher Raum elementar verwandt sein.

3 *Plücker's Line Geometry*

Julius Plücker (1801–68) was the other leading German representative of the analytic approach to geometry in the first half of the nineteenth century. It is well known that Plücker's influence on the development of group theory was less direct than that of Möbius. On the other hand, his influence on the development of geometry was far more decisive than that of Möbius.

Plücker's close relations with mathematicians in France and (especially) in England, and the high esteem that he won largely through his contributions to physics, had a very beneficial effect on the development of geometry in Germany. On the other hand, his relations with German mathematicians were ambivalent. He was the target of attacks launched by followers of Steiner and the "synthetic" school, who enjoyed the support of Jacobi. By design or coincidence, Plücker resumed work on geometry in 1863, the year of Steiner's death, and between 1863 and 1868 he developed the essentials of line geometry. Plücker, and F. Klein when he was his assistant, prepared the ground for the superimposition of explicit (permutational) group theory on geometry.

In spite of Plücker's significant contributions, direct and indirect, to the development of geometry, few of his wide-ranging contributions to mathematics and physics are of direct group-theoretic interest. Those few are associated with his contribution to the evolution of the coordinate concept.

Plücker's "triangular" and "tetrahedral" coordinates are essentially identical to Möbius's barycentric coordinates. Plücker published his work [560] on these coordinates in 1829, two years after the publication of Möbius's *Der barycentrische Calcul.* But in spite of this, and in spite of the fact that Grassmann, Hamilton, and others relied upon the work of Möbius, it was through Plücker that these coordinates entered mathematics. This is probably because the Plücker version of the new coordinates was labeled a mathematical tool, whereas the mathematical content of the Möbius variant was hidden by motivations and definitions derived from mechanics. Also, beginning in 1826, Plücker specifically favored the viewpoint of homogeneous coordinates by selecting as a starting point, what he called the "fundamental theorem on homogeneous functions," (Fundamentaltheorem über homogene Funktionen), that is, what we would call today the defining equation

$$F(xt, yt, zt) = t^{\alpha}F(x, y, z)$$

of a homogeneous function of degree α. The link between a broader view of the coordinate concept and the transition to homogeneous coordinates played an important role in the controversy over analytic as against synthetic geometry, and assured the eventual triumph of the analytic approach in the sixties and seventies.

Plücker's line geometry turned out to be extraordinarily important for the subsequent group-theoretic formulation of the "geometry problem." Like his contemporaries, Plücker was deeply influenced by the discussion of the duality principle. Following the precedent set by Poncelet and Gergonne, this principle was as a rule used heuristically, but it had become a kind of mysterious *deus ex machina* encumbered with mystical connotations. This is where Plücker stepped in.[20] He managed to strip away the mysticism and to clarify the principle's essential mathematical content, which is that point and line are equivalent starting elements of plane geometry and that point and plane are equivalent starting elements of 3-dimensional geometry. Plücker's search for determining elements that could serve as suitable coordinates of a line in the plane and of a plane in space led him, in the years between 1827 and 1830, to his line and plane coordinated,[21] respectively—the first step on the road to line geometry.

Let $u_1x + u_2y + u_3z + u_4 = 0$ be the equation of a plane in parallel coordinates. Then the u_i are the homogeneous coordinates of the plane in space; this is Plücker's usage of 1829. A corresponding statement holds for the coordinates of a line in the plane. But in that very year Plücker admitted the line as a possible basic element in the plane *and* in space. Thus it can be said that, in a sense, Plücker's line geometry already existed in 1829. But I shall follow accepted usage and designate 1846, the year of publication of Plücker's *System der Geometrie des Raumes*, as the year of the definitive emergence of line geometry, since it was then that Plücker defined the coordinates of a line in space. (The task was far from easy. Due to his involvement in physical research Plücker found it difficult to catch up with developments in algebra.)

Nowadays we make the following definition. Let X and Y be points in 3-dimensional (real or complex) projective space with coordinates (x_i) and (y_i), $i = 1, 2, 3, 4$. Then by the (Plücker) line coordinates of the line determined by X and Y we mean the six determinats $p_{12}, p_{13}, p_{14}, p_{34}, p_{42}, p_{23}$ where $p_{ik} = x_iy_k - x_ky_i$. The first hint of this approach is found in No. 258 of Plücker's *Geometrie des Raumes* of 1846. Our present knowledge of source materials gives firm support to the view that until 1846 Plücker knew nothing of Möbius's parallel development of new coordinate systems. As a result of contact with British scientists, Plücker resumed investigations in this area in 1864. In 1868 appeared the first volume of his *Neue Geometrie des Raumes, gegründet auf die Betrachtung der geraden Linie als Raumelement*, with its systematic line-geometric approach. The second volume, edited by F. Klein, appeared in 1869, after Plücker's death.

There is a feature in Plücker's line geometry of which he himself was well aware and that helped in the systematic classification of the most varied "geometries." It concerns the rudiments of n-dimensional geometry. Plücker's influence must be highly rated, since it endowed the n-dimensional manifolds of Grassmann—hitherto viewed as empty generalizations—with compelling geometric substance and thus moved them out of the realm of formal algebraic constructs. More specifically, the Plücker line coordinates are determined up to a proportionality constant and are connected by the relation $p_{12}p_{34} + p_{13}p_{42} + p_{14}p_{23} = 0$; this means that they form a kind of 4-dimensional manifold. Plücker wanted these coordinates to be thought of as elements of a kind of 4-dimensional geometry. Thus line geometry, built around a set depending on four parameters, furnishes a model for a theory of 4-dimensional objects in ordinary space whose dimension is 3 as long as we regard points as its basic elements. Plücker added that by choosing a basic set depending on sufficiently many parameters it is possible to study manifolds of arbitrarily high dimension without leaving 3-dimensional space: the multidimensional remains embedded in the 3-dimensional. This showed that, in the mathematical sense, a manifold does not have a natural dimension. Also—and this cannot be regarded highly enough—this lent support to the conceptual separation of objectively existing (3-dimensional) space from the mathematical construct of a manifold (of arbitrary dimension). A proper, that is, materialist, interpretation of this separation paved the way for the ideas of Riemann and Hilbert and for the theory of relativity and was a stepping stone toward the concept of a group of transformations on n-dimensional space.

4 The Role of Synthetic Geometry in the Development of Group Theory

Synthetic geometers also pursued the search for ordering principles in geometry. For example, Steiner formulated the aim of his principal work, "Systematische Entwicklung der Abhängigkeit geometrischer Gestalten von einander," (1832) in these words:

* The present work contains the final results of a prolonged search for fundamental spatial properties which contain the germ of all theorems, porisms and problems of geometry so freely bequeathed to us by the past and the present. It must be possible to find for this host of disconnected properties a leading thread and common root that would give us a comprehensive and clear overview of the theorems and a better insight into their distinguishing features and mutual relationships. ... The present work is an attempt to discover the organic connection [*Organismus*] between the most varied manifestations in the world of space. There are a few very simple fundamental relations that supply the pattern for the evolution of the remaining mass of theorems in a manner orderly and free from any difficulties. By properly ordering these few fundamental relations we become rulers of the whole discipline. Order supplants chaos and we see how all parts inter-

(* see p. 46)

connect in a natural way, how they line up in the most beautiful order, and how related ones combine in well-defined groups. [683, p. 233]

Steiner's aim was the systematic development of the whole of projective geometry. This goal was to be achieved by projectively generating successively higher geometric forms from "fundamental forms" (in the plane—line, pencil of rays, plane; in space—plane pencil of rays, pencil of planes, bundle of rays, bundle of planes, space). But apart from the fact that the "systematische Entwicklung" foundered after the first of its proposed five volumes, Steiner's method could not, in principle, encompass more than a part of geometry. For example, due to his failure to adopt Möbius's principle of orientation—that is, the use of signs—his method did not provide for the inclusion of the imaginary in geometry. On the other hand, as a result of Steiner's fundamental idea of successive projective generation, the role of projective correspondence as a method was increasingly overshadowed by its role as a geometric correspondence. This shift can be clearly seen in Chasles's "Aperçu historique" of 1837. (Incidentally, Chasles called a projective correspondence of figures a "homography" (Homographie), a term used in Germany until the sixties and seventies.)

The emphasis on projective correspondence among followers of the synthetic approach tended to obliterate the gradations of geometric correspondences elaborated by the "analytic" school. This meant that the synthetic approach did not yield group-theoretic viewpoints for the mastery of the geometry problem. The logical organization of the results of synthetic geometry demanded the use of analytic means; purist synthetic geometry had reached the end of its possibilities once it had developed its system without utilizing metric ideas. Later, the principal notion of synthetic geometry was absorbed in the fruitful concept of a transformation group.

(* see p. 45)
Das vorliegende Werk enthält die Endresultate mehrjähriger Forschungen nach solchen räumlichen Fundamentaleigenschaften, die den Keim aller Sätze, Porismen und Aufgaben der Geometrie, womit uns die ältere und neuere Zeit so freigebig beschenkt hat, in sich enthalten. Für dieses Heer von auseinander gerissenen Eigenthümlichkeiten musste sich ein leitender Faden und eine gemeinsame Wurzel auffinden lassen, von wo aus eine umfassende und klare Uebersicht der Sätze gewonnen, ein freierer Blick in das Besondere eines jeden und seiner Stellung zu den übrigen geworfen werden kann. ... Gegenwärtige Schrift hat es versucht, den Organismus aufzudecken, durch welchen die verschiedenartigsten Erscheinungen in der Raumwelt mit einander verbunden sind. Es giebt eine geringe Zahl von ganz einfachen Fundamentalbeziehungen, worin sich der Schematismus ausspricht, nach welchem sich die übrige Masse von Sätzen folgerecht und ohne alle Schwierigkeit entwickelt. Durch gehörige Anordnung der wenigen Grundbeziehungen macht man sich zum Herrn des ganzen Gegenstandes; es tritt Ordnung in das Chaos ein, und man sieht, wie alle Theile naturgemäss in einander greifen, in schönster Ordnung sich in Reihen stellen, und verwandte zu wohlbegrenzten Gruppen sich vereinigen.

Thus the return to computational methods became the driving force behind the further development of geometry. Beyond doubt, this assured the triumph of the analytic approach to geometry in the sixties. But it remained to determine the most effective algebraic methods. Thus it soon became apparent that efforts (by Jacobi and Hesse) to fashion an effective tool by combining the calculus of determinants and the use of homogeneous coordinates were doomed to failure. Progress was achieved, however, by the English school around Boole, Cayley, Sylvester and others, who went back to the theory of forms, originally a number-theoretic discipline. In the form of invariant theory, this was to offer the first means of classifying different geometries (see III.1).

3 Implicit Group Theory in the Domain of Number Theory: The Theory of Forms and the First Axiomatization of the Implicit Group Concept

As in the analysis of the development of geometry, so too in the analysis of the development of number theory the historian encounters (implicitly) the idea of a group, and, in addition, very many group-theoretic arguments involved in the concrete number-theoretic material. In contrast to the situation in geometry, the initial occurrence of group-theoretic thinking in number theory is not at all an expression of a need to unify far-ranging developments; it grew directly out of the computations—more precisely, out of their penetration of the inner structure of the number systems under investigation.

What follows is an illustration rather than an exhaustive presentation[22] of the implicit group-theoretic material in number theory. At the time, developments in this area did not directly contribute to the evolution of an equivalent of the group concept. Later, however, when the equivalence of the abstract content of several basic modes of reasoning common to the already developed group theory ("group" here meaning permutation group) and to number theory was recognized, these developments gave an essential impulse to the shaping of the abstract group concept, and its central role in mathematics was thereby emphasized. This occurred in the 1880s,[23] when it became apparent that the theory of permutation groups permitted the conceptual mastery of large parts of number theory.

1 Euler's Paper on the Theory of Power Residues

It is well known that number theory, after a long pause following the work of P. de Fermat (1601–65), became again a serious and useful discipline, worthy of study, largely through the efforts of Leonhard Euler (1707–83).[24] Euler's 1761 paper on power residues, "Theoremata circa residua ex divisione potestatum relicta" [193], proved a rich source of early implicit group-theoretic thinking.[25]

Euler here considers the remainders obtained on division of powers a^ν, ν a natural number, by a prime p. He assumes that a is not divisible by p; then a^ν is obviously never a multiple of p. In §3[26] Euler writes: "I decided to investigate carefully the remainders of the terms of this geometric sequence upon division by p" ("Ich habe mir vorgenommen, die Reste, welche bei der Division der Terme dieser geometrischen Reihe durch p entstehen, aufmerksam zu betrachten") [628, p. 111].

Of decisive significance for the sequel is Euler's insight that what matters is not just the remainder r such that $0 < r < p$, but rather that "all remainders $r + np$" (n a natural number) "can be regarded as the same remainder r" ("alle Reste $r \pm np$" (n natürliche Zahl) "als ein und derselbe Rest r angesehen werden können") [628, p. 111]; that, as he puts it, all these remainders are "equivalent" (gleichbedeutend).[27] Since there are no more than $p - 1$ different nonequivalent remainders, it follows that a number of terms of the infinite sequence a^ν must leave the same remainder (§4). In particular, infinitely many of the a^ν leave the remainder 1 (§12). If a^λ is the lowest power, other than a^0, which leaves the remainder 1, then all powers a^μ that leave the remainder 1 are of the form $a^{\lambda m}$, where m is a natural number (§20). Then the remainders of the powers $1, a, a^2, \ldots, a^{\lambda-1}$ are all distinct (§27). Also, it suffices to investigate the remainders of the powers $1, a, a^2, \ldots, a^{\lambda-1}$, since, as Euler puts it, the "segment" (Abteilung) $a^\lambda, \ldots, a^{2\lambda-1}$ and all further segments $a^{2\lambda}, \ldots, a^{3\lambda-1}$, and so on, leave the same remainders in the same order (§27). Thus the remainder left by a^σ, where $\sigma = n\lambda + \mu$, is equal to that left by a^μ (§30).

With a^λ as defined there are just λ different remainders. If $\lambda < p - 1$, then certain numbers never turn up as remainders (§33).

* Thus if one looks for different remainders, then it may happen that all the powers of a leave just one remainder, or just two remainders, or just three, and so on. There can never be more than $p - 1$ remainders. Whatever the number of actual remainders, the number 1 is always among them. [628, p. 119]

Now Euler considers separately two cases: (1) All the numbers from 1 to $p - 1$ turn up as remainders, so that a^{p-1} is the least power that leaves the remainder 1; and (2) the number of remainders is less than $p - 1$.

The second case presents a clear example of group-theoretic thinking, namely, in modern terms, the decomposition of a group into a subgroup and its cosets. Specifically, Euler proves the following theorem:

** When the number of remainders resulting from the division of the powers $1, a, a^2, a^3, a^4, a^5$ and so on by a prime p is less than $p - 1$, then there are at least as many numbers which are not remainders as there are remainders. [628, §37, p. 120]

---

* Wenn man also auf die Verschiedenheit der Reste achtet, so kann es vorkommen, dass bei allen Potenzen von a nur ein Rest oder nur zwei Reste oder nur drei usw. auftreten. Mehr als $p - 1$ kann es aber niemals geben. Welches auch die Zahl der vorkommenden Reste ist, immer findet sich die Einheit darunter.

** Wenn die Anzahl der Reste, welche bei der Division der Potenzen $1, a, a^2, a^3, a^4, a^5$ usw. durch die Primzahl p entstehen, kleiner ist als $p - 1$, dann gibt es mindestens ebenso viele Zahlen, die nicht Reste sind, als es Reste gibt.

This implies that the total number of remainders and of nonremainders is at least 2λ. Since this number cannot exceed $p - 1$ (§38), $\lambda = (p - 1)/2$ or $\lambda < (p - 1)/2$ (§39).

Thus in the second case, in which $\lambda < p - 1$, there are two subcases: (a) $\lambda = (p - 1)/2$ or (b) $\lambda < (p - 1)/2$:

* If $\lambda < (p - 1)/2$, then the exponent λ cannot exceed $(p - 1)/3$. Thus $\lambda = (p - 1)/3$ or $\lambda < (p - 1)/3$. [628, §41, p. 122]

Euler proves this as follows:

** Since a^λ is the lowest power which leaves the remainder 1 upon division by p it follows that there are λ different numbers among the remainders of the sequence $1, a, a^2, a^3, a^4, \ldots, a^{\lambda-1}$. Since $\lambda < p - 1$, there are just $p - 1 - \lambda$ numbers that are not remainders. If r is one such number, then we saw [in the proof of an earlier theorem, in §37] that the numbers

$$r, ar, a^2r, a^3r, a^4r, \ldots, a^{\lambda-1}r,$$

when reduced by division by p, do not appear among the remainders of the powers of a. This excludes λ numbers as possible remainders. Since $\lambda < (p - 1)/2$, we have $\lambda < p - 1 - \lambda$, so that there are still other numbers that are not remainders of the powers of a. Let s be a number which is neither a remainder nor a nonremainder belonging to the above sequence. Then all the numbers

$$s, as, a^2s, a^3s, a^4s, \ldots, a^{\lambda-1}s$$

are nonremainders and, as we showed in the previous proof [in the proof of the theorem in §37] they are all distinct. Also, none of these numbers appears ... in the previous sequence of nonremainders, that is, we never have an equality $a^\mu s = a^\nu r$ .... Thus when $\lambda < (p - 1)/2$, then there are at least λ numbers that are nonresidues, so that in addition to the λ remainders we have at least 2λ nonremainders. Since all these 3λ numbers are less than p, their number cannot exceed $p - 1$, that is, we cannot have $\lambda > (p - 1)/2$. It follows that $\lambda = (p - 1)/3$ or $\lambda < (p - 1)/3$, under the assumption that $\lambda < (p - 1)/2$ and that p is a prime. [628, §41, pp. 122–123]

Subcase (b) again splits into two, and the partitioning continues until we reach an equality of the form $\lambda = (p - 1)/n$:

---

* Nun sei $\lambda < (p - 1)/2$, dann kann dieser Exponent λ nicht grösser als $(p - 1)/3$ sein; entweder wird also $\lambda = (p - 1)/3$ oder es ist $\lambda < (p - 1)/3$.

** Da a^λ die niedrigste Potenz von a ist, welche bei der Division durch p den Rest 1 ergibt, so kommen nur λ verschiedene Zahlen unter den Resten dieser Reihe

$$1, a, a^2, a^3, a^4, \ldots, a^{\lambda-1}$$

vor; da $\lambda < p - 1$, so gibt es genau $p - 1 - \lambda$ Zahlen, die nicht Reste sind; sei r eine solche, dann sahen wir, dass alle diese Zahlen:

$$r, ar, a^2r, a^3r, a^4r, \ldots, a^{\lambda-1}r,$$

wenn wir sie durch Division mit p reduzieren, nicht unter den Resten der Potenzen von a vorkommen. Hiermit sind λ Zahlen von den Resten ausgeschlossen; da nun $\lambda < (p- 1)/2$, so wird $\lambda < p - 1 - \lambda$ und es gibt also neben diesen Zahlen noch weitere Zahlen, welche nicht Reste der

(** continued on p. 51)

* Quite generally, when it is known that $\lambda < (p-1)/n$, then we can show in the same way that we cannot have $\lambda > (p-1)/(n+1)$, and that therefore $\lambda = (p-1)/(n+1)$ or $\lambda < (p-1)/(n+1)$.
This shows that the number of numbers that cannot be remainders is 0 or λ or 2λ or some multiple of λ. For if there are more than $n\lambda$ such numbers, then a single such number gives rise to λ new ones, and if not all nonremainders are thereby included, then we immediately obtain λ new ones. [628, §46, pp. 124–125]

This result of Euler is most remarkable. In group-theoretic terms, the last formulation amounts to the theorem that the order of a subgroup is a divisor of the order of the group. Also, Euler used this result to give a quick new proof of the "little theorem" of Fermat; this proof was superior to his earlier one, based on the series expansion of $(a + b)^n$, in the sense that it established a number-theoretic result by number-theoretic means. In this sense, the new proof was, as Euler put it,[28] "more natural" ("natürlicher") than the old one.

The theory of power residues was considerably advanced at the end of the eighteenth century. Thus Euler discovered the law of quadratic reciprocity by inductive means,[29] and A.-M. Legendre (1752–1833) in his "Recherches d'analyse indéterminée" [452] gave the law its modern form, namely,

$$\left(\frac{p}{q}\right)\left(\frac{q}{p}\right) = (-1)^{\frac{p-1}{2}\cdot\frac{q-1}{2}}$$

This was a central result of his textbook *Essai sur la théorie des nombres* [453] of 1798. Only three years later the *Disquisitiones arithmeticae* [236] of C. F. Gauss (1777–1855) surpassed Legendre's textbook in content and, even more decisively, in method.[30]

(** continued from p. 50)
Potenzen von a sind. Es sei s eine solche Zahl, die weder Rest noch einer der Nichtreste der obigen Reihe ist. Dann werden auch alle diese Zahlen

$$s,\ as,\ a^2s,\ a^3s,\ a^4s,\ \ldots,\ a^{\lambda-1}s$$

Nichtreste sein und diese Zahlen werden, wie wir im vorangegangenen Beweis gezeigt haben, untereinander verschieden sein. Aber es kommt auch keine einzige dieser Zahlen ... in der vorigen Reihe der Nichtreste vor, d. h. es ist niemals $a^{\mu}s = a^{\nu}r$ .... Wenn also $\lambda < (p-1)/2$, dann gibt es noch mindestens λ Zahlen, die Nichtreste sind, und so erhalten wir zu den λ Resten noch mindestens 2λ Nichtreste. Da alle diese 3λ Zahlen kleiner als p sind, so kann ihre Anzahl nicht grösser als $p-1$ sein, d. h. es wird nicht $\lambda > (p-1)/2$. Daher wird entweder $\lambda = (p-1)/3$ oder $\lambda < (p-1)/3$, vorausgesetzt, dass $\lambda < (p-1)/2$ und dass p eine Primzahl ist.
* Allgemein, wenn bekannt ist, dass $\lambda < (p-1)/n$, dann zeigt man auf dieselbe Weise, dass gewiss nicht $\lambda > (p-1)/(n+1)$, es wird daher entweder $\lambda = (p-1)/(n+1)$ oder $\lambda < (p-1)/(n+1)$.
Hieraus erhellt, dass die Anzahl aller Zahlen, welche nicht Reste sein können, entweder $= 0$ oder $= \lambda$ oder gleich 2λ oder irgend ein Vielfaches von λ ist; wenn es nämlich mehr als $n\lambda$ solcher Zahlen gibt, dann liefert eine einzige solche schon λ neue; und wenn hierin noch nicht alle Nichtreste enthalten sind, so kommen wiederum zugleich λ neue hinzu.

2 *Implicit Group Theory in Gauss's Work*

The theory of power residues is also an essential part of the innovative *Disquisitiones arithmeticae*. By the nature of things, Gauss was led to the same partitioning[31] of the remainders of the terms of the sequence $a^0, a^1, a^2, \ldots$, described above in connection with Euler's 1761 paper and that is also found in J.-L. Lagrange's "Réflexions sur la résolution algébrique des équations" [447] of 1770–71. Nevertheless, using his concept of congruence, Gauss could express himself far more succinctly than Euler:

* If p is a prime that does not divide a, and a^t is the least power of a congruent to 1 modulo p, then the exponent t either equals $p - 1$ or is a divisor of that number. [237, p. 31]

To avoid repetition, I refrain from giving Gauss's proof of this theorem—notwithstanding its importance for the decomposition of a group with respect to a subgroup. Instead, I present Gauss's treatment of higher-order congruences, which have additional group-theoretic interest.

Quite generally, Gauss investigates the congruence

$$x^n \equiv a \pmod{m}$$

with $(a, m) = 1$. Extending Euler's terminology, Gauss refers to a as an nth-power residue or nonresidue according as this equation does or does not have a solution. First he proves (article 45) that if $(a, m) = 1$, then there is always an exponent t with $0 < t < m$ such that $a^t \equiv 1 \pmod{m}$. Later, in article 53, Gauss says that a belongs to the exponent $t \pmod{m}$ if no power of a with positive exponent less than t is congruent to 1 mod m. In article 54 Gauss proves the important theorem that if d is a divisor of $p - 1$, then there are just $\varphi(d)$ numbers belonging to d; here $\varphi(x)$ is the Euler φ-function. In particular, this implies the existence of a primitive root mod p, that is, a number g belonging to the exponent $p - 1$ such that $g^{p-1} \equiv 1 \pmod{p}$ and $g^m \not\equiv 1 \pmod{p}$ if $1 \leqslant m < p - 1$. It follows that, apart from order, the remainders of the powers

$$g^0, g^1, g^2, \ldots, g^{p-2}$$

are $1, 2, \ldots, p - 1$. Put differently, this says that the powers of g form a (finite) abelian, and even cyclic, group. Further, for every choice of a, with $a \not\equiv 0 \pmod{p}$, and for every choice of a primitive root g we can always find an α such that

$$a \equiv g^\alpha \pmod{p}.$$

---

* Ist p eine Primzahl, welche in a nicht aufgeht, und ist a^t die niedrigste Potenz von a, welche nach dem Modul p der Einheit congruent ist, so ist der Exponent t entweder gleich $p - 1$ oder ein aliquoter Teil dieser Zahl.

In group-theoretic terms, this is the simplest case of a basis representation of a cyclic abelian group.

What is fascinating about Gauss's work is that it contains clear prototypes of modern algebraic proofs and concepts. While not formulated abstractly—they are tied to concrete mathematical material—they exhibit very great profundity of thought. The fascinating task of confronting the concrete investigations of Gauss with the corresponding parts of modern algebra was undertaken (see [238]) by a number of authors, under the editorship of H. Reichardt, on the occasion of the 100th anniversary of Gauss's death.[32] I therefore limit myself to presenting a few of Gauss's group-theoretic ideas.

Gauss's theory of cyclotomic equations, "De aequationibus circuli sectiones definientibus," forms part 7 (articles 335–366) of *Disquisitiones arithmeticae*. This theory is a strikingly beautiful example of implicit group-theoretic methods that are in parts altogether independent of the special nature of the group elements. Another distinction of this theory is that it promoted the development of the theory of algebraic equations and the associated evolution of the permutation-based group concept.

Gauss shows that it is possible to reduce the study of regular polygons, without loss of generality, to the study of the equation

$$X = \frac{x^n - 1}{x - 1} = x^{n-1} + x^{n-2} + \cdots + x + 1 = 0,$$

where n is an odd prime. He proceeds to investigate the structure of the set Ω of the $n - 1$ roots of $X = 0$ and finds that, together with the number 1, they form a finite cyclic group:

* If ... r is any root in Ω, then $1 = r^n = r^{2n}$ and so on. Quite generally, $r^{en} = 1$ for every integral, positive or negative, value of e. It follows that if λ and μ are integers congruent modulo n, then $r^\lambda = r^\mu$. On the other hand, if λ and μ are incongruent modulo n, then r^λ and r^μ are unequal for, in that case, we can find an integer ν such that $(\lambda - \mu)\nu \equiv 1 \pmod{n}$, that is $r^{(\lambda-\mu)\nu} = r$. But then $r^{\lambda-\mu}$ is definitely not equal to 1. Also, it is clear that every power of r is likewise a root of the equation $x^n - 1 = 0$. Since the numbers $1\ (= r^0)$, r, r^2, ..., r^{n-1} are distinct, it follows that they represent all the roots of the equation $x^n - 1 = 0$. This means that the roots $r, r^2, r^3, \ldots, r^{n-1}$ coincide with Ω. More generally, if e is any positive or negative integer not divisible by n, then Ω coincides with the complex of numbers $r^e, r^{2e}, r^{3e}, \ldots, r^{(n-1)e}$. [237, p. 401]

* Wenn ... r irgendeine Wurzel aus Ω ist, so ist $1 = r^n = r^{2n}$ usw., und allgemein $r^{en} = 1$ für jeden ganzen, positiven oder negativen Wert von e; hieraus ist ersichtlich, dass, wenn λ, μ nach dem Modul n congruente ganze Zahlen sind, $r^\lambda = r^\mu$ wird. Sind dagegen λ, μ nach dem Modul n incongruent, so werden r^λ und r^μ ungleich sein; denn in diesem Falle kann man eine ganze Zahl ν derart annehmen, dass $(\lambda - \mu)\nu \equiv 1$ (mod. n) wird, woraus $r^{(\lambda-\mu)\nu} = r$ und daher $r^{\lambda-\mu}$ sicher
(* continued on p. 54)

This is a genuinely group-theoretic argument, and a comparison with its modern counterpart in abstract group theory is of interest. Gauss deduces the existence of a number ξ with $r^{\xi} = 1$ from the assumption that r is a root of unity. Abstractly, the existence of ξ is deduced from the assumption that if an element A generates a finite group, then there must exist a power of A with least exponent $n + 1$ that coincides with a preceding power A^m of A. But then $m = 1$, and so $A^n = 1$.

Of even greater interest is Gauss's investigation, in articles 344ff. of *Disquisitiones arithmeticae*, of the "periods" (Perioden).[33] In modern terms, this is the investigation of the subgroups of the Galois group of the cyclotomic equation. To judge from Abel's testimony, it deeply influenced the development by Abel and Galois of the theory of algebraic equations.

Gauss chooses a primitive root g mod n (where n is an odd prime). Then, apart from order, the set of $n - 1$ powers

$$r^{\lambda}, r^{\lambda g}, r^{\lambda g^2}, \ldots, r^{\lambda g^{n-2}},$$

where λ is any integer not divisible by n, coincides with Ω. Since $g^{n-1} \equiv 1 \pmod{n}$, $r^{\lambda g^{\mu}}$ and $r^{\lambda g^{\nu}}$ are equal or unequal according as μ and ν are congruent or incongruent modulo $n - 1$.[34] Gauss shows that this does not depend on the choice of the primitive root g.

In modern terms, Gauss's assertions are equivalent to the theroem that states that the Galois group of the cyclotomic equation of degree $n = p$ has order $p - 1$, is cyclic and isomorphic to the group of relatively prime residue classes modulo p, and is generated by the automorphism which takes r to r^g.

Gauss turns next to the construction of the periods. Let e be a divisor of $n - 1$, so that, for some f, $n - 1 = e \cdot f$. Then Gauss shows that it is possible to partition Ω into e disjoint classes of f roots each, and this partitioning is independent of the choice of the primitive root g mod n. Specifically, Gauss puts $g^e = h$ and forms he f numbers $r, r^h, r^{h^2}, \ldots, r^{h^{f-1}}$; more generally, the numbers $r^{\lambda}, r^{\lambda h}, r^{\lambda h^2}, \ldots, r^{\lambda h^{f-1}}$.

* The aggregate $r^{\lambda} + r^{\lambda h} + \cdots + r^{\lambda h^{f-1}}$, which[35] [as Gauss showed earlier (Wussing)] must be regarded as independent of g since it does not change if we take g to be any other primitive root, will be denoted by (f, λ), and the complex of these roots will be called the *period* (f, λ); the order of the roots is immaterial. [237, p. 406]

(* continued from p. 53)
nicht $= 1$ folgt. Ferner ist klar, dass jede Potenz von r ebenfalls Wurzel der Gleichung $x^n - 1 = 0$ ist; daher werden, da die Grössen $1\ (= r^0), r, r^2, \ldots, r^{n-1}$ sämtlich verschieden sind, diese Grössen sämtliche Wurzeln der Gleichung $x^n - 1 = 0$ darstellen und somit die Wurzeln $r, r^2, r^3, \ldots, r^{n-1}$ mit Ω identisch sein. Man schliesst hieraus leicht allgemeiner, dass Ω mit dem Complex der Grössen $r^e, r^{2e}, r^{3e}, \ldots, r^{(n-1)e}$ übereinstimmt, wenn e irgend eine durch n nicht teilbare positive oder negative ganze Zahl ist.
(* see p. 55)

Thus defined, each period consists of f numbers, each of which is the g^eth power of its predecessor. For $\lambda = 1$ we obtain the period that contains r^1.[36] If we regard each of the summands of the corresponding "aggregate" as a single element, then this period is a cyclic group of order f with identity r^1. [The implicit operation for this group is $r^a \odot r^b = r^{ab}$, where ab is the product of a and b modulo n (Translator).] Thus the formation of the Gaussian periods corresponds to the theorem of abstract group theory to the effect that for every divisor a of the order α of a cyclic group there is a cyclic subgroup of order a.

These examples of group-theoretic thinking in number theory are intended only to shed some light on this area as it was at the beginning of the century. It is possible to give many similar examples. One such example, related to the work of Gauss, is P. G. Lejeune-Dirichlet's (1805–59) theory of characters [454, §3] of 1839. In II.3, I shall take up additional instances—for example, the theory of so-called Galois imaginaries—of internal links between group theory and number theory at the time when group theory was being created (see II.3).

3 *Gauss's Theory of Composition of Forms*

The above and other examples of the preconceptual application of the group structure of sets of numbers lent only indirect support to the elaboration of the group concept. But this is not true of the theory of binary quadratic forms and forms of higher degree in more variables. Early in the 1870s, the algebraic invariant theory of the 1850s and 1860s acted as a liberating, and thus direct, factor in the extension of the group concept from permutation groups to transformation groups; it will be treated, in this connection, in suitable detail (see III.1–2).

A direct line of development leads from the theory of binary quadratic forms—the earliest instance of a complete theory of forms—and, in particular, from Gauss's theory of composition of forms in *Disquisitiones arithmeticae*, to the implicit, and then the explicit, axiomatization of an (abelian) group (without, admittedly, any conceptual fixing of the resulting structure by means of the word "group" or some synonym). This historically and logically uninterrupted line of development was brought to completion in 1870 by L. Kronecker (1823–91), long after the group concept had been generally accepted in connection with the Galois theory of algebraic equations, and at a time when it was beginning to prove itself

---

(* see p. 54)
Das Aggregat von f solchen Wurzeln $r^\lambda + r^{\lambda h} + \ldots + r^{\lambda h^{f-1}}$, welches als unabhängig von g zu betrachten ist, da es sich nicht ändert, wenn man für g eine andere primitive Wurzel nimmt, werden wir mit (f, λ) bezeichnen und den Complex derselben Wurzeln die **Periode** (f, λ) nennen, wobei keine Rücksicht auf die Anordnung der Wurzeln genommen wird.

as a far-reaching concept in many diverse areas of mathematics. The remarkable result of this temporal coincidence was (see III.4) that Kronecker's axiomatization of the (in his work) implicit group concept immediately prompted that concept's elaboration. Thus Gauss's theory of forms represents more than just another instance of implicit group-theoretic thinking.

It is commonly agreed [238, p. 45] that a genuine theory of forms originated in 1773, when J.-L. Lagrange published the first part of his "Recherches d'Arithmétique" [448]. With number-theoretic ends in mind, Lagrange systematically applied transformations of variables to binary quadratic forms. Basing his inquiry on the work of others as well as on his own special results, Lagrange set himself the comprehensive objective of finding all the integers representable by the form $Bt^2 + Ctu + Du^2$ with given coefficients B, C, D [448, p. 695]. One of his principal results was, in modern terms, the complete proof that the number of non-equivalent classes of forms with fixed determinant is finite—a result that became the starting point of Gauss's theory of "composition of forms."

At this point (and again later, in connection with algebraic invariant theory; see III.1) we require another result of Lagrange and a related methodological approach of the utmost importance. Lagrange applies (I use his symbolism) to the indeterminates y, z of the form $py^2 + 2qyz + rz^2$, where p, q, and r are positive or negative integers, a linear transformation

$$y = Ms + Nx, \qquad z = ms + nx,$$

with arbitrary M, N, m, and n, and finds that, under certain conditions, the negative discriminant $a = pr - q^2$ of the initial form is equal to that of the transformed form:

* In effect, these substitutions yield the form

$$Ps^2 + 2Qsx + Rx^2$$

with

$$P = pM^2 + 2qMm + rm^2$$

$$Q = pMN + q(Mn + Nm) + rmn$$

$$R = pN^2 + 2qNn + rn^2,$$

and all one need do is see whether one can determine numbers M, N, m, n so that

$$PR - Q^2 = a \dots .$$

[448, pp. 723–724]

* En effet ces substitutions donneront une transformée de cette forme

$$Ps^2 + 2Qsx + Rx^2,$$

(* continued on p. 57)

Lagrange finds that the required relation is

$$PR - Q^2 = (pr - q^2)(Mn - Nm)^2,$$

so that the discriminants are equal if $(Mn - Nm)^2 = 1$. In the language of the algebraic invariant theory of Boole, Cayley, and Sylvester (developed about seventy years later), the discriminant of a binary quadratic form is an "invariant" under this substitution. Incidentally, this result of Lagrange also occupies a central position in the part of Legendre's *Essai sur la théorie des nombres* (1798) dealing with quadratic forms.

But despite the work of Lagrange and Legendre, the consistent development and systematization of the theory of quadratic forms is due to Gauss. The theory of composition of forms is presented in part 5, "Von den Formen und unbestimmten Gleichungen zweiten Grades," of the *Disquisitiones arithmeticae*, in particular in articles 223–225 and 234–244.

Gauss considers (article 153) the general binary quadratic form

$$ax^2 + 2bxy + cy^2 \qquad (a, b, c \text{ integers}),$$

and denotes it by (a, b, c). Following Gauss's usage in part 5, I shall refer to binary quadratic forms simply as "forms." It turns out that, from the formal algebraic point of view, Gauss's use of $2b$ as the coefficient of the middle term of the form—as against Legendre's b—is most advantageous. The quantity $D = b^2 - ac$, which Gauss called the determinant of the form and which we call its discriminant, plays a crucial role in the theory.

The behavior of forms under transformations (Article 157) is decisive for the composition of forms. Gauss speaks of transformation of forms and substitution of indeterminates as follows:

* If a form F with indeterminates x, y can be transformed into a form F' with indeterminates x', y' by means of substitutions

$$x = \alpha x' + \beta y', \qquad y = \gamma x' + \delta y',$$

wehre $\alpha, \beta, \gamma, \delta$ are integers, then we say that the former *contains* the latter or that the latter *is contained* in the former. [237, p. 115]

(* continued from p. 56)
dans laquelle on aura

$$P = pM^2 + 2qMm + rm^2$$
$$Q = pMN + q(Mn + Nm) + rmn$$
$$R = pN^2 + 2qNn + rn^2$$

et il ne s'agira que de voir si l'on peut déterminer les nombres M, N, m, n en sorte que l'on ait

$$PR - Q^2 = a, \ldots.$$

(* see p. 58)

The connection between the discriminants D of $F = ax^2 + 2bxy + cy^2$ and D' of $F' = a'x'^2 + 2b'x'y' + c'y'^2$ is given by the formula $D' = D(\alpha\delta - \beta\gamma)^2$.

A number M is said to be "represented" by a form $ax^2 + 2bxy + cy^2$ if there exist integers m, n such that $M = am^2 + 2bmn + cn^2$. The number-theoretic aim of part 5 is to investigate all possible representations of numbers by means of forms:

* Given a form, find out whether a given number can be represented by it and obtain all such representations. [237, p. 117]

We need not here concern ourselves with Gauss's problem. To solve it, Gauss surveys all possible forms. He orders the forms by equalitiy of discriminant and then further classifies them in terms of what we would call an equivalence relation. Thus let F and F' be two forms such that F contains F'. Then $D' = D(\alpha\delta - \beta\gamma)^2$. Hence D divides D' and both have the same sign.

** If, in addition, the form F' can be transformed by a similar substitution into the form F, that is, if F' is contained in F as well as F in F', then the determinants [that is, discriminants (Translator)] of the forms are equal and $(\alpha\delta - \beta\gamma)^2 = 1$. Then we say that the forms are *equivalent*. [237, p. 115]

Gauss calls a substitution of the above type proper or improper according as $\alpha\delta - \beta\gamma > 0$ or $\alpha\delta - \beta\gamma < 0$ and speaks of proper ("eigentlich") and improper ("uneigentlich") containment of forms. Finally, Gauss calls two forms properly or improperly equivalent according as $\alpha\delta - \beta\gamma = 1$ or -1. In a footnote, Gauss provides for the inclusion of forms with discriminant 0 but states that the equation $(\alpha\delta - \beta\gamma)^2 = 1$ "must not be extended to that case" ("nicht auf diesen Fall ausgedehnt werden darf").

It is clear that equality of the discriminants of two forms does not imply their equivalence. Since this is so, it is proper to ask about the number of classes of inequivalent forms with the same discriminant D. After a great deal of preliminary work, Gauss finally proves that the number of such classes is finite, but he is forced

---

(* see p. 57)

Wenn eine Form F, deren Unbestimmten x, y sind, in eine andere Form F', deren Unbestimmten x', y' sind, durch Substitutionen von der Form

$$x = \alpha x' + \beta y', \qquad y = \gamma x' + \delta y',$$

in denen α, β, γ, δ ganze Zahlen sind, übergeführt werden kann, so werden wir sagen, dass die erstere die letztere **enthalte** oder dass die letztere unter der ersteren **enthalten** sei.

* Wenn irgend eine Form gegeben ist, so soll man finden, ob eine gegebene Zahl durch sie dargestellt werden kann, und soll alle Darstellungen angeben.

** Wenn daher überdies die Form F' durch eine ähnliche Substitution in die Form F verwandelt werden kann, d. h. wenn sowohl F' unter F als auch F unter F' enthalten ist, werden die Determinanten der Formen einander gleich und $(\alpha\delta - \beta\gamma)^2 = 1$ sein. In diesem Falle nennen wir die Formen äquivalent.

to discuss several cases of values of D. Before that, in articles 158 and 159, he had proved all the properties of the division of forms into classes induced by his definition of equivalence.

In articles 223ff. Gauss resumes the discussion of forms on a deeper level.

* We showed earlier (Articles 175, 195, 211) that, given a (positive or negative) integer D, it is possible to find a finite number of forms $F, F', F'', \ldots$ with determinant D such that every form is directly equivalent to just one of these forms. Using these forms we can subdivide the (infinitely many) forms with determinant D into classes by making one class out of the forms equivalent to the form F, another out of the forms equivalent to the form F', and so on.

Any member of a class of forms with given determinant D can serve as a *representative* form (representative) of that class. [237, pp. 212–213]

After partitioning the forms into classes of equivalent forms, and classes of forms into "orders"—an investigation without relevance to the present context—Gauss defines in article 235 a composition rule for forms, and so broaches "a very important subject hitherto not dealt with by anyone." ("einen sehr wichtigen, bisher noch von Niemand berührten Gegenstand") [237, p. 229].

** When the form

$$AX^2 + 2BXY + CY^2 = F$$

goes over into the product of the forms

$$ax^2 + 2bxy + cy^2 = f, \qquad a'x'^2 + 2b'x'y' + c'y'^2 = f'$$

under a substitution of the form

$$X = pxx' + p'xy' + p''yx' + p'''yy'$$

$$Y = qxx' + q'xy' + q''yx' + q'''yy',$$

..., then we shall simply say that F is *transformable* into ff'; if this transformation is such that the six numbers

$$pq' - qp',\ pq'' - qp'',\ pq''' - qp''',\ p'q'' - q'p'',\ p'q''' - q'p''',\ p''q''' - q''p'''$$

have no common divisor, then we shall say that the form F is composed of the forms f and f'. [237, pp. 231–232]

---

* Schon oben (Artikel 175, 195, 211) haben wir gezeigt, dass, wenn irgend eine ganze (sei es positive, sei es negative) Zahl D gegeben ist, eine endliche Anzahl von Formen $F, F', F'', \ldots$ mit der Determinante D von der Beschaffenheit angegeben werden kann, dass jede beliebige Form mit der Determinante D irgend einer von jenen und nur einer einzigen eigentlich äquivalent ist. Somit können sämtliche Formen mit der Determinante D (deren Anzahl unendlich gross ist) nach jenen Formen in Klassen geteilt werden, indem man nämlich aus der Gesamtheit aller Formen, welche der Form F eigentlich äquivalent sind, die erste Klasse, aus den Formen, welche der Form F' eigentlich äquivalent sind, die zweite Klasse, u.s.w. bildet.

Aus den einzelnen Klassen der Formen mit der gegebenen Determinante D kann irgend eine Form ausgewählt und gleichsam als **repräsentierende** Form (Repräsentant) der ganzen Klasse betrachtet werden.

(** see p. 60)

At this very point—in a footnote—Gauss states that the transformation $F = ff'$ is in a sense commutative; interchanging f and f' produces just certain permutations of the coefficients p and q, respectively.

* With this notation, we must pay attention to the order of the coefficients p, p' ... as well as the forms f, f'. However, it is easy to see that if we interchange the order of the forms f, f', then we need only interchange the coefficients p', q' and p'', q'' respectively, leaving each of the remaining coefficients in its place. [237, p. 231]

In the next article (No. 235) Gauss proves that the reverse procedure, that is, the construction of the composed form F from the component forms f and f', is always possible. He states this result as a problem:

** Find the form composed of two given forms whose determinants [i.e., discriminants (Translator)] are equal or, failing that, are in the ratio of two squares. [237, p. 238]

The fundamental theorem on the composition rule, which, among other things, provides the link with the above partition into classes, is proved in article 239. This theorem states that if the forms f and g are (directly) equivalent, and the same is true of f' and g', then the forms $F = f \cdot f'$ and $G = g \cdot g'$ are also (directly) equivalent. (The existence and constructibility of the forms F and G are guaranteed by article 236.) In other words, the result of the composition of forms is independent

(** see p. 59)
Wenn die Form

$$AX^2 + 2BXY + CY^2 = F$$

übergeht in das Product zweier Formen:

$$ax^2 + 2bxy + cy^2 = f, \qquad a'x'^2 + 2b'x'y' + c'y'^2 = f'$$

durch eine Substitution von der Form

$$X = pxx' + p'xy' + p''yx' + p'''yy'$$
$$Y = qxx' + q'xy' + q''yx' + q'''yy',$$

..., so werden wir einfach sagen, die Form F sei **transformierbar** in ff'; ist diese Transformation überdies so beschaffen, dass die sechs Zahlen

$$pq' - qp', pq'' - qp'', pq''' - qp''', p'q'' - q'p'', p'q''' - q'p''', p''q''' - q''p''',$$

keinen gemeinschaftlichen Teiler haben, so werden wir die Form F aus den Formen f, f' **zusammengesetzt (componiert)** nennen.

* Bei dieser Bezeichnung hat man also auf die Reihenfolge sowohl der Coefficienten p, p' ... als auch der Formen f, f' wohl zu achten. Man sieht aber leicht, dass, wenn man die Reihenfolge der Formen f, f' derart ändert, dass die erste zur zweiten wird, die Coefficienten p', q' mit p'', q'' zu vertauschen sind, jeder der übrigen aber an seinem Platze bleibt.

** Wenn zwei Formen gegeben sind, deren Determinanten entweder gleich sind oder wenigstens in dem Verhältnis zweier Quadratzahlen zu einander stehen, so soll man die aus jenen zusammengesetzte Form finden.

of the choices of representatives of the classes of forms:

* If the form G is composed of the forms g and g' in the same way as F is composed of f and f', and if the forms g, g' are directly equivalent to the forms f, f' respectively, then the forms F, G are likewise directly equivalent. [237, p. 243]

This defines composition of classes of forms; the closure of the finite set of classes of forms with repect to this composition is proved by articles 236 and 239. Commutativity had been proved earlier. Associativity of the composition is proved by Gauss in article 240, where the result is stated as follows:

** Let the form F be composed of the forms f and f'; the form $\mathfrak{F}$ of F and f''; the form F' of f and f''; and the form $\mathfrak{F}'$ of F' and f'. Then the forms $\mathfrak{F}$, $\mathfrak{F}'$ are directly equivalent. [237, p. 244]

A few pages later Gauss adds to this result the following explanation:

*** The previous article shows that the order of composition of three forms is immaterial. [237, p. 248]

In modern terms, this proves that the finite set of classes of forms constitutes a (finite) abelian group under the composition rule. The role of the unit element is played by the "principal class" (Hauptklasse), that is, the class of forms equivalent to the "principal form" (Hauptform) $x^2 - Dy^2$.

Later (1869), E. Schering (1833–97) investigated the structure of this abelian group (see [637]). Schering had known Gauss when he was a student in Göttingen and was the first editor of his works; he was deeply familiar with Gauss's ideas.[37] Schering found certain fundamental classes from which all classes of forms could be obtained by multiplication. In group-theoretic terms, Schering found a basis of the abelian group of classes of forms.

4 *Kronecker's Axiomatization of the Implicit Group Concept*

Gauss's theory of cyclotomic polynomials and his composition of forms led directly to the study of algebraic number fields. Especially interesting for the attempts to elaborate the group-theoretic germ of the investigation of algebraic number fields is the contribution of E. E. Kummer (1810–93). It is well known that when the young Kummer attempted to prove Fermat's last theorem, Lejeune-

---

* Wenn die Form G aus den Formen g, g' in derselben Weise zusammengesetzt ist, wie F aus f, f' respective, und die Formen g, g' den Formen f, f' eigentlich äquivalent sind, so werden auch die Formen F, G eigentlich äquivalent sein.

** Wenn aus den Formen f, f' die Form F, aus F, f'' die Form $\mathfrak{F}$, aus f, f'' die Form F' und aus F', f' die Form $\mathfrak{F}'$ zusammengesetzt ist, so sind die Formen $\mathfrak{F}$, $\mathfrak{F}'$ eigentlich äquivalent.

*** aus dem vorigen Artikel geht hervor, dass es gleichgültig ist, in welcher Reihenfolge die drei Formen componiert werden.

Dirichlet called his attention to the fact that, in general, unique factorization does not hold in algebraic number fields. To construct a multiplicative composition of algebraic numbers, Kummer then introduced the concept of an "ideal complex number" (ideale complexe Zahl) [440]. With justified pride, Kummer says that from the theorems connected with this concept

* it follows that the introduction of ideal primes made computation with complex numbers into an analog of computation with integers and real integer primes. [440, p. 323]

Already in this paper of 1845 Kummer realized that

** The whole theory of quadratic forms in two variables can be regarded as the theory of complex numbers of the form $x + y\sqrt{D}$, a fact that then leads necessarily to ideal complex numbers of the same form. But the latter are classified in terms of the ideal factors that are necessary and sufficient to make them into genuine complex numbers of the form $x + y\sqrt{D}$. In this way, in agreement with *Gauss's* classification, the ideal complex numbers explain Gauss's theory.

There is a very close analogy between the general study of the ideal complex numbers and Gauss's very difficult section *De compositione formarum*. Also the main results that *Gauss* ... proved for quadratic forms hold for the composition of general ideal complex numbers. [440, pp. 324–325]

In the same paper Kummer defined an equivalence of ideal complex numbers, and in this way obtained a partition of ideal numbers into classes. In view of the difficulties involved in the investigation of ideal classes, Kummer restricted himself at first to cyclotomic fields. He published the results in 1847 [441].

As early as 1850 Kummer, starting from Lejeune-Dirichlet's analytic method of determining the number of classes of binary quadratic forms with a given determinant, computed [442] the class number h of ideals in the cyclotomic field of pth roots of unity for an odd prime p. Kummer's result states that h can be written as a product of two numbers, called the "first" and "second factor of the class

---

* geht hervor, dass die Rechnung mit complexen Zahlen durch Einführung der idealen Primfactoren ganz dieselbe geworden ist, wie die Rechnung mit den ganzen Zahlen und den ganzzahligen realen Primfactoren.

** Die ganze Theorie der Formen vom zweiten Grade, mit zwei Variablen, kann nämlich als Theorie der complexen Zahlen von der Form $x + y\sqrt{D}$ aufgefasst werden, und führt dann nothwendig zu idealen complexen Zahlen derselben Art. Diese classificiren sich aber eben so nach den idealen Multiplicatoren, welche nöthig und hinreichend sind, um sie zu wirklichen complexen Zahlen von der Form $x + y\sqrt{D}$ zu machen. Mit der ***Gauss***ischen Classification übereinstimmend, erschliessen diese so den wahren Grund derselben.

Die allgemeine Untersuchung über die idealen complexen Zahlen hat die grösste Analogie mit dem bei ***Gauss*** sehr schwierig behandelten Abschnitte: ***De compositione formarum***, und die Hauptresultate, welche ***Gauss*** für die quadratischen Formen ... bewiesen hat, finden auch für die Zusammensetzung der allgemeinen idealen complexen Zahlen Statt.

number" ("erster" bzw. "zweiter Faktor der Klassenzahl"), respectively. Kummer investigated the divisibility of these factors by 2, and in 1870 he gave a lecture on these matters in the Berlin Academy [443].

At this point Kronecker entered the picture. His lecture, entitled "Auseinandersetzung einiger Eigenschaften der Klassenzahl idealer complexer Zahlen" [432], followed Kummer's lecture,[38] and cited it directly:

* One of the main theoretical results of the paper just presented is the theorem that the second factor of the class number of ideal numbers constructed form the λth roots of unity is divisible by *two* only if the first factor is also divisible by *two*. When my friend *Kummer* communicated this theorem to me some time ago ... I tried, first of all, to deduce the properties of both factors of the class number contained in that theorem directly from their definition or, at least, without using the expansions of the two factors employed by *Kummer* in his proof. [432, p. 273]

Kronecker's attempt to prove Kummer's theorem by purely arithmetical means, and to extend it to arbitrary number fields, was not entirely successful. It stemmed from his general attitude to questions concerning the foundations of mathematics. It is well known that Kronecker aimed at total arithmetization. He wished to "set aside" all extensions of the number concept beyond the natural numbers and, in particular, to banish irrational numbers from arithmetic.[39]

In the realm of algebraic number fields, arithmetization meant the development of an arithmetical theory of algebraic magnitudes and the reduction of algebra to number theory, that is, the study of "rational magnitudes" (rationale Grössen), "entire magnitudes" (ganze Grössen), and of "divisibility" (Teilbarkeit) in the broadest sense of the term. In 1881, on the occasion of the fiftieth anniversary of the awarding of a doctoral degree to Kummer, Kronecker dedicated to his "teacher and friend Kummer" ("Lehrer und Freunde Kummer") a paper [435] containing such a theory. In the introduction Kronecker states that since 1853 he had "emphasized the arithmetical viewpoints in algebra" in his publications as well as in his university lectures, and had "frequently applied arithmetical methods to algebraic problems" ("in gedruckten Publicationen wie in meinen Universitätsvorlesungen die arithmetischen Gesichtspunkte in der Algebra besonders hervorgehoben und auch vielfach die arithmetischen Methoden auf einzelne algebraische Fragen angewendet") [435, p. 245]. In his paper "Auseinander-

* Eines der hauptsächlichsten theoretischen Resultate in der soeben vorgetragenen Abhandlung ist der Satz, dass der zweite Factor der Klassenzahl idealer aus λ^{ten} Wurzeln der Einheit gebildeter Zahlen nur dann durch *Zwei* theilbar sein kann, wenn auch der erste Factor durch *Zwei* theilbar ist. Als mir mein Freund *Kummer* vor einiger Zeit diesen Satz mittheilte ..., bemühte ich mich zuvörderst, die in dem Satze enthaltenen Eigenschaften der beiden Factoren der Klassenzahl unmittelbar aus deren Definition herzuleiten, oder wenigstens ohne, wie es in dem *Kummer*'schen Beweise geschieht, die entwickelten Ausdrücke der beiden Factoren zu benutzen.

setzung einiger Eigenschaften der Klassenzahl idealer complexer Zahlen" [432], Kronecker tried to generalize Kummer's result using an arithmetical—that is, in this case, a number-theoretic—approach. Inevitably, this meant a return to Gauss's *Disquisitiones arithmeticae*, and counteracted what Kronecker considered an alien methodology introduced by Lejeune-Dirichlet. To quote Kronecker:

* In articles 305 and 306 of the *Disquisitiones arithmeticae*, *Gauss* ordered the different classes of quadratic forms by means of the theory of composition, and Herr *Schering* has recently advanced the subject in a paper ... devoted to the relevant construction of fundamental classes.[40] The very simple principles underlying *Gauss's* method are applied not only in the context indicated but also frequently elsewhere—even in the elementary parts of number theory. This shows, and it is otherwise easy to see, that these principles belong to a more general and more abstract realm of ideas. It is therefore proper to free their development from all inessential restrictions, thus making it unnecessary to repeat the same argument when applying it in different cases. This advantage appears in the development itself. Also, when stated with all admissible generality, the presentation gains in simplicity and, since only the truly essential features are thrown into relief, in transparency. [432, pp. 274–275]

This opening passage from §1 of Kronecker's paper shows the author's sure sense of the advantages of a firmly based formal position. Here are presented, with complete clarity, the foundations of a viewpoint that is part of the essence of modern algebra. This is immediately followed by laws governing an abstract operation on a set of undefined "elements" (Elemente) and forming a complete axiom system for a finite abelian group. The laws are closure of the finite set under the operation; associativity and commutativity of the operation; and uniqueness of the inverse of an element under this operation whenever such an inverse exists. It is well known that, under the assumption of the finiteness of the number of elements, these axioms imply the existence of a unit element and of an inverse for each element.

---

* In den Artikeln 305 und 306 der ‚*Disquisitiones arithmeticae*' hat *Gauss* eine Anordnung der verschiedenen Klassen quadratischer Formen auf die Theorie der Composition gegründet und Hr. *Schering* hat neuerdings der weiteren Ausführung dieses Gegenstandes eine Arbeit gewidmet, welche ... eine sachgemässe Aufstellung von ‚Fundamentalklassen' zum Zwecke hat. Die überaus einfachen Principien, auf denen die *Gauss*'sche Methode beruht, finden nicht blos an der bezeichneten Stelle, sondern auch sonst vielfach und zwar schon in den elementarsten Theilen der Zahlentheorie Anwendung. Dieser Umstand deutet darauf hin, und es ist leicht sich davon zu überzeugen, dass die erwähnten Principien einer allgemeineren, abstrakteren Ideensphäre angehören. Deshalb erscheint es angemessen die Entwickelung derselben von allen unwesentlichen Beschränkungen zu befreien, sodass man alsdann einer Wiederholung derselben Schlussweise in den verschiedenen Fällen des Gebrauchs überhoben wird. Dieser Vortheil kommt sogar schon bei der Entwickelung selbst zur Geltung und die Darstellung gewinnt dadurch, wenn sie in der zulässig allgemeinsten Weise gegeben wird, zugleich an Einfachheit und durch das deutliche Hervortreten des allein Wesentlichen auch an Uebersichtlichkeit.

* Let

$$\theta', \theta'', \theta''', \ldots$$

be finitely many elements such that with any two of them we can associate a third by means of a definite procedure. Thus if $\mathfrak{f}$ denotes the procedure and θ' and θ'' are two (possibly equal) elements, then there exists a θ''' equal to $\mathfrak{f}(\theta', \theta'')$.
Furthermore, we require that

$$f(\theta', \theta'') = \mathfrak{f}(\theta'', \theta')$$

$$\mathfrak{f}(\theta', f(\theta'', \theta''')) = \mathfrak{f}(\mathfrak{f}(\theta'\theta''), \theta''')$$

and if θ'' is different from θ''', then

$\mathfrak{f}(\theta', \theta'')$ is different from $\mathfrak{f}(\theta', \theta''')$.

Once this is assumed we can replace the operation $\mathfrak{f}(\theta', \theta'')$ by multiplication $\theta' \cdot \theta''$ provided that instead of equality we employ equivalence.[41] Thus using the usual equivalence symbol "$\sim$" we define the equivalence

$$\theta' \cdot \theta'' \sim \theta'''$$

by means of the equation

$$\mathfrak{f}(\theta', \theta'') = \theta'''.$$

[432, pp. 275–276]

From the finiteness of the set of elements, whose number he denotes by n, Kronecker deduces the same consequences that Euler and Gauss deduced from particular realizations.

---

* Es seien

$$\theta', \theta'', \theta''', \ldots$$

Elemente in endlicher Anzahl und so beschaffen, dass sich aus je zweien derselben mittels eines bestimmten Verfahrens ein drittes ableiten lässt. Demnach soll, wenn das Resultat dieses Verfahrens durch $\mathfrak{f}$ angedeutet wird, für zwei beliebige Elemente θ' und θ'', welche auch mit einander identisch sein können, ein θ''' existieren, welches gleich: $\mathfrak{f}(\theta', \theta'')$ ist. Ueberdies soll:

$$\mathfrak{f}(\theta', \theta'') = \mathfrak{f}(\theta'', \theta')$$

$$\mathfrak{f}(\theta', \mathfrak{f}(\theta'', \theta''')) = \mathfrak{f}(\mathfrak{f}(\theta', \theta''), \theta''')$$

und aber, sobald θ'' und θ''' von einander verschieden sind, auch

$\mathfrak{f}(\theta', \theta'')$ nicht identisch mit $\mathfrak{f}(\theta', \theta''')$

sein. Dies vorausgesetzt, kann die mit $\mathfrak{f}(\theta', \theta'')$ angedeutete Operation durch die Multiplikation der Elemente $\theta'\,\theta''$ ersetzt werden, wenn man dabei an Stelle der vollkommenen Gleichheit eine blosse Aequivalenz einführt. – Macht man von dem üblichen Aequivalenzzeichen: $\sim$ Gebrauch, so wird hiernach die Aequivalenz:

$$\theta' \cdot \theta'' \sim \theta'''$$

durch die Gleichung

$$\mathfrak{f}(\theta', \theta'') = \theta'''$$

definirt.

* I. Certain powers of an element θ are equivalent to the unit element. The exponents of these powers are multiples of one of them and—as I shall say—θ belongs to it.[42]

II. If a certain θ belongs to the exponent ν, then there exist elements belonging to every divisor of ν.

III. If the elements θ' and θ'' belong, respectively, to the relatively prime exponents ϱ and σ, then the product $\theta' \cdot \theta''$ belongs to the exponent $\varrho\sigma$.

IV. If n_1 is the smallest number divisible by all the exponents to which the n elements θ belong, then there are also elements that belong to n_1 itself. In fact, if $p^\alpha q^\beta r^\gamma$ ... is the prime factorization of n, then, by II., there exist elements θ' belonging to p^α, elements θ'' belonging to q^β, elements θ''' belonging to r^γ, and so on. Then, by III., $p^\alpha \cdot q^\beta \cdot r^\gamma$ ... belongs to n_1. [432, p. 276]

This abstract investigation yields a "basis theorem" (Basissatz)—in group-theoretic terms, the basis theorem for finite abelian groups. Specifically, one obtains

** a "fundamental system" of elements $\theta_1, \theta_2, \theta_3, \ldots$, such that the expression

$$\theta_1^{h_1}\theta_2^{h_2}\theta_3^{h_3} \cdots \qquad (h_i = 1, 2, 3, \ldots, n_i)$$

represents each element θ, up to equivalence, just once. The numbers $n_1, n_2, n_3, \ldots$, to which, respectively, $\theta_1, \theta_2, \theta_3, \ldots$ belong, are such that each is divisible by its successor, the product $n_1 n_2 n_3 \cdots$ is equal to the totality n of elements θ, and thus n has no prime factors other than those in n_1. [432, p. 278]

Once the preliminary abstract work has been done, the elements are given concrete interpretations. Then the advantages of the formal standpoint are obvious.

---

* I. Unter den verschiedenen Potenzen eines Elementes θ giebt es stets solche, die der Einheit äquivalent sind. Die Exponenten aller dieser Potenzen sind ganze Vielfache eines derselben, zu welchem – wie ich mich ausdrücken werde – das betreffende θ gehört.

II. Gehört irgend ein θ zum Exponenten ν, so gehören auch zu jedem Theiler von ν gewisse Elemente θ.

III. Wenn in beiden Exponenten ϱ und σ, zu denen resp. die Elemente θ' und θ'' gehören, relative Primzahlen sind, so gehört das Produkt $\theta' \cdot \theta''$ zu Exponenten $\varrho\sigma$.

IV. Ist n_1 die kleinste Zahl, welche die sämmtlichen Exponenten als Theiler enthält, zu denen die n Elemente θ gehören, so giebt es auch Elemente, welche zu n_1 selbst gehören. Denn, wenn n_1 in seine Primfaktoren zerlegt gleich: $p^\alpha q^\beta r^\gamma$... ist, so giebt es nach II. Elemente θ', die zu p^α, ferner Elemente θ'', die zu q^β, Elemente θ''', die zu r^γ etc. gehören, und das Produkt: $\theta' \cdot \theta'' \cdot \theta'''$ ... gehört alsdann nach III. zu: $p^\alpha \cdot q^\beta \cdot r^\gamma$ ... d. h. zu n_1.

** ein ‚Fundamentalsystem' von Elementen: $\theta_1, \theta_2, \theta_3, \ldots$, welches die Eigenschaft hat, dass der Ausdruck:

$$\theta_1^{h_1}\theta_2^{h_2}\theta_3^{h_3} \ldots \qquad (h_i = 1, 2, 3, \ldots n_i)$$

im Sinne der Aequivalenz sämtliche Elemente θ und zwar jedes nur ein Mal darstellt. Dabei sind die Zahlen $n_1, n_2, n_3, \ldots$, zu denen resp. $\theta_1, \theta_2, \theta_3, \ldots$ gehören, so beschaffen, dass jede derselben durch jede folgende theilbar ist, das Produkt: $n_1 n_2 n_3$ ... ist gleich der mit n bezeichneten Anzahl sämmtlicher Elemente θ, und diese Zahl n enthält demnach keine anderen Primfactoren als diejenigen, welche auch in n_1 enthalten sind.

* If the elements θ are thought of as a system of inequivalent ideal numbers or inequivalent composable arithmetical forms, then the above representation of each element θ as a product of powers of selected elements $\theta_1, \theta_2, \theta_3, \ldots$ is identical with that in the above-mentioned paper of Herr *Schering*. [432, p. 278]

In this way Kronecker proved a fundamental theorem connecting ideal classes in arbitrary algebraic number fields and their multiplicative representation: The ideal classes of a field always include m classes $A_1, A_2, \ldots, A_m$, with the property that every ideal class A can be uniquely represented as a product

$$A = A_1^{q_1} A_2^{q_2} \cdots A_m^{q_m}$$

where $0 \leqslant q_i \leqslant h_i - 1$, $i = 1, 2, \ldots, m$, and h, the number of ideal classes in the field, equals the product $h = h_1 h_2 \cdots h_m$.

Kronecker's general and abstract theorem can be interpreted as the basis theorem for finite abelian groups. Strangely enough, Kronecker did not interpret his theorem in group-theoretic terms or apply it in group theory, although one knows that he had been familiar with Galois theory since the midfifties and that he himself had made fundamental contributions to field theory (see II.4). The application of his theorem to group theory was made only in 1882, by Kronecker's student and direct follower E. Netto (1846–1919) in his *Substitutionentheorie* [541]. The fact that by the end of the nineteenth century the basis theorem for (finite) abelian groups was deduced without using the concept of a permutation group indicates that, by that time, the evolution of the abstract group-theoretic viewpoint had been completed.

The historical developments that revolutionized the content of the group concept will be treated in detail in part III. But first we must describe the basic problem that for the first time gave rise, explicitly, to the group concept: the problem of solvability of algebraic equations.

* Wenn man unter den Elementen θ ein System von nicht äquivalenten idealen Zahlen oder ein System von nicht äquivalenten zusammensetzbaren arithmetischen Formen versteht, so fällt die hier entwickelte Darstellung sämmtlicher Elemente θ durch ein Product von Potenzen ausgewählter Elemente $\theta_1, \theta_2, \theta_3, \ldots$ vollständig mit derjenigen zusammen, welche sich in der oben erwähnten Abhandlung des Hrn. *Schering* angegebenen findet.

II Evolution of the Concept of a Group as a Permutation Group

1 Discovery of the Connection between the Theory of Solvability of Algebraic Equations and the Theory of Permutations[43]

1 Lagrange and the Theory of Solvability of Algebraic Equations

With his extensive "Réflexions sur la théorie algébrique des équations" [447] of 1770–71, Joseph-Louis Lagrange (1736–1813) initiated a far-reaching development. In particular, this work shows how difficult it was to abandon the traditional patterns of thought of contemporary algebra in favor of a radically different kind of algebraic thinking—more specifically, to effect the transition from the computation of the roots of an equation to the study of their structure.

Lagrange tried to determine *why* the solutions of cubic and quartic equations, given by Sc. del Ferro (1465?–1526), N. Tartaglia (1500?–1557), G. Cardano (1501–76), L. Ferrari (1522–1565) and, later, J. Hudde (1628–1704), E. W. von Tschirnhaus (1651–1708), L. Euler (1707–83), E. Bezout (1730–83), and others, worked. He arrived at the conclusion that this *kind* of solution was *bound* to fail in the case of (general) equations of higher degree. His analysis yielded the insight that in order to treat equations of higher degree it was necessary to develop an entirely new approach, "a kind of combinational calculus"—to use Lagrange's term for the vague anticipations of the subsequent permutation-theoretic and then group-based theory of solvability of algebraic equations.

In the introduction to the "Réflexions," Lagrange makes a few historical remarks,[44] subsequently elaborated in the text, and describes his aim as follows:

* In this paper I propose to examine the different methods found thus far for the algebraic solution of equations, to reduce them to general principles,[45] and to show *a priori* why these methods work for degrees three and four and fail for higher degrees. [447, p. 206]

This statement is taken from the introduction. In the summary Lagrange says,

** Here, if I am not mistaken, are the true principles for the solution of equations and for the best analysis of its conduct. As one can see, everything reduces to a kind of combinatorial calculus by which one finds *a priori* the results to be expected. It may seem reasonable to apply these principles to equations of the fifth and higher degrees, of which the solution is at present unknown,

---

* Je me propose dans ce Mémoire d'examiner les différentes méthodes que l'on a trouvées jusqu'à présent pour la résolution algébrique des équations, de les réduire à des principes généreaux, et de faire voir *à priori* pourquoi: des méthodes réussissent pour le troisième et le quatrième degré, et sont en défaut pour les degrés ultérieurs.
** see p. 72

but such application requires too large a number of examinations and combinations whose success is very much in doubt. Thus, for the time being, we must give up this work. We hope to be able to return to it at another time. At this point we shall be satisfied with having set forth the foundations of a theory which appears to us to be new and general. [447, p. 403]

Although Lagrange did not reach the decisive insight that algebraic equations of degree higher than four are in principle unsolvable by radicals—an insight arrived at by Gauss a short time afterward[46]—and although he did not return in later papers on the theory of equations[47] to his promised study based on the "combinatorial calculus," he did draft, in the "Réflexions," "the foundations of a new and general theory."

Of greatest interest to us are the parts of the "Réflexions" that foreshadow, in germinal form, the future permutation-theoretic treatment of algebraic equations.

In the first section Lagrange analyzes the solution of the general cubic. He begins with the equation

$$x^3 + mx^2 + nx + p = 0$$

and uses the familiar procedure to reduce it to the form

$$x^3 + nx + p = 0.$$

In Lagrange's words, "the most natural method" for dealing with this equation is to use the substitution due to J. Hudde, who put the root x equal to the sum of two indeterminates: $x = y + z$. Lagrange makes this substitution, restricts z—as Hudde had done—by the condition $3yz + n = 0$, and obtains the resolvent (reduite[48])

$$y^6 + py^3 - n^3/27 = 0.$$

This reduces the solution of a cubic to the solution of a quadratic linked to a pure cubic. Of course, Lagrange shows that the six roots of the resolvent coincide in pairs. We are thus led to Cardano's formulas for the roots of a cubic.

Next Lagrange analyzes the procedures for the solution of a cubic. He denotes

(** see p. 71)

Voilà, si je ne me trompe, les vrais principes de la résolution des équations et l'analyse la plus propre à y conduire; tout se réduit, comme on voit, à une espèce de calcul des combinaisons, par lequel on trouve *à priori* les résultats auxquels on doit s'attendre. Il serait à propos d'en faire l'application aux équations du cinquième degré et des degrés supérieurs, dont la résolution est jusqu'à présent inconnue; mais cette application demande un trop grand nombre de recherches et des combinaisons, dont le succès est encore d'ailleurs fort douteux, pour que nous puissions quant à présent nous livrer à ce travail; nous espérons cependant pouvoir y revenir dans un autre temps, et nous nous contenterons ici d'avoir posé les fondements d'une théorie qui nous paraît nouvelle et générale.

the roots of the cubic $x^3 + mx^2 + nx + p = 0$ by x', x'', x''', and the three third roots of unity by $1, \alpha, \beta = \alpha^2$. The result of his analysis is an assertion about the different forms of the roots of the resolvents associated with the different solution procedures.

* Our analysis shows that these methods are basically the same, since they consist in finding the resolvents whose roots are represented quite generally by $x' + \alpha x'' + \alpha^2 x'''$ or by $(x' + \alpha x'' + \alpha^2 x''')^3$, or, equivalently, by quantities proportional to these. If the root of the resolvent is of the form $x' + \alpha x'' + \alpha^2 x'''$, then the resolvent is of degree six but can be solved like a quadratic because it contains only the third and sixth powers of the unknown. ... In the case where the root of the resolvent is $(x' + \alpha x'' + \alpha^2 x''')^3$, the resolvent is necessarily a quadratic. [447, p. 243]

Thus the solution of a cubic reduces to the investigation of the expression $R = (x' + \alpha x'' + \alpha^2 x''')$ whose outstanding ("très remarquable") property is that it takes on just two different values under the six permutations (permutations) of the roots x', x'', and x'''. Since the properties of the third roots of unity play an important role in this investigation, it is natural for the first section to conclude with remarks about roots of unity in general.

The second section is devoted to analyzing the solutions of the quartic given by Ferrari, Tschirnhaus, Bezout, Euler, and others. Lagrange shows that all of these procedures lead to resolvents whose roots are three-valued functions of the roots $x', x'', x''', x^{\text{iv}}$ of the given quartic, for example $x'x'' + x'''x^{\text{iv}}$ or $x' + x'' - x''' - x^{\text{iv}}$. Such functions can hardly be said to be transparent, and so the proofs of their three valuedness are consistently complicated—all the more because Lagrange must do without the calculus of permutations. His method of proof relies on the gradual restriction of the possible 24 values. This is particularly clear in article 43 of the "Réflexions" where it is said that the function is not changed if we permute x' and x'', which means that the function takes on at most 12 rather than 24 values. Similarly, the function is not changed if we permute x''' and x^{iv}, which leaves just six possible values. The function also remains unchanged if we permute x' and x''' and, simultaneously, x'' and x^{iv}. Since these new permutations are also independent of the previous ones ("comme ces nouvelles permutations sont aussi indépendantes des précédentes") [447, p. 289], we are left with just three possible functional values. While it is true that the function does not change if we permute x' and x^{iv}

* Par l'analyse que nous venons d'en faire il est visible que ces méthodes reviennent toutes au même pour le fond, puisqu'elles consistent à trouver des réduites dont les racines soient représentées en général par $x' + \alpha x'' + \alpha^2 x'''$, ou par $(x' + \alpha x'' + \alpha^2 x''')^3$, ou bien, ce qui est la même chose, par des quantités proportionelles à celles-ci. Dans le cas où la racine de la *réduite* est $x' + \alpha x'' + \alpha^2 x'''$, cette réduite est du sixième degré, résoluble à la manière du second parce qu'elle ne renferme que la troisième et la sixième puissance de l'inconnue ... Dans l'autre cas, où la racine de la *réduite* est $(x' + \alpha x'' + \alpha^2 x''')^3$, cette réduite ne peut être que du second degré ....

as well as x' and x''', this permutation can be ignored since it is subsumed under the preceding ones ("mais il ne dois pas entrer en ligne de compte, parce qu'il est déjà renfermé dans les précédents") [447, p. 289].

At the end of the second section Lagrange summarizes the results of his analysis of the (general) quartic as follows:

* We have not only brought together these different methods and shown their connection and interdependence, but we have also—and this is the main point—given the reason, *a priori*, why some of them lead to resolvents of degree three and others to resolvents of degree six that can be reduced to degree three. This is due, in general, to the fact that the roots of these resolvents are functions of the quantities $x', x'', x''', x^{\text{iv}}$, which, like the function $x'x'' + x'''x^{\text{iv}}$, take on just three values under all the permutations of these four quantities; or which, like the function $x' + x'' - x''' - x^{\text{iv}}$, take on six values that are pairwise equal but have opposite signs; or which take on six values that can be separated into three pairs such that the sum or product of the values in each pair is not changed by the permutations of the quantities $x', x'', x''', x^{\text{iv}}$. ... The general solution of quartics depends uniquely on the existence of such functions. [447, p. 304]

The third section, "De la résolution des équations du cinquième degré et de degrés ultérieurs," extends these arguments to equations of higher degree. Guided by the earlier insight into the dependence of solvability on the existence of resolvents, Lagrange investigates the likelihood of finding the corresponding resolvents.

Lagrange thinks that the (general) equation of degree higher than four may be solvable or, as he puts it, that nothing proves its unsolvability.[49] But he knows of only two approaches, the one used by Tschirnhaus in 1683 [699] and the one used by Euler and Bezout in 1765 [192, 41], that in his opinion, offer some prospect of success. Their presumed advantage is that the solution of the cubic and quartic by these methods rests on uniform principles, and that these may therefore yield a precedent ("préjugé") for equations of higher degree. Even for equations of degrees five and six, however, the necessary computations are too extensive and complicated—"so long and complicated that they can discourage the most intrepid calculators" ("si longs et si compliqués, que le plus intrépide calculateur peur en

* Non-seulement nous avons rapproché ces méthodes les unes des autres, et montré leur liaison et leur dépendance mutuelle; nous avons encore, ce qui était le point principal, donné la raison *a priori* pourquoi elles conduisent, les unes à des *réduites* du troisième degré, les autres à des réduites du sixième, mais qui peuvent s'abaisser au troisième; et l'on a dû voir que cela vient en général de ce que les racines de ces *réduites* sont des fonctions des quantités $x', x'', x''', x^{\text{IV}}$, telles, qu'en faisant toutes les permutations possibles entre ces quatre quantités, elles ne peuvent recevoir que trois valeurs différentes comme la fonction $x'x'' + x'''x^{\text{IV}}$, ou six valeurs, mais deux à deux égales et de signes contraires, comme la fonction $x' + x'' - x''' - x^{\text{IV}}$, ou bien six valeurs telles, qu'en les partageant en trois couples et prenant la somme ou le produit des valeurs de chaque couple, ces trois sommes ou ces trois produits soient toujours les mêmes, quelque permutation qu'on fasse entre les quantités $x', x'', x''', x^{\text{IV}}$, ... C'est uniquement de l'existence de telles fonctions que dépend la résolution générale des équations du quatrième degré.

être rebuté") [477, p. 305]. In his analysis of the methods under investigation, Lagrange finally comes to the following conclusion.

* We conclude from these reflections that it is very doubtful that the methods we have just discussed can give a complete solution of equations of degree five and, all the more so, of higher degrees. This uncertainty, combined with the length of the calculations required by these methods, will discourage in advance all those who might be tempted to use them to solve one of the most celebrated and important problems of algebra. [447, p. 307]

In view of the "doubtfulness" of the methods of Tschirnhaus, Euler, and Bezout, Lagrange thinks that he must develop criteria that make it possible to decide—as he puts it—*a priori* whether these methods lead to the desired goal. In this he follows the approach developed in analyzing the solution methods for the cubic and quartic.

** It would be highly desirable to be able to judge *a priori* the success that one could expect from the application of these methods to degrees higher than the fourth. We shall try to accomplish this by an analysis similar to that used until now in connection with the known methods for the solution of equations of degrees three and four. [447, p. 307]

Before discussing the starting points selected by Lagrange for a new and progressive theory of equations, we must comment on the relevant contribution of A. Vandermonde (1735–96).

2 *Vandermonde and the Theory of Solvability of Algebraic Equations*

The papers of Bezout and Euler analyzed by Lagrange served as the starting point for another significant paper on the theory of solvability of algebraic equations, likewise published in 1771. This was the "Mémoire sur la résolution des équations" [701] of A. Vandermonde, submitted to the Paris Academy of Sciences in 1770. Since it originated before the "Réflexions" of Lagrange, and was obviously of great importance, Kronecker characterized it as the beginning of "the new revival of algebra" ("des neuen Aufschwungs der Algebra").[50]

Vandermonde became acquainted with the "Réflexions" of Lagrange and the *Meditationes algebraicae* [711] of Waring (1734–93), which was devoted

* Il résulte de ces réflexions qu'il est très-douteux que les méthodes dont nous venons de parier puissent donner la résolution complète des équations du cinquième degré, et à plus forte raison celle des degrés supérieurs; et cette incertitude, jointe à la longueur des calculs que ces méthodes exigent, doit rebuter d'avance tous ceux qui pourraient être tentés d'en faire usage pour résoudre un des Problèmes les plus célèbres et les plus important de l'Algèbre.

** Il serait donc fort à souhaiter que l'on pût juger *à priori* du succès que l'on peut se promettre dans l'application de ces méthodes aux degrés supérieurs au quatrième; nous allons tâcher d'en donner les moyens par une analyse semblable à celle dont nous nous sommes servis jusqu'ici à l'égard des méthodes connues pour la résolution des équations du troisième et du quatrième degré.

to similar problems, only between 1770 and the date of publication of his own paper. Lagrange, for his part, learned of Vandermonde's contribution to the solvability theory of algebraic equations only after the appearance of his "Réflexions," and later expressed his appreciation of Vandermonde's procedure, which he termed "more direct" than his own.[51] While Vandermonde's "direct" method for determining the nature of the roots of algebraic equations shows remarkable acuteness, it is less significant than Lagrange's procedure from a group-theoretic point of view. This is borne out by a comparison of the two papers.

Vandermonde attributes some of the difficulties associated with the solvability of higher-degree algebraic equations to the nature of the computing schemes used up to that time. He states his own idea as follows:

* One requires the simplest general values that can simultaneously satisfy an equation of a given degree. [701, p. 365]

As stated, his objective is admittedly very vague, but it becomes more understandable through his examples involving quadratic and cubic equations: One is to find algebraic functions (valeurs générales) of the sum of the roots, of the sum of their products taken two at a time, and of the sum of their products taken three at a time (that is, algebraic functions of the elementary symmetric functions of the roots) that take on as values *each* of the roots. For example, in the case of the quadratic

$$x^2 - (a + b)\,x + ab = (x - a)\,(x - b)$$

with roots a and b, we must find a function satisfying simultaneously the conditions $a = \text{function}[(a + b), ab]$ and $b = \text{function}[(a + b), ab]$, that is a two-valued (ambigue) function. The function given by Vandermonde is

$$\tfrac{1}{2}\left(a + b + \sqrt{(a + b)^2 - 4ab}\right).$$

Such functions are necessarily unchanged by a permutation of the roots.

Vandermonde achieved his objective for quadratic, cubic, and quartic equations, but his attempts to treat equations of higher degree in terms of his formulation of the problem landed him in tremendous computational difficulties. For example, he used square and cube roots to form a number of six-, eight-, and nine-valued functions only to find that they could not be used to form a root of the equation of the corresponding degree. The only positive outcome of these computations, for all their multitude of abbreviations and symbols, was the proof of the existence of a resolvent of degree six for the equation of degree five and of resolvents of

* On demande les valeurs générales les plus simples, qui puissent satisfaire concurrement à une Equation d'un degré déterminé.

degree ten and fifteen for the equation of degree six. Vandermonde went on to find the resolvents of his resolvents, but discontinued these efforts when he realized that he was not making significant progress.

We see that Vandermonde's "direct" method also led him to the connection between solvability of equations and combinatorics. Both Vandermonde and Lagrange established a connection between the degree of the resolvent and the number of values taken on by functions of the roots of the original equation when the roots are permuted in all possible ways. But in this area Lagrange made significantly greater progress than Vandermonde.

3 *Beginnings of a Group-Theoretic Treatment of Algebraic Equations in the Work of Lagrange*

It is only in the fourth section of his book ("Conclusion des réflexions précédentes, avec quelques remarques générales sur la transformation des équations, et sur leur réduction ou abaissement à un moindre degree") that Lagrange spells out with full clarity his fundamental insights into algebraic equations of all degrees. The roots of a resolvent θ of an equation—obtained by means of his rule—are functions of the roots of that equation. The degree of the resolvent is the same as the number m of different values that can be taken on by a root of that resolvent when the roots of the given equation are permuted in all possible ways. More specifically, if $f_i(x', x'', \ldots, x^{(\mu)})$ is a root of the resolvent θ, which is a function of the μ roots of the given equation of degree μ, and f_i takes on just m values $f_1, f_2, \ldots, f_m$ under the $\mu!$ permutations of the μ roots, then the Lagrange resolvent is of the form

$$\theta = \prod_{i=1}^{m} (t - f_i),$$

and m divides $\mu!$.

Strictly speaking, Lagrange proves an even more general theorem. He denotes by f a rational function, by $f[(x)]$ a rational function of x, and by $f[(x), (y)]$ a rational function of x and y. If a rational function of x and y is unchanged when x and y are permuted, that is, if $f[(x), (y)] = f[(y), (x)]$, then Lagrange writes for brevity $f[(x, y)]$. Lagrange's theorem is to the effect that

* In general, the function

$$f[(x', x'', x''', \ldots, x^{(\alpha)}), (x^{(\alpha+1)}, x^{(\alpha+2)}, x^{(\alpha+3)}, \ldots, x^{(\alpha+\beta)}), (x^{(\alpha+\beta+1)}, \ldots) \ldots]$$

yields an equation $\theta = 0$, where the quantity θ is a power with exponent

$$1 \cdot 2 \cdot 3 \cdots \alpha \times 1 \cdot 2 \cdot 3 \cdots \beta \times 1 \cdot 2 \cdot 3 \cdots.$$

---

* see p. 78

Hence this equation can be reduced to one of degree

$$\frac{1 \cdot 2 \cdot 3 \cdot 4 \cdots \mu}{1 \cdot 2 \cdot 3 \cdots \alpha \times 1 \cdot 2 \cdot 3 \cdots \beta \times 1 \cdot 2 \cdot 3 \cdots}.$$

[447, pp. 371–372]

This connection between the degree of the resolvent and the number of values of a rational function leads Lagrange in the fourth section to consider the number of values that can be taken on by a rational function of μ variables. His conclusion is that the number in question is always a divisor of $\mu!$. Lagrange proves this only for two-valued functions and says that an analogous proof works in other cases.

Lagrange saw the "metaphysics" of the procedures for the solution of algebraic equations by radicals in this connection between the degree of the resolvent and the valuedness of rational functions [447, p. 357]. His discovery was the starting point of the subsequent development due to Ruffini, Abel, Cauchy, and Galois. But Lagrange himself never returned to this subject despite an explicit announcement in the "Réflexions" that he would do so.

For all that, it is remarkable to see in Lagrange's work the germ, in admittedly rudimentary form, of the group concept.[52] In spite of the fact that Lagrange always speaks of "permutations" (permutations) without ever using a calculus of permutations, the idea of a group arises quite naturally from his question about the number of values of rational functions of μ variables. In the fourth section, in an apparently casual way, Lagrange introduces the concept of "similar functions":

* Given a number of functions of the same quantities, we call those of them which simultaneously vary or stay the same under the same permutations of their component quantities *similar*. Thus such functions can be designated in analogous ways. [477, pp. 358–359]

(* see p. 77)

En général, la fonction

$$f[(x', x'', x''', \ldots, x^{(\alpha)}), (x^{(\alpha+1)}, x^{(\alpha+2)}, x^{(\alpha+3)}, \ldots, x^{(\alpha+\beta)}), (x^{(\alpha+\beta+1)}, \ldots), \ldots]$$

donnera une équation $\Theta = 0$, où la quantité Θ sera une puissance qui aura pour exposant le nombre

$$1 \cdot 2 \cdot 3 \ldots \alpha \times 1 \cdot 2 \cdot 3 \ldots \beta \times 1 \cdot 2 \cdot 3 \ldots,$$

de manière que cette équation s'abaissera au degré

$$\frac{1 \cdot 2 \cdot 3 \cdot 4 \ldots \mu}{1 \cdot 2 \cdot 3 \ldots \alpha \times 1 \cdot 2 \cdot 3 \ldots \beta \times 1 \cdot 2 \cdot 3 \ldots}.$$

* Si l'on a plusieurs fonctions des mêmes quantités, on appelera fonctions *semblables* celles qui varient en même temps ou demeurent les mêmes lorsqu'on y fait les mêmes permutations entre les quantités dont elles sont composées, de manière qu'elles puissent être désignées d'une manière analogue.

As examples of "similar" functions Lagrange mentions the symmetric functions, which even in his time had long been known.

Of course, Lagrange concentrates on the reproduction of the function and not, to use a later term, on the totality of permutations admitted by the function. Therefore he does not attempt to single out those permutations of a certain order that leave the function invariant or simultaneously change it. Lagrange does not consider the composition of permutations and certainly not the question of the closedness of a system of permutations.

In spite of these and other limitations, and in spite of the fact that the "Réflexions" marked, so to speak, the beginning of the road, we must rate very highly its effect on progress toward the solvability theory of algebraic equations. Quite apart from the formal advantages of this detailed paper—its transparency, the clear formulation of its aim, and the consistent realization of that aim—it set forth fundamental questions and approaches. Lagrange showed that all known solutions of equations of degree not exceeding four defined *rational* functions of the roots that take on a small number of different values as a result of all the permutations of the roots. What remained unclear, and was thus revealed as a problem, was whether this was an accident or a manifestation of deeper relations, as yet unknown, whose discovery—of this Lagrange was sure—would provide starting points for the treatment of equations of higher degree. Also, Lagrange's insight lacked a clear conceptual formulation of the term "rationally determined"—in other words, an elaboration of the fundamental concept of the domain of rationality (field) of an equation. Nevertheless, Lagrange provided the assurance that one could expect progress if one worked in the area set out by him, that is, if one systematically investigated functions of the roots of an equation with a view to determining the number of distinct values that they take on as a result of permutations of the roots. (To be sure, given the lack of knowledge of the fundamental features of the later calculus of permutations, the proper method of investigation was hardly discernible.) Finally, in view of all the unsuccessful efforts to solve higher-degree equations by radicals, Lagrange's conclusion that equations of degree higher than four cannot be solved by radicals by means of the known methods for the solution of equations of degree up to four, strengthened the conviction of mathematicians that no such solution is possible. But as yet this conjecture, though slowly gaining ground, could not be proved.

On the face of it, Lagrange's "Réflexions" was intended as the conclusion of a prior development. But its content shows that it began a new phase of the theory of solvability of algebraic equations, a phase that culminated in Galois theory.[53] The path to that outcome proved long indeed and posed great conceptual difficulties.

4 Ruffini's Proof of the Unsolvability of the General Quintic

At first, Lagrange's results were accepted very slowly. Thus no edition till the fifth (1797) of A.-Cl. Clairaut's *Eléments d'algèbre* [157] contained any excerpts from the "Réflexions"—and that edition had only a few. The next significant step in the theory of solvability of algebraic equations, decidedly group-theoretic in nature, was taken about 30 years after the publication of the "Réflexions," by the many-sided Italian P. Ruffini (1765–1822).

Ruffini was the first to assert—and to attempt a proof of—the unsolvability by radicals of algebraic equations of degree higher than four. His two-volume textbook of algebra, *Teoria generale delle equazioni* [592], published in 1799, bore in its subtitle "in cui si dimostrata impossibile la soluzione algebraica delle equazioni generali di grado superiore al quarto" ("in which is demonstrated the impossibility of the algebraic solution of general equations of degree greater than four"). Incidentally, at that time the term "algebraic solution" meant a solution by radicals.

In later years there appeared six versions of Ruffini's proof. They were, in part, a reaction to the objections raised in 1804 by G.-F. Malfatti (1731–1807), who, as a representative of the older generation, could not come to terms with the idea of the nonexistence of a conventional "solution." The details are as follows. Ruffini's textbook appeared in 1799. In 1802 he published a paper, dated 21 October 1801 and bearing the title "Della soluzione delle equazioni algebraiche determinate particolari di grado superiore al quarto" [593], in which he investigated the conditions under which the algebraic solution of an equation could be reduced to the solution of equations of lower degree. In 1802, in response to these two works of Ruffini, his friend P. Abbati (1768–1842) wrote a letter, published in 1803 [1], in which he accepted the proof of nonsolvability and suggested a number of improvements. This in turn prompted Ruffini to publish, in 1803, the paper "Della insolubilità delle equazioni algebraiche generali di grado superiore al quarto" [594]. It was dated 18 December 1802, and in it Ruffini, using Abbati's suggestions, once more gave a detailed presentation of his proof. Malfatti stated a number of objections to Ruffini's proofs in a paper [501] published in 1804. These objections were based on his earlier—and by this time thoroughly discredited—attempt [500] to prove the solvability of the quintic. Malfatti's objections gave Ruffini a reason for yet another detailed presentation of his proof. This appeared in 1805 under the title "Riposta di Paolo Ruffini ai dubbi propostigli dal socio Gian-Francesco Malfatti sopra la insolubilità algebraica dell' equazioni di grado superiore a quarto" [596]. Malfatti died in 1807 without having responded to Ruffini's 1805 paper. Ruffini returned to the topic of unsolvability of equations of higher degree in a number of subsequent papers,[54] including a minor contribution, published in 1806 [597], a closely related paper of 1807 [598] dealing with general properties of (algebraic) functions, and, finally, the paper "Riflessioni intorno alla soluzione delle equazioni algebraiche generali" [599], appearing in 1813, in which Ruffini uses much of Lagrange's "Réflexions" in stating his initial position.

The above-mentioned papers of Ruffini, the supplements of Abbati, and the objections of Malfatti have been analyzed by H. Burkhardt in his thorough and detailed work "Die Anfänge der Gruppentheorie und Paolo Ruffini" [69] of 1892, in which special attention is given to permutation-theoretic group theory.[55] Apart from minor points, later works dealing with this topic have completely confirmed Burkhardt's analysis.[56] For this reason, as well as for the sake of overall continuity, I shall devote relatively little space to the discussion of Ruffini's work.

Ruffini went beyond the mere recognition that there exists a connection between the solvability of algebraic equations and permutations. In his work the theory of permutations no longer plays the role of a mere computing device but is rather a structural component of solvability theory.

He begins with Lagrange's program of systematic investigation of permutations from the point of view of their effect on algebraic functions of n variables (a permutation can fix or change such a function). In addition to many particular references to Lagrange, Ruffini states explicitly in the introduction to his *Teoria generale delle equazioni* that

> * the immortal LAGRANGE, with his marvellous reflections about equations, published in the Proceedings of the Berlin Academy, has supplied the foundation for my proof. Therefore, it was only natural that I should, for the sake of greater clarity, preface my proof with a summary of these reflections. [590, p. 3]

Already his *Teoria* of 1799 [592], which was intended as a survey of the theory of equations, shows significant progress relative to its predecessors. To be sure, Ruffini treats (algebraic) equations of degrees two through four in the conventional manner; but he prefaces his treatment of fifth- and sixth-degree equations in chapter 13 ("Riflessioni intorno alla soluzione generale delle equazioni") with a classification of permutations. Apart from a different terminology, the modern permutation-theoretic group concept appears in that chapter with full clarity. Not only is Ruffini—like Lagrange—concerned with permutations that leave

---

* L'immortale de LA GRANGE con le sublimi sue riflessioni intorno alle equazioni, inserite negli Atti dell'Accademia di Berlino, ha somministrato il fondamento alla mia dimostrazione: conveniva dunque premettere a questa, per la maggiore sua intelligenza, un ristretto di simili riflessioni.

a rational function of the roots invariant; he deals also with the totality of such permutations and their properties. He calls such a set of permutations "permutazione."[57] Thus Ruffini's "permutazione" coincides with what Cauchy later called a "system of conjugate substitutions" and Galois called a (permutation) "group."

It is remarkable that Ruffini used consistently the fact that his "permutazione" is closed, both in connection with the composition of the permutations reproducing the function, and even for the purpose of classifying groups in connection with the question of generators of the "permutazione." Ruffini speaks of "permutazione semplice" and "permutazione composta" according as the "permutazione" is generated by, respectively, one or more than one "sostituzione" (permutation). The groups of the first kind are conceptually identical with (finite) cyclic groups. Ruffini divides the "composite" groups into three classes:

* The composite "permutazioni" can be divided into three classes. The first class consists of those in which none of the roots existing in any one of the composing "permutazioni" can pass through the roots, or in the place occupied by the roots of another one.

The second class contains those in which the roots of one composing "permutazione" can pass all to occupy the place already first occupied by the roots of another one, but without it happening that the roots of the first one intermingle with those of the second.

And, finally, the third class contains the "permutazioni" in which the roots of one of the composing ones can pass to mix among the roots of the other. [590, p. 163]

Thus the "composite" groups of the first kind are the intransitive groups, those of the second kind are the transitive imprimitive groups, and those of the third kind are the transitive primitive groups.

There is another difference between Ruffini's terminology and modern usage. While unimportant in itself, this was significant in the sense that—as pointed out already by Burkhardt [69, p. 134]—it involved an inaccuracy that barred Ruffini from discovering the fundamental property of normal subgroups. If a function of n quantities takes on m distinct values under the $n!$ possible permutations and if it is invariant under the permutations of a "permutazione" (group), then—says Ruffini—all of these distinct functional values are left fixed by the same "permutazione." Of course, the truth of the matter is that the different values admit the permutations of a class of conjugate groups.

---

* *Le permutazioni composte distinguonsi in tre generi. Il genere primo comprende quelle nelle quali niuna delle radici esistenti in una qualunque delle permutazioni componenti può passare tra le radici, o nel luogo occupato dalle radici di un'altra.*

Il secondo abbraccia quelle nelle quali le radici di una permutazioni componente possono passare tutte ad occupare il luogo già prima occupato dalle radici di un'altra, senza però che le radici della prima si framischino a quelle della seconda.

Il terzo finalmente comprende le permutazioni, in cui le radici di una delle componenti possono passare a mescolarsi tra le radici di un'altra.

Ruffini calls the number p of permutations that leave invariant a given function of n roots of an equation the "degree of equality" (grado di uguaglianza). Thus this concept coincides with that of the "order" of a (permutation) group. In this connection Ruffini quotes Lagrange's result to the effect that p divides $n!$.

Ruffini goes on to determine the value of p for all groups that occur in connection with five quantities, the roots of a quintic, and thus for the simple and composite groups of all three kinds. In terms of content, this investigation comes down to an (almost)[58] complete determination of all subgroups of the symmetric group S_5. In this way Ruffini obtains the main result, formulated in article 275, to the effect that p can never be 15, 30, or 40, that is, that there are no (rational) functions of five quantities that take on 8, 4, or 3 different values when these quantities are permuted in all possible ways. On the basis of this correctly proved group-theoretic result, Ruffini gives a proof—with some gaps (see II.3)—of the unsolvability of the general quintic in radicals. This "proof" that the general equations of degrees six and higher are unsolvable is based on the same principles and has additional gaps in the group-theoretic part as well.

Ruffini's 1802 paper [593] treats two other important problems. At one point he considers a function of the roots with the property that a permutation changes its form but (because of the special values of the roots) not its numerical value. The paper's central concern, however, is a proof that the group of permutations associated with an irreducible equation is transitive. In this he went far beyond Lagrange.

Abbati starts from the *Teoria*. According to Burkhardt, his letter [1] contains the following simplification of Ruffini's work:

* The simplification consists in the replacement of the many separate investigations of all possible subgroups by results of a more general nature. For example, he uses the following simple argument to prove the main result that a rational function of five quantities cannot take on just three or four values ...: A group of index less than 5 must contain all cyclic permutations of five letters and at least one of the six permutations of three letters other than the identity, that is, a transposition or a cyclic permutation of three letters. But Ruffini had proved that a cyclic permutation of five letters and a transposition generate the symmetric group, and a cyclic permutation of five letters and a cyclic permutation of three letters generate the alternating group. The proof that there exists no eight-valued function of five magnitudes is simplified in a similar way. Abbati's generalization, omitted by Ruffini, is the explicit proof that there are no three- or four-valued functions of more than five magnitudes. [69, pp. 140–141]

---

* Die Vereinfachung besteht vor allem darin, dass er die grosse Anzahl von Einzeluntersuchungen aller möglichen Untergruppen, die bei Ruffini einen so breiten Raum einnahm, durch Entwickelungen allgemeineren Charakters ersetzt. So beweist er den Hauptsatz, dass eine rationale Function von fünf Grössen nicht drei oder vier Werte haben kann, durch folgende einfache Überlegung ...: Eine Gruppe, deren Index kleiner als 5 sein soll, muss jede cyclische Permutation von (continuation see p. 84)

In his paper of 1803 [594], Ruffini used Abbati's simplifications to reformulate the group-theoretic part of his proof.

This is all that I shall say here about the high points of Ruffini's work in group theory. The reader will find more detailed explanations of the content of Ruffini's other papers in Burkhardt [69]. But, in view of the gaps in his proof, I shall return to Ruffini in my discussion of N. H. Abel (II.3), where I point out those elements in Abel's proof of the unsolvability by radicals of the quintic that were essential for the development of the permutation-theoretic group concept.

The meritorious and detailed work of Burkhardt [69] puts Ruffini's and Abbati's contributions to the development of (permutation-theoretic) group theory in proper perspective. In particular, it defines Ruffini's place vis-à-vis Cauchy, who previously had been given credit for most of the results cited above. Obviously, Ruffini's flawed presentation was in large part responsible for the slight impact of his results. Cauchy began his group-theoretic work in 1815, and at that time he made a vague reference to Ruffini as his predecessor in this area. On the other hand, in his comprehensive account, given in 1844 [108, pp. 151–252], Cauchy mentioned neither Ruffini nor other authors.

(* continuation from p. 83)
fünf Buchstaben enthalten; ferner aber von den sechs Substitutionen, welche irgend drei Buchstaben unter sich versetzen, ausser der Identität mindestens noch eine, also entweder eine Transposition oder eine cyclische Permutation von drei Buchstaben. Dass aber eine cyclische Permutation von fünf Buchstaben mit einer Transposition die symmetrische, mit einer cyclischen Permutation von drei Buchstaben die alternirende Gruppe erzeugt, hatte bereits Ruffini dargethan. Auch der Beweis, dass keine achtwertigen Functionen von fünf Grössen existiren, wird in ähnlicher Weise vereinfacht. Die Verallgemeinerung Abbati's besteht darin, dass er ausdrücklich zeigt – was Ruffini unterlassen hatte –, dass auch von mehr als fünf Grössen keine drei- oder vierwertigen Functionen existiren.

Klügel's mathematical dictionary of 1808 contains neither the cue word "permutation" nor the cue word "substitution" (in the sense of permutation). Permutations are discussed solely in the context of a few articles such as the one on "combinatorics" and the one honoring the contributions of K. F. Hindenburg (1741–1828), head of the German school of combinatorics. While it is true that at the end of eighteenth century permutation theory boasted significant results (such as E. Waring's theorem of 1762 on the representability of a symmetric function as a polynomial with integer coefficients in the elementary symmetric functions), and that these results were available through systematic expositions of combinatorics (such as Hindenburg's treatise [284] of 1781), this theory was "lifted out" of combinatorics only by the contributions of Lagrange, Vandermonde, and Ruffini. At the beginning of the nineteenth century, in view of the applications of permutations to the theory of equations and the determination of the valuedness of rational functions of the roots, permutation theory became an independent field of mathematical research.

This happened before the group-theoretic formulation of the problem of solvability of algebraic equations by radicals by Abel and Galois brought about a deeper understanding of the connection between the theory of equations and permutation theory. Viewed in historic terms, the fact that permutation theory acquired independence so quickly led to a temporary weakening of its connection with the theory of algebraic equations, whose development paralleled (in date) that of permutation theory (see II.3). In these areas the transition to permutation-theoretic group theory during the thirties and fourties occurred without mutual conceptual interaction.

It was Galois who obtained what was in principle the group-theoretic formulation of the problem of solvability of algebraic equations. In the fifties and sixties Galois's work brought about the reunification and amalgamation of two areas that until then had appeared entirely distinct. This was a deliberate merger, based on a fresh recognition of the depth of their connection. The perfected permutation theory supplied an advanced calculus for the amalgamation and computational elaboration of Galois's ideas and, in turn, the group-theoretic formulation of the solvability theory by Abel and Galois lent support to the tendency, which arose independently in permutation theory, to study permutation groups. The shift from independence to amalgamation of these two directions of investigation, with their genetic links and their connections (not at all obvious, yet extremely close) of subject matter, provided a basis for the eventual independence of permutation-theoretic group theory and for the rise of all aspects of permutation theory to a central position.

This anticipation of the outcome of the historical analysis of the period from about 1815 to about 1870 allows us to study separately the development of permutation theory and the progress toward the solution of algebraic equations with a view to reconstructing their actual historical evolution. Only then can we study the amalgamation of these two widely separated areas—a process that led to the eventual refinement and independence of group theory on a permutation-theoretic basis.

1 The Systematic Development of Permutation Theory by Cauchy

Undoubtedly, A.-L. Cauchy (1789–1857) played a central role in the shaping of permutation theory. It is true that Ruffini's work anticipated many of Cauchy's pioneering papers, but the fact remains that it was Cauchy's activity that consolidated permutation theory as an independent mathematical discipline. For a variety of reasons, Cauchy's papers in this area were destined to have an incomparably greater influence on the mathematical public than Ruffini's. Cauchy systematized the results of his predecessors and used the outcome as a basis for a multitude of decisive general theorems. Finally, after some notational indecision, he elaborated a consistent terminology, which was adopted, by and large, by his successors.

Cauchy's activity in the area of permutation theory can be divided into two distinct periods. The first began in 1815 with a paper [98] (written in 1812) on the number of negative roots of algebraic equations of arbitrary degree. In 1815 Cauchy published two papers on permutation theory, respectively entitled "Sur les Nombre des Valeurs qu'une Fonction peut acquérir, lorsqu'on y permute de toutes les manières possibles les quantités qu'elle renferme" [99] and "Sur les Fonctions que ne peuvent obtenir que deux valeurs égales et de signes contraires par suite des transpositions opérées entre les variables qu'elles renferment" [100]. The second of these had been submitted to the Academy already in 1812. In 1815, 1820, and 1837 Cauchy published four additional papers [101, 102, 103, 104] on roots of equations. These were traditional in the sense that they did not go beyond pre-Galois ideas. Cauchy's second creative period in the area of permutation theory began in 1844, twelve years after Galois's death and two years before the publication of decisive writings from his *Nachlass*. First to be published (1844) was a comprehensive exposition of permutation theory, *Mémoire sur les arrangements que l'no peut former avec des lettres données* [108], which formed part of the third volume of *Exercises d'analyse et de physique mathématique*. Between 1844 and 1846 Cauchy published a great many[59] memoirs in the Paris *Comptes rendus* that explored the topics delineated in the *Mémoire sur les arrangements* of 1844 and contained excerpts or supplements of various themes in that comprehensive work. Of this series of papers I mention just two of decidedly group-theoretic character: "Mémoire sur un nouveau calcul qui permet de simplifier et d'étendre la théorie des permutations" [122] and "Note sur une théorème fondamental rélatif à deux systémes de substitutions conjuguées" [128]. Cauchy's first papers on permutation theory ([99, 100] of 1815) are noteworthy in the sense that they are devoted not to the presentation of special results but to a fundamental development of the concepts and results whose focus is the valuedness of rational functions of n quantities.

Cauchy had worked out the content of "Sur les Fonctions" [100] in 1812. In a sense, this paper is an analogue of Waring's result on elementary symmetric functions. It deals with functions of n quantities whose absolute values are unchanged by the permutations of the full symmetric group. Cauchy calls such a function "fonction symétrique alterné" [100, p. 30]; he calls a function which does not change at all "fonction symétrique permanente" [100, p. 30]. In the first part of the paper Cauchy presents the then widely known results on symmetric and alternating functions (the terminology developed in that part of the paper is discussed below). In the second part he treats frequently-appearing alternating functions, primarily discriminants and resultants.[60]

"Sur le Nombre" [99], the second of the papers published in 1815, is much shorter than "Sur les Fonctions" [100]. In it Cauchy continues his 1812 study of the number of values taken on by a function of n quantities under the full symmetric group S_n. Cauchy refers to Lagrange, Vandermonde, and Ruffini as follows:

> * I believe that Messrs. LAGRANGE and VANDERMONDE are the first to have considered the number of values taken on by functions of several variables as a result of permutations of these variables. They obtained many interesting theorems on this subject, theorems which appeared in two papers published in Berlin and Paris respectively. Since then, a number of Italian geometers successfully pursued this issue. Here we must single out Mr. *Ruffini*, the results of whose research are found in volume XII of the Memoirs of the Italian Society and in his Theory of Numerical Equations. One of the most remarkable consequences of the work of these geometers ist that it is not always possible to form a function of a given number of letters that takes on a specified number of values. [99, p. 9]

---

* MM. LAGRANGE et VANDERMONDE sont, je crois, les premières qui aient considéré les fonctions de plusieurs variables relativement au nombre de valeurs qu'elles peuvent obtenir, lorsqu'on substitue ces variables à la place les unes des autres. Ils ont donné plusieurs théorèmes intéressans relatifs à ce sujet, dans deux mémoires imprimés en 1771, l'un à Berlin, l'autre à Paris. Depuis ce temps, quelques géomètres italiens se sont occupés avec succès de cette matière, et particulièrement M. *Ruffini*, qui a consigné le résultat de ses recherches dans la tome XII des Mémoires de la Société italienne, et dans sa Théorie des équations numériques. Une des con-
(continuation see p. 88)

Ruffini had proved that it is not possible to form a four-valued function of five quantities.[61] Cauchy went far beyond Ruffini by proving the following theorem:

* *The number of different values of a nonsymmetric function of n quantities cannot be less than the largest odd prime p which divides n.* [99, p. 9][62]

Cauchy's long summary of permutation theory, *Mémoire sur les arrangements* [108], published in 1844, is noteworthy in that it represents permutation theory as an independent theory and completes the transition to a theory of permutation groups.[63]

By an "arrangement" (arrangement) Cauchy means an ordered string of quantities (which may be represented by numbers). A "permutation" (permutation) or "substitution" (substitution) denotes a transition from one arrangement to another.

In §1 of the work devoted to generalities ("Considérations générales"), Cauchy defines arrangements, the permutation notation—for example $\binom{xzy}{xyz}$—the product (produit) of an arrangement and a permutation and the product of two permutations. Two permutations are said to be commutative (permutables entre elles) if their product is independent of the order of the factors ("lorsque leur produit sera indépendant de l'ordre dans lequel se suivront les deux facteurs") ([108], p. 154). He introduces the power notation, puts the zeroth power of a permutation equal to 1 and calls it the unity (l'unité). He calls the lowest positive power of a permutation that is equal to the unit permutation its degree or order (le degré or l'ordre), "order" being the term he prefers. Then he introduces the cycle notation and calls a cycle $(x, y, \ldots, w)$ a cyclic substitution (substitution circulaire). He proves that the order of a 1-cycle is 1, and follows this with the theorem about the order of a permutation that is the product of cycles of different orders. In particular, he defines a regular (régulière) permutation as the product of cycles of equal length (and therefore order).

In §2 ("Extensions des notions adoptées dans le premier paragraphe. Substitutions semblables entre elles"), Cauchy introduces symbolic powers p^{-1}, p^{-2}, ... of permutations, defines the inverse of a permutation, and, finally, defines similar permutations:

(* continuation from p. 87)
séquences les plus remarquables des travaux de ces divers géomètres, est qu'avec un nombre donné de lettres on ne peut pas toujours former une fonction qui ait un nombre déterminé de valeurs.

* *Le nombre des valeurs différentes d'une fonction non symétrique de* n *quantités, ne peut s' abaisser au-dessous du plus grand nombre premier* p *contenu dans* n, *sans devenir égal à* 2.

* Two different substitutions are said to be similar if the cycles of either can be paired off with the cycles of the other in such a way that corresponding cycles have the same number of letters, that is, the same order. [108, p. 165]

This paragraph culminates in the theorem that states that

** *every substitution similar to a given substitution P is the product of three factors, of which the extreme ones are inverse to one another and the middle one is precisely P.* Conversely, *every product of three factors of which the extreme ones are inverse to one another and the middle one is P is a substitution similar to P.* [108, p. 169]

Of the problems in §§3–5 there is one (in §5) that is of special interest to us. Cauchy states that the "symbolic equation" $RP = QR$ connecting two similar permutations P and Q admits of two interpretations, which can be stated as two problems. One is, Given similar permutations P and Q to find R such that $RP = QR$; and the other is, To find all permutations $Q = RPR^{-1}$ similar to P.

In the very next section, "Sur les dérivées d'une ou de plusieurs substitutions, et sur les systèmes de substitution conjuguées," Cauchy proceeds to the theory of permutation groups. His terminology corresponds to our modern concepts of generator and group:

*** Given one or more substitutions involving some or all of the n letters $x, y, z, \ldots$ I call the products of these substitutions, by themselves or by one another, in any order, *derived* substitutions. The given substitutions, together with the derived ones, form what I call a *system of conjugate substitutions.* The order of the system is the number of substitutions in the system, including the substitution with two equal terms, that reduces to unity. [108, p. 183]

It is remarkable that Cauchy defines the concept of a permutation group generated by certain elements—a starting point used later by A. Cayley (see III.4). After pointing to the cyclic group (the term is not used by Cauchy) generated

---

* Deux substitutions distinctes ... seront dites *semblables* entre elles, quand elles offriront le même nombre de facteurs circulaires et le même nombre de lettres dans les facteurs circulaires correspondants, en sorte que les facteurs circulaires, comparés deux à deux, soient de même ordre.
** *toute substitution semblable à P sera le produit de trois facteurs dont les deux extrêmes seront inverses l'un de l'autre, le facteur moyen étant précisement la substitution donnée P.* Réciproquement, *tout produit de trois facteurs dont les deux extrêmes seront deux substitutions inverses l'une de l'autre, le facteur moyen étant la substitution P, sera une substitution semblable à P.*
*** Etant données une ou plusieurs substitutions qui renferment les n lettres $x, y, z, \ldots,$ ou du moins plusieurs d'entre elles, je nommerai substitutions *dérivées* toutes celles que l'on pourra déduire des substitutions données, multipliées une ou plusieurs fois les unes par les autres, ou par elles-mêmes, dans un ordre quelconque; et les substitutions données, jointes aux substitutions dérivées, formeront ce que j'appellerai un *système de substitutions conjuguèes.* L'ordre de ce système sera le nombre total des substitutions qu'il présente, y compris la substitution qui offre deux termes égaux et se réduit à l'unité.

by a single permutation and the symmetric group with $N = 1 \cdot 2 \cdots n$ elements, Cauchy proves the key theorem:

* The order of a system of conjugate substitutions on n variables always divides the number N of arrangements that can be formed from these variables. [108, p. 184]

Cauchy gives the now common proof of this theorem. He then proves that the order of each element divides the order of the system of conjugate permutations.

At the end of this section Cauchy proves that if the generating permutations of a group commute in pairs, then its order is the product of the orders of the generators. In §10 ("Sur les systèmes de substitutions permutables entres eux"), in another context, Cauchy gives a second proof of this theorem (which is stated here in modern terms) using the concept of "commuting groups" and the elementary properties of commuting permutations—specifically, the connections between their cycle representations—deduced in §9.

In §10 Cauchy considers two systems of conjugate permutations with respective orders a and b:

** (1) $1, P_1, P_2, \ldots, P_{a-1}$,

(2) $1, Q_1, Q_2, \ldots, Q_{b-1}$.

We say that the two systems commute if any product of the form

$$P_h Q_k$$

can also be written

$$Q_k P_h. \quad [108, p. 227]$$

The following remarks, bearing on this definition, specifically delimit the normal subgroup property. It is remarkable that this is done without any reference to the already available concept of "décomposition propre" (see II.3) —Galois's contribution to the elaboration of normal subgroups. Cauchy only says,

---

* L'ordre d'un système de substitutions conjuguées relatives à n variables est toujours un diviseur du nombre N des arrangements que l'on peut former avec ces variables.

** (1) $1, P_1, P_2, \ldots, P_{a-1}$

(2) $1, Q_1, Q_2, \ldots, Q_{b-1}$

Nous dirons que les deux systèmes sont *permutables* entre eux, si tout produit de la forme

$$P_h Q_k$$

est en même temps de la forme

$$Q_k P_h.$$

* Furthermore, it may happen that the indices h and k do not change upon transition from the first form to the second, so that

$$P_h Q_k = Q_k P_h;$$

or that the indices h and k change, so that

$$P_h Q_k = Q_{k'} P_{h'},$$

where h', k' are new indices related in some manner to h and k. In the first case, each substitution in (1) commutes with each substitution in (2). On the other hand, in the second case two substitutions P_h and Q_k, in general, will not commute, although the system of substitutions of the form P_h commutes with the system of substitutions of the form Q_k. [108, p. 227]

In view of the implicit group-theoretic treatment of number-theoretic problems (see I.3), which preceded Cauchy's 1844 presentation of permutation theory, and the later development (in the sixties) that emphasized the central position of permutation theory in all of mathematics (see II.5), I wish to make an additional comment on the last two sections of Cauchy's paper [108]. Both sections deal with a kind of anticipation of representation theory, namely, with what Serret and Jordan would later call "analytic representation of substitutions" and with what would occupy much space in Jordan's *Traité des substitutions* of 1870.

The main concepts of that theory, which is equivalent to the study of the additive group of residue classes with respect to an arbitrary modulus when interpreted in number-theoretic terms, are the concepts of "arithmetic substitution" and "geometric substitution." Both concepts are defined by Cauchy in §11 ("Des substitutions arithmétique et des substitutions géométriques"). Here given variables $x, y, z, \ldots$ are represented by indices $0, 1, 2, \ldots, n-1$, and the variables themselves are now denoted

$$(1) \qquad x_0, x_1, x_2, \ldots, x_{n-1}.$$

The variables are to be considered equal if their indices are congruent modulo

---

* Il pourra d'ailleurs arriver, ou que les indices h et k restent invariables dans le passage de la première forme à la seconde, en sorte qu'on ait

$$P_h Q_k = Q_k P_h;$$

ou que les indices h et k varient dans ce passage, en sorte qu'on ait

$$P_h Q_k = Q_{k'} P_{h'},$$

h', k' étant de nouveaux indices, liés d'une certaine manière aux nombres h et k. Dans le premier cas, l'une quelconque des substitutions (1) sera permutable avec l'une quelconque des substitutions (2). Dans le second cas, au contraire, deux substitutions de la forme P_h, Q_k, cesseront d'être généralement permutables entre elles, quoique le système des substitutions de la forme P_h soit permutable avec le système des substitutions de la forme Q_k.

n; Cauchy puts

$$x_l = x_{l+n} = x_{l+2n} = \cdots = x_{l-n} = x_{l-2n} = \cdots$$

for integer l. The addition of a number h to each of the indices in the sequence (1), and the multiplication of each of the indices by a number coprime with n, induce permutations referred to by Cauchy as "rapport arithmétique" and "rapport géométrique," respectively. To quote Cauchy:

* For these reasons, supposing, as above, that the given variables are represented by a single letter and subscripts

$$0, 1, 2, \ldots, n-1,$$

we call the substitution that consists in replacing every term in the sequence (1) by the corresponding term of the sequence $x_h, x_{h+1}, x_{h+2}, \ldots, x_{h+n-1}$ an *arithmetic substitution*, and, on the other hand, we call the substitution that consists in replacing every term in the sequence (1) by the corresponding term of the sequence $x_0, x_r, x_{2r}, \ldots, x_{(n-1)r}$ *a geometric substitution*. [108, p. 233][64]

This remark, and the earlier one that normal subgroups appear in Cauchy's work only in veiled form independent of Galois's "décomposition propre," show that, at the time, the evolution of permutation theory, essentially due to Cauchy, was definitely not inspired directly by the contemporary group-theoretic formulation of the problem of solvability of algebraic equations. This conclusion is reinforced by an investigation of the monographs that preceeded and followed Cauchy's comprehensive presentation [108] of 1844.

In the first of the many[65] monographs on permutation theory published in 1845 and 1846, namely, [109] of 1845, Cauchy explained that he was publishing abstracts of results obtained in his 1844 paper. While the content of [109] is of no interest to us, the following statement is:

** For more than thirty years I have been concerned with the theory of permutations, and especially with the problem of the number of values that can be taken on by functions. Lately, as I shall explain in the next installment, Mr. Bertrand has added a few new theorems to those previously established and to those obtained by myself.[66] But as regards Lagrange's theorem, according to which the number of values taken on by a function of n letters divides the product $1 \cdot 2 \cdots n$, one has until now obtained almost exclusively theorems concerning the impossibility of obtaining

---

* Pour ce motif, en supposant, comme ci-dessus, que les variables données sont représentées par une seule lettre successivement affectée des indices

$$0, 1, 2, \ldots, n-1,$$

nous appellerons *substitution arithmétique* la substitution qui consiste à remplacer chaque terme de la série (1) par le terme correspondant de la série $x_h, x_{h+1}, x_{h+2}, \ldots, x_{h+n-1}$, et nous appellerons, au contraire, *substitution géométrique* la substitution qui consiste à remplacer chaque terme de la série (1) par le terme correspondant de la série $x_0, x_r, x_{2r}, \ldots, x_{(n-1)r}$.
** see p. 93

functions that take on certain prescribed numbers of values. In a new work I have attacked directly the following two questions: 1° What is the number of values that can be assumed by a function of n letters? and 2° How can one effectively form functions that take on the permissible numbers of values? Incidentally, research in this area has led me to new formulas in the theory of sequences that are not without interest. I plan to publish in *Exercises d'Analyse et de Physique mathématique* the results of my work, including all the details that seem to me to be useful. I shall only ask the Academy for permission to publish, in *Compte rendus*, abstracts indicating some of the most remarkable propositions at which I have arrived. [109, pp. 593–594]

This statement by Cauchy clarifies once more the starting point[67] of his investigations in the area of permutation theory. His own assessment is not quite correct, in the sense that Cauchy did not recognize the ramifications of the group-theoretic viewpoint that he had attained;[68] he speaks only of a "Théorie des suites." It should be noted that [109], the first of his series of monographs, leads into the theory of permutation groups, for its definition (p. 605) of a system of conjugate permutations is identical with the definition Cauchy gave in [108] in 1844.

The series of monographs published in 1845–46 marked the end of Cauchy's activity in the area of permutations. In 1846, possibly in response to Cauchy's results, Liouville published essential parts of the *Nachlass* of Galois. Galois's group-theoretic formulation of problem of solvability of algebraic equations provided the study of permutation groups with a deeper context and restored its relevance not as a derivative of permutation theory but as a necessary object of study in the theory of equations.

This historical survey shows that until 1846 the connection between permutation theory and the theory of solvability of algebraic equations was loose and indirect, but that after that date the two theories began to influence one another directly. After 1846, the development of the theory of permutations and permutation

---

(** see p. 92)

Je m'étais déja occupé, il y a plus de trente années, de la théorie des permutations, particulièrement du nombre des valeurs que les fonctions peuvent acquérir; et dernièrement, comme je l'expliquerai plus en détail dans une prochaine séance, M. Bertrand a joint quelques nouveaux théorèmes à ceux qu'on avait précédemment établis, à ceux que j'avais moi-même obtenus. Mais à la proposition de Lagrange, suivant laquelle le nombre des valeurs d'une fonction de n lettres est toujours un diviseur du produit $1.2.3 \ldots n$, on avait jusqu'ici ajouté presque uniquement des théorèmes concernant l'impossibilité d'obtenir des fonctions qui offrent un certain nombre de valeurs. Dans un nouveau travail, j'ai attaqué directement les deux questions qui consistent à savoir: 1° quels sont les nombres de valeurs que peut acquérir une fonction de n lettres; 2° comment on peut effectivement former des fonctions pour lesquelles les nombres de valeurs distinctes soient les nombres trouvés. Mes recherches sur cet objet m'ont d'ailleurs conduit à des formules nouvelles relatives à la théorie des suites, et qui ne sont pas sans intérêt. Je me propose de publier, dans les *Exercises d'Analyse et de Physique mathématique*, les résultats de mon travail avec tous les développements qui me paraîtront utiles; je demanderai seulement à l'Académie la permission d'en insérer des extraits dans le *Compte rendu*, en indiquant quelques-unes des propositions les plus remarquables auxquelles je suis parvenu.

groups was influenced, to an ever growing extent, by the work of Galois (see II.4). Thus we must study next the development that, beginning with the achievements of Lagrange and Ruffini, led to the group-theoretic formulation of the problem of solvability of algebraic equations (see II.3).

2 *History of the Fundamental Concepts of the Theory of Permutations*

The development of the theory of permutations and of permutation groups and the evolution of abstract group theory were accompanied by the formulation of an appropriate terminology. As a kind of appendix to a discussion of Cauchy, to whom we owe so many fundamental concepts in this area of mathematics, I give a short historical survey of the basic concepts of the theory of permutation groups. To some extent, this survey anticipates my later exposition.

The papers most relevant to Cauchy's contribution to the development of permutation-theoretic terminology are "Sur le Nombre des Valeurs qu'une Fonction peut acquérir ..." [99] of 1815, the systematic presentation of permutation theory [108] of 1844, and the paper "Mémoire sur les substitutions permutables entre elles" [113] of 1845.

At first, in 1815, Cauchy called an ordered n-tuple $a_1, a_2, \ldots, a_n$ "une permutation" [99, p. 3]. In 1844 he changed "permutation" to "arrangement" and used "permutation" as well as "substitution" (the latter dating back to 1815 [99, p. 4]) to denote the transition from one ordered n-tuple, or arrangement, to another. Cauchy used his permutation notation—in which the arrangements are written one below the other and both are enclosed in parantheses—for the first time in 1815 [99, p. 10]. At first Cauchy wrote the original arrangement in the top line, but after 1844 he shifted it to the bottom line. Already in 1815 Cauchy denoted a permutation by means of a single letter [99, p. 4] and used "product"—"produit" [99, p. 10]—for the composition of two permutations. The term "identity permutation"—"permutation identique"—appeared in 1815 [99, p. 10], but the symbol 1 was not used until 1844 ([108], p. 155). The term "inverse permutation"—"substitution inverse"—was used only from 1844 [108, p. 165; 113, p. 780]. The notation $S, S^2, S^3, \ldots$ for powers of a permutation appeared already in 1815 [99, p. 11]. In 1815 Cauchy used "degré" for the order of a permutation [99, p. 13], but from 1845 he used both "degré" and "ordre"—the latter, possibly, as a result of Abel's using it in 1826 [6, p. 18]. The use of "substitution circulaire" [100, p. 39] and "permutation circulaire" [99, p. 17] for a cyclic permutation goes back to 1815, but the representation of a permutation as a product of disjoint cycles and the cycle notation $(ab \cdots d)$ appear first in 1844 [108, pp. 157, 159]. The term "similar permutations"—"substitutions semblables"—[108, p. 168] for permutations that have analogous cycle representations, and thus are obtainable from one another by conjugation with a third permutation, dates back to 1844 [108, p. 168]; E. Betti used "derivata" only in 1852 [34, p. 55] and C. Jordan used "transformée par" in 1867 [309, p. 110]. A special word for commutativity of multiplication of permutations was first used by Cauchy in 1844. Cauchy used the term "permutable" [108, p. 154], while Jordan used "échangeable" [809, p. 117].

Cauchy's elaboration of the terminology for the concepts which we now call group, order of a group, index of a group, and subgroup is of special interest. In 1815 Cauchy used no special term for a set[69] of permutations closed under multiplication, but called the number of its elements "diviseur indicatif" [99, p. 7], following Lagrange's theorem that the order of a group on n letters divides $n!$. Thus it is natural that he should have used the word "index" [99, p. 6], in its present sense, just in connection with the symmetric group. We saw that it was only in 1844 [108, p. 183] that Cauchy introduced the term "système des substitutions conjuguées" as a synonym for "group," and, in this connection, the term "ordre" [108, p. 183] for the order of a group. Cauchy had no synonym for our "subgroup"; Galois used "diviseur" [231, p. 58] and this was echoed by the German "Teiler," used, for example, by G. Frobenius and L. Stickelberger [207, p. 220]. The German term "Untergruppe" seems to derive from S. Lie [427, p. 536].

The concept of transitivity goes back to Ruffini; it was reintroduced by Cauchy in 1845 [109, p. 669]. The division of transitive groups into primitive and imprimitive is actually found in the work of Ruffini (see II.1). For the term "imprimitive" in connection with groups, Cauchy used "fonction transitive complexe" in 1845 [109, p. 731], and E. Betti used "gruppo a lettero congiunte" in 1852 [34, p. 37]. The German "imprimitiv" was (first?) used by Netto in his *Substitutionentheorie* of 1882 [54, p. 77]. As for the term "primitive group," Ruffini (see II.1) used "permutazione composta di 3ª specie" and Galois used "groupe primitif" in 1831 [231, p. 58].

3 The Group-Theoretic Formulation of the Problem of Solvability of Algebraic Equations

This chapter discusses the contributions made to the theory of equations by C. F. Gauss (1777–1855), N. H. Abel (1802–1829), and, above all, E. Galois (1811–1832) and the assimilation of Galois's ideas by the succeeding generations of mathematicians.

Gauss's contribution to the solution of algebraic equations has been analyzed in many detailed works; a recent account is contained in the superb *Gauss-Gedenkband* [238][70] of 1957. Gauss's implicit group-theoretic methods relating to the theory of solvability of algebraic equations were discussed in I.3.2. I shall therefore limit myself here to mentioning Gauss's proofs of the fundamental theorem of algebra, his theory of the cyclotomic equation, and his work on what amounted to the preliminaries[71] of the theory of finite fields.

1 *Abel and the Theory of Solvability of Algebraic Equations*

While Lagrange did not rule out, in principle, the solution by radicals of equations of degree higher than four, Gauss did so in 1801.[72] The mastery of this problem involved a chain of ideas stretching from Ruffini to Abel, including the first stage of its group-theoretic formulation. Moreover, in N. H. Abel's papers on the theory of equations—which he started working on in 1823, at the latest—we see with complete clarity the new view of the purpose, scope, and nature of the solvability theory of algebraic equations. In this sense, the essential algebraic component of the impulse first given to that theory by Lagrange reached fruition in the work of Abel.

The work of Lagrange and Gauss provided Abel with a starting point for his solvability theory of algebraic equations. Abel read Lagrange when he was still a high school student [545, p. 70], and studied Gauss's *Disquisitiones arithmeticae* thoroughly as a first-year student at the university of Oslo [545, p. 110]. Next, he made himself familiar with Cauchy's results of 1815 on permutations. At first, however, he had no knowledge of Ruffini's work.

Gauss's treatment of the cyclotomic equation, an equation of arbitrary degree whose solution by radicals is used as an example to this day, made an extremely strong impression on Abel [545, p. 110]. In this connection, Abel was very sharply confronted with the number-theoretic consequences, and with the patterns of implicit group-theoretic thinking embedded within them, of the *Disquisitiones* (see I.3). This is clear from his abortive attempt, in the summer of 1823, to prove

Fermat's last theorem. In connection with the problem of algebraic solvability—to use the contemporary terminology—of the general algebraic equation of arbitrary degree, that is, its solution by radicals, the then student Abel began with the mistaken idea that he had found the solution of the general quintic. After reading Lagrange and Gauss, Abel realized that what was required was a proof of unsolvability. Lagrange was skeptical about the possibility of proving solvability—at least with contemporary methods. Gauss was convinced of the unsolvability but did not prove it. For Abel the proof of unsolvability was a test of his unfolding genius. In 1824, before his European study trip, he published his first proof, "Mémoire sur les équations algébriques, où l'on démontre l'impossibilité de la résolution de l'équation générale du cinquième degré" [5].

It should be noted that Abel's approach to this complex problem, a problem that had been discussed in vain for centuries, was inspired by a clear insight into the new mathematics of his time. Abel found it often necessary—partly because he wished to justify departures from his travel route in reports to Oslo—to make projections of his scientific plans. These personal programs invariably embodied the developmental tendencies of contemporary mathematics.[73]

From the many documents that show Abel consciously preparing the way for the new view of the nature of mathematical method, I wish to quote a passage that is directly related to the solvability theory of algebraic equations. It was set down only in 1828 and formed the introduction to Abel's last paper [12] on algebraic equations solvable by radicals. This paper appeared posthumously in 1839. In it Abel describes the state of the theory of equations. In spite of the contributions of outstanding scientists, including Lagrange, the failure of all efforts suggested that a solution of general (algebraic) equations of degree higher than the fourth is impossible:

* In this paper I shall treat the problem of the algebraic solution of equations in a very general form. The first, and if I am not mistaken the only one, who before me had tried to show the impossibility of the algebraic solution of general equations is the geometer [that is, mathematician (Translator)] *Ruffini.* But his paper is so complicated that it is very difficult to decide the correctness of his reasoning. It seems to me that his reasoning is not always satisfactory. I believe that the proof I gave in the first volume of this journal [this paper was also intended for *Crelles Journal der reinen und angewandten Mathematik,* the first issue of which contained Abel's paper on the unsolvability of the quintic (Wussing)] leaves nothing to be desired as regards correctness, but it is not as simple as it could be. While looking for a solution of a more general problem I came upon another proof, based on the same principles but simpler.

One knows that every algebraic expression can satisfy an equation of higher or lower degree according to the particular nature of the expression. Thus there exists an infinity of particular

* see p. 98

equations that are algebraically solvable. From this there derive in a natural manner the following two problems, whose complete solution comprises all the theory of the algebraic solution of equations.

1. To find all equations of any given degree that are algebraically solvable.
2. To decide whether a given equation is algebraically solvable or not. The object of this paper is the consideration of these two problems.

Although we shall not give their complete solution, we shall nevertheless indicate reliable methods for reaching it. [3, pp. 218–219]

This passage is noteworthy in several ways. First there is a clear demand for existence proofs, a demand that Cauchy made fashionable at just about the same time in analysis and, in particular, in the theory of differential equations.

Then there is Abel's position vis-à-vis Ruffini. Abel learned of Ruffini's ideas for a proof of unsolvability not later than the early summer of 1826, during his stay in Vienna. The source of Abel's information was a summary, by anonymous authors, of Ruffini's main ideas [545, p. 125]. During his stay in Paris, Abel published in volume 6 (1826) of the *Bulletin* of de Férussac an anonymous note[74] [9] in which he presented a detailed discussion of his second paper on the theory of equations, the "Démonstration de l'impossibilité de la résolution algébrique des équations générales qui passent le quatrième degré" [7], published in *Crelles Journal* in the same year. Among other things, Abel here discussed and corrected a decisive gap in Ruffini's argument.

(* see p. 97)

Dans ce mémoire je vais traiter le problème de la résolution algébrique des équations, dans toute sa généralité. Le premier, et, si je ne me trompe, le seul qui avant moi ait cherché à démontrer l'impossibilité de la résolution algébrique des équations générales, est le géomètre *Ruffini*; mais son mémoire est tellement compliqué qu'il est très difficile de juger de la justesse de son raisonnement. Il me paraît que son raisonnement n'est pas toujours satisfaisant. Je crois que la démonstration que j'ai donnée dans le premier cahier de ce journal, ne laisse rien à désirer du côté de la rigueur; mais elle n'a pas toute la simplicité dont elle est susceptible. Je suis parvenu à une autre démonstration, fondée sur les même principes, mais plus simple, en cherchant à résoudre un problème plus général.

On sait que toute expression algébrique peut satisfaire à une équation d'un degré plus cu moins élevé, selon la nature particulière de cette expression. Il y a de cette manière un infinité d'équations particulières qui sont résolubles algébriquement. De là dérivent naturellement les deux problèmes suivans, dont la solution complète comprend toute la théorie de la résolution algébrique des équations, savoir:

1. Trouver toutes les équations d'un degré déterminé quelconque qui soient résolubles algébriquement.
2. Juger si une équation donnée est résoluble algébriquement, ou non.

C'est la considération de ces deux problèmes qui est l'objet de ce mémoire, et quoique nous n'en donnions pas la solution complète, nous indiquerons néanmoins des moyens sûrs pour y parvenir.

The fact is that Ruffini's proof of the unsolvability by radicals of the general equation of degree higher than four contains flaws that Ruffini was not able to eliminate from the various versions of his proof.[75] The main difficulty in Ruffini's proof is due to the fact that he lacked the concept of a domain of rationality, that is, a field.

In the second paper on the theory of equations [7] published in 1826 in the first issue of *Crelles Journal*, Abel closed one of Ruffini's obvious gaps. In terms current at that time, Ruffini assumed without proof that the radicals involved in the solution of an equation can be expressed rationally in terms of its roots. That is why Abel pays special attention to this point in the announcement [9]. He points out that the author of [7]—that is, Abel himself—had proved the theorem that

* *if an algebraic equation is solvable, then one can always write each root so that the component algebraic expressions are expressible in terms of rational functions of the root of the given equation.* [2, p. 10]

Beyond that, until his first paper on the theory of equations [7] Abel's work on the theory of solvability of algebraic equations (and his study of elliptic functions) went hand in hand with efforts to make precise the terms "algebraically solvable" and "rationally expressible." He advanced furthest in these efforts in the introduction, set down in 1828, to the paper "Sur la résolution algébrique des équations" [12], quoted earlier.[76]

The passage quoted above is also noteworthy in that Abel here develops a program for the theory of solvability of algebraic equations. Only a few years later, Galois formulated the solvability problem in group-theoretic terms. This brought him to the second of Abel's problems, and he proceeded to solve that problem in principle. Abel himself worked on the first of his two problems, that is, the problem of constructing[77] all algebraically solvable equations of a given degree. This is where Abel went beyond Lagrange, Ruffini, and Cauchy in the application of permutations in the theory of equations, and, rather naturally, investigated commutative (permutation) groups.

Actually, Abel had indicated earlier, after his arrival in Berlin in the late fall of 1825, that he intended to work in the problem of constructing all equations of a given degree that are solvable by radicals.[78] This is made clear in his paper "Mémoire sur une classe particulière d'équations résoluble algébriquement" [11], dated 29 March 1828 and published in volume 4 of *Crelles Journal* (1829). The paper

* *Si une équation algébrique est résoluble algébriquement, on peut toujours donner à la racine une forme telle, que toutes les expressions algébriques dont elle est composée pourront s'exprimer par des fonctions rationnelles des racines de l'équation proposée.*

begins as follows:

* While the algebraic solution of equations is not possible in general, there are particular equations of all degrees that admit of such solution. Such are, for example, the equations $x^n - 1 = 0$. The solution of these equations is based on certain relations among the roots. I have tried to generalize this method under the assumption that two roots of the given equation are so related that one is rationally expressible in terms of the other. [2, p. 478]

To clarify matters Abel considers the special case when all the roots can be calculated by iteration of the same rational function of one of the roots. Specifically, if x is the root and θx is a rational function of x, then Abel uses the notation $\theta^2 x, \theta^3 x, \ldots$ for functions of the same form as θx ("des fonctions de la même forme que θx, prise deux fois, trois fois, etc.") [2, p. 478], with $\theta^n x = x$. The connection between the roots of the cyclotomic equation is of this type. Abel quotes the example of the cyclotomic equation here as well as elsewhere in the text, and mentions specifically its source—Gauss's *Disquisitiones arithmeticae*. Abel states the main result proved in this paper as follows:

** If the roots of an equation of a certain degree are related so that *all* of them are rationally expressible in terms of one, which we designate as x, and if, furthermore, for any two of the roots θx and $\theta_1 x$ we have

$$\theta\theta_1 x + \theta_1\theta x,$$

then the equation is algebraically solvable. Similarly, if the equation is assumed to be irreducible and its degree is

$$\alpha_1^{\nu_1}\alpha_2^{\nu_2} \cdots \alpha^{\nu_\omega},$$

where $\alpha_1, \alpha_2, \ldots, \alpha_\omega$ are different primes, then one can reduce its solution to that of ν_1 equations of degree α_1, ν_2 equations of degree α_2, ν_3 equations of degree α_3, and so on. [2, p. 479]

---

* Quoique la résolution algébrique des équations ne soit pas possible en général, il y a néanmoins des équations particulières de tous les degrés qui admettent une telle résolution. Telles sont par exemple les équations de la forme $x^n - 1 = 0$. La résolution de ces équations est fondée sur certaines relations qui existent entre les racines. J'ai cherché à généraliser cette méthode en supposant que deux racines d'une équation donnée soient tellement liées entre elles, qu'on puisse exprimer rationnellement l'une par l'autre ....

** Si les racines d'une équation d'un degré quelconque sont liées entre elles de telle sorte, que *toutes* ces racines puissent être exprimées rationnellement au moyen de l'une d'elles, que nous désignerons par x; si de plus, en désignant par $\theta x, \theta_1 x$ deux autres racines quelconques, on a

$$\theta\theta_1 x = \theta_1\theta x,$$

l'équation dont il s'agit sera toujours résoluble algébriquement. De même, si l'on suppose l'équation irréductible, et son degré exprimé par

$$\alpha_1^{\nu_1}\alpha_2^{\nu_2} \ldots \alpha_\omega^{\nu_\omega},$$

où $\alpha_1, \alpha_2, \ldots, \alpha_\omega$ sont des nombres premiers différens, on pourra ramener la résolution de cette équation à celle de ν_1 équations du degré α_1, de ν_2 équations du degré α_2, de ν_3 équations du degré α_3 etc.

There is a direct connection, in terms of method, between Abel's idea for the proof of this theorem and Gauss's discussion of the cyclotomic equation.[79] In modern terms, Abel argues as follows: Let $\varphi(x)$ be an irreducible equation of degree n $(=\alpha_1^{\nu_1}\alpha_2^{\nu_2}\cdots\alpha_\omega^{\nu_\omega})$ with n roots $x_1, x_2, \ldots, x_n$. By assumption, it is possible to express all the roots as rational functions of one, say x_1. Then

$$x_1 = \theta_1(x_1), \qquad x_2 = \theta_2(x_1), \;\ldots,\; x_n = \theta_n(x_1),$$

where θ_1 is the identity map. Abel proves that if in the rational functions θ_k, $k = 1, \ldots, n$, we replace the root x_1 with another root x_i, where i is an arbitrary one of the integers $1, \ldots, n$, then $\theta_k(x_i)$, $k = 1, \ldots, n$, are again the roots of our equation in possibly different order. Thus the roots $x_i = \theta_1(x_i), \theta_2(x_i), \ldots, \theta_n(x_i)$ are a permutation of the roots $x_1, x_2, \ldots, x_n$. The number of such permutations of the roots is n. Abel goes on to prove—and this is the main point—that the nature of these permutations is crucial. In modern terms, Abel shows that if the group of permutations of the roots is commutative, then the solution of the equation $\varphi(x) = 0$ is reducible to the solution of auxiliary equations of lower degree, and these, in turn, are solvable by radicals. In the case of the cyclotomic equation, the group in question is even cyclic.

Abel does not explicitly construct this permutation group. He merely uses the assumed commutativity of the rational functions that express the roots in terms of the root x; he writes $\theta\theta_1 x = \theta_1\theta x$.

Following French usage, Abel sometimes speaks of "ordering the roots of the equation in groups." Thus in [2, p. 482], we find "les racines de l'équation seront partagées en plusieurs groupes."

We find the same spontaneous recourse to the nontechnical use of "le groupe" in the early work of Galois, who, however, later used the term in its present technical sense.

A cursory glance at the preceding few items bearing on Abel's contribution to the solution of algebraic equations shows the fundamental importance of the long passage set down by Abel in 1828 and quoted above. It is an objective historical document expressing the historical evolution of solvability theory. A few years later, Galois's group-theoretic formulation of the solvability problem completed this development. Given the aim of our investigation, Galois's contribution calls for a detailed analysis.

2 *Galois's Group-Theoretic Formulation of the Problem of Solvability of Algebraic Equations*

Today we possess all of Galois's mathematical manuscripts [221, 222], including sketches of papers and notes of various length. On the other hand, his critical social statements are available only in expurgated form or have otherwise been completely suppressed.

Galois could not abide the rule of mediocrity in science or in public life. He pilloried the self-seeking ways of those who do not hesitate to use science as a means of attaining personal glory or lucrative posts. He searched textbooks and papers in vain for the great, leading ideas. He was convinced that the failure to emphasize these ideas allows the intellectual core of the subject to be smothered by a mass of theorems. He goes so far as to say that the papers and textbooks of his contemporaries are "largely the work of people with no knowledge of the science they profess" ("la plupart composés par des hommes qui n'ont pas l'intelligence de la science qu'ils professent") [222, p. 28].

Galois's papers and notes [222], edited by J. Tannery, show that his mathematical investigations went hand in hand with his reflections on the nature of mathematics and, above all, its method. Galois represents a fascinating mixture of arrogance, true superiority, self-insight, and sense of mission. He was convinced that he knew the true, new road in mathematics. Nevertheless, he was aware that his tendency to avoid formalisms and computational schemes made it difficult to understand his papers.

Careful study of Galois's notes shows how he was able to complete the work of Lagrange, Ruffini, Gauss, Abel and Cauchy—the group-theoretic formulation of the theory of solvability of algebraic equations—and discloses the inner forces of the theory that would trigger the development of the (permutation-theoretic) group concept in the following decades. More than in the case of Abel, whose last paper on the theory of equations (1828) attests his spiritual kinship with Galois, Galois's own work stems from an explicit new methodology, and from his deliberate policy of thinking in terms of structures.[80] Incidentally, Galois's notes are among the few contemporary documents that attest to an awareness of the beginning of structural changes in mathematics.[81]

Of the greatest interest are Galois's thoughts, set down in October 1831 in the Saint-Pélagie prison in the introduction to two planned papers on pure analysis. Here Galois tries to determine the role of calculi in mathematics. His approach is historical. He claims that it was only since the time of Euler that perfected calculi became indispensable tools of progress; before that, he says, little was gained by translation of verbal statements of results into algebraic formulas. In his own time, the continued formation of calculi and of mathematical algorithms had produced a sudden shift in their roles. Once supreme aids, they now became supreme obstacles:

* From the beginning of this century, the algorithm has attained such levels of complexity that all progress by this means had become impossible . ... [222, p. 25]

At a time when the perfection of calculational mathematics was triumphant, Galois declared the self-liquidation of calculation as a method. In essence, he posed the question how mathematical thinking should be externalized. His own position was that, while useful at certain periods, symbolism and masses of formulas are incidental attributes of mathematics. When attributes become ends in themselves and thus a hindrance, then, like wraps, they must be discarded, and mathematics itself must be permitted to come forward.

As a first step in this direction Galois noted the remarkable tendency of contemporary mathematicians toward "elegance." Its essence, he said, was a certain simplicity and clarity, a gracefulness and lightness of presentation that enabled "the mind to grasp quickly and at one time a large number of operations" ("l'esprit saisit promptment et d'un seul coup un grand nombre d'opérations") [222, p. 25]. This is Galois's characteristic turn of phrase. He is concerned with insight into principles. As for elegance, its only possible and reasonable aim is "intellectual simplification" ("simplification intellectuelle").[82]

Galois thinks, quite rightly, that the power of the "elegant method" is limited:

** I think that the simplifications produced by the elegance of a calculus ... have their limit. [222, p. 25]

What then is to take the place of barren algorithmization and of an intrinsically limited elegance? His answer is that the road, already instinctively entered upon, on which one can simultaneously survey a large number of opérations, must be consistently followed and ad hoc approaches must become methods:

*** I think that the mission of future geometers will be to group operations, to classify them according to their difficulty and not according to their form. [222, p. 26]

Galois dispels the misunderstanding that he defends "the affectation that some people have of seemingly avoiding all calculi, an affectation that makes them render by means of very long sentences that which can be expressed very

* Dès le commencement de ce siècle, l'algorithme avait atteint un degré de complication tel que tout progrès était devenu impossible par ce moyen ....

** Or je crois que les simplifications produites par l'élégance des calculs ... ont leur limite.

*** Grouper les opérations, les classer suivant leurs difficultés et non suivant leurs formes; telle est, suivant moi, la mission des géométres futurs.

briefly in terms of algebra" ("l'affectation que certaines personnes ont d'éviter en apparence toute espèce de calcul, en traduisant par de phrases fort longues ce qui s'exprime très brièvement par l'algèbre") [222, p. 26]. His concern is "the analysis of analysis" ("l'analyse de l'analyse"):

* Here the most advanced calculi ... executed until now are considered as special cases .... [222, p. 26]

Thus, at least in the sense of the transition from the spontaneous to the conscious, Galois stands on the threshold of modern mathematics. Galois put before nineteenth-century mathematicians the problem of "the analysis of analysis" (which D. Hilbert referred to at the end of the nineteenth century as "Metamathematik"), of finding the abstract core of various domains and methods. This problem led to the development of the axiomatic method and to the study of mathematical structures.

Nevertheless, one must not overlook certain anarchistic features of Galois's mathematical activity, tendencies quite analogous to the anarchistic component of his republican politics. His famous toast to the king, his attitude to contemporary scientists—to Cauchy, Poisson, and especially the professors of the *Ecole Polytechnique*—as well as his contempt for the traditional presentation of mathematics, cannot be explained by his youth and character alone. On the basis of his experiences, he broke consciously with both the political and the mathematical worlds of his time.

He refused to admit that he belonged to a mathematical tradition. His almost aphoristic mode of presentation increased the level of difficulty of his intrinsically difficult papers. The result was that only the next generation of mathematicians could appreciate his work—and then only after it had been presented in the traditional way. But the truth of the matter is that, whether he liked it or not, Galois *was* part of a mathematical tradition. The study of Lagrange's works provided him with an algebraic education. He was familiar with, and quoted, Cauchy's papers and most of Abel's. He valued Abel as the person with the best grasp of the solvability of algebraic equations [222, p. 24]. He was familiar with Gauss's papers, used Gauss's congruence symbol $\equiv$, and, in his famous letter to Auguste Chevalier, dated 29 May 1832—the evening before his fatal duel—he suggested Gauss and Jacobi as judges of the significance of his theories.

Galois adopted from others not only the problem of the solvability of algebraic equations but also the means of its solution, namely, the theory of permutations as well as the by then considerably advanced theory of equations. His discovery

* Ici les calculs les plus élèves ... exécutés jusqu'à présent sont considérés comme des cas particuliers ....

of the inner connections between the two aspects of the problem enabled him to meet Lagrange's demand for *a priori* criteria for the solvability of equations by radicals. The structure of the roots of an equation could now be read off from the structure of a certain group of permutations associated with the equation.

The beauty and depth of the theory of solvability of algebraic equations, the theory later named for Galois, together with Galois's tragic fate, his long-deferred recognition as a mathematician, and the fact that he must be credited with the introduction of "group" (le groupe) as a technical term into mathematics—all this has tended to produce distorted evaluations of his contribution to the part of the theory of permutation groups not connected with the solvability of equations. Compared with the contributions of his predecessors—Lagrange, Ruffini, and Cauchy—Galois's momentous discovery is not the idea of a group but the insight into the structure of the group uniquely associated with an equation, and, in particular, the discovery of the role of certain of its subgroups, namely, the normal subgroups.

In line with the concern of this book, I shall show how Galois's primary objective, the creation of an effective theory of algebraic equations, led to the concept of "le groupe"—our "group"—and how Galois came upon the decisive discovery of the normal subgroup property. This is the extent to which shall investigate Galois's theory of equations. As much as possible, I shall analyze his papers and notes in the order in which they were written rather than in their order of publication.

In September 1830 Galois wrote the short (two-page) "Discours préliminaire" [230].[83] According to his friend A. Chevalier, he meant this to serve as an introduction to a long, future paper on the theory of equations, which was to replace a lost manuscript that had been submitted to the Academy. Previously he had published, in addition to two short papers on periodic continued fractions [224] and on analysis [228], three small papers on the theory of equations: *Analyse d'une Mémoire sur la résolution algébrique des équations* [225] in April 1830; *Note sur la résolution des équations numériques* [226] in June 1830; and *Sur la théorie des nombres* [227], also in June 1830.

The first of these papers is very short. It consists of two pages of results, stated without proof, dealing for the most part with the solvability of primitive equations of degree p^ν. Here Galois follows Gauss. He designates equations that can be decomposed into m factors of degree n as nonprimitive or Gaussian of degree $m \cdot n$. He also states partial results on the general modular equation. Galois does not give even an indication of a proof. Two thirds of the way through the paper, at the end of the results on the solvability of primitive equations, one finds the laconic remark that "all of these results have been deduced from the theory of permutations" ("Toutes ces propositions ont été déduites de la théorie des permutations") [221, p. 12]. The next sentence is a rather overbearing introduction to further results dealing with the modular equation: "Here are additional results that follow from my theory" ("Voici d'autres résultats qui découlent de ma [sic (Wussing)] théorie") [221, p. 12].

The content of the second paper is traditional. Following Legendre, Galois investigates the problem of determining the numerical values of the roots of an algebraic equation that can be put in the form $f(x) = x$. But the third paper shows conclusively that Galois had new ideas. He definitely wanted these to be regarded as a basic contribution to an original general theory of algebraic solvability of equations. That is why the title *On the theory of numbers* in de Ferussac's *Bulletin* was accompanied by the footnote, "This paper is part of the investigations of M. Galois on the theory of permutations and algebraic equations" ("Ce Mémoire fait partie des recherches de M. Galois sur la théorie des permutations et des équations algébriques") [221, p. 15]. This paper became the starting point of the theory of finite (commutative) fields,[84] known also as the theory of Galois fields.

Galois investigates the solutions of the congruence $F(x) \equiv 0\ (p)$, where $F(x)$ is an algebraic function and p is a prime. In contrast to the traditional, number-theoretic statement of the problem, Galois's approach admits "incommensurable," that is, irrational solutions. He assumes that $F(x)$ is irreducible, that is, that there are no three functions $\varphi(x)$, $\psi(x)$, and $\chi(x)$[85] such that

$$\varphi(x) \cdot \psi(x) = F(x) + p \cdot \chi(x).$$

Then, in Galois's words, the equation has neither integral nor irrational roots of lower degree. Here one must bear in mind that a root is of degree ν if it satisfies an equation (or congruence) of degree ν. Galois has this to say about such roots of congruences:

* It is thus necessary to view the roots of this congruence as a variety of imaginary symbols with none of the properties of whole numbers, symbols whose use in calculations will often be as fruitful as the use of the imaginary $\sqrt{-1}$ in ordinary analysis. [221, p. 15]

According to Galois, the aim of the paper is

** The classification of these imaginaries and the reduction of their number to the least possible. [221, p. 15]

---

* Il faut donc regarder les racines de cette congruence comme des espèces de symboles imaginaires, puisqu'elles ne satisfont pas aux questions des nombres entiers, symboles dont l'emploi, dans le calcul, sera souvent aussi utile que celui de l'imaginaire $\sqrt{-1}$ dans l'analyse ordinaire.
** la classification de ces imaginaires, et leur réduction au plus petit nombre possible, qui va nous occuper.

Galois acquired[86] a complete understanding of the structure of the solutions of the congruence $F(x) \equiv 0\ (p)$: All its roots are roots of equations of the form $x^{p^\nu} = x$, and all the nonzero roots are powers of a single one; in short, there exists a primitive root. In addition to this result, the paper is notable for two specific modes of reasoning, and for Galois's first use of the term "le groupe."

The first noteworthy passage contains the proof of the fact that, in modern terms, the number of elements of a Galois field of characteristic p is a power of p. Here Galois consciously adopts methods from the theory of power residues and arrives necessarily at Euler's decomposition process (see I.3). Galois considers a congruence $F(x) \equiv 0$ of degree ν and designates one of its roots by i. Then he considers the expression

(A) $$a_0 + a_1 i + a_2 i^2 + \cdots + a_{\nu-1} i^{\nu-1}$$

with integer coefficients. Modulo p, the expression (A) can take on only p^ν values,

* which have, as I will show, the same properties as the natural numbers in *the theory of power residues*. [221, p. 16]

Now let α be one of the $p^\nu - 1$ nonzero values of (A). Its powers are necessarily also of the form (A) and thus cannot all be distinct. Hence $\alpha^n = 1$ for some natural number n. We assume that n is the least natural number with this property. Thus

** one will have a set of n expressions

$$1, \alpha, \alpha^2, \alpha^3, \ldots, \alpha^{n-1}$$

completely different from each other. Let us multiply these n expressions by another expression β of the same form.[87] We shall obtain a new group of quantities completely different from the first and from each other. [221, p. 16]

Now the argument follows the familiar pattern. If the above two sequences do not exhaust the expressions (A), then we again multiply the powers of α by a γ in (A) which has not yet turned up, and so on. Thus n must turn up among the $p^\nu - 1$ quantities (A). But then

$$\alpha^{p^\nu - 1} = 1 \text{ and therefore } \alpha^{p^\nu} = \alpha.$$

---

* qui jouissent, ainsi que je vais le faire voire, des mêmes propriétés que les nombres naturels dans la *théorie des résidus des puissances*.

** On aura un ensemble de n expressions, toutes différentes entre elles

$$1, \alpha, \alpha^2, \alpha^3, \ldots, \alpha^{n-1}.$$

Multiplions ces n quantités par une autre expression β de la même forme. Nous obtiendrons encore un nouveau groupe de quantités toutes différentes des premières et différentes entre elles.

In proving the existence of a primitive root, Galois makes only a passing reference to number theory.

The last-quoted excerpt from Galois's paper includes his first use of the term "le groupe" in a mathematical context. It is clear that the term as used does not yet have the definite modern sense, is not yet a technical mathematical term. Like Abel, Cauchy, and others before him, Galois uses "le groupe" here in the sense of everyday speech as a synonym for a set or a collection. The term is used mentally to bracket related quantities. We shall see that Galois very often used "le groupe" and verb forms of "grouper" in their everyday senses even when "le groupe" was beginning to take on for him the sense of a "group" in the mathematical sense of a group of permutations.

A second passage from *Sur la théorie des nombres* throws a great deal of light on the mental process associated with the evolution of "le groupe" from its everyday sense to its technical, mathematical sense. At the same time, that passage indicates clearly that by June 1830 Galois's ideas on a theory of solvability of equations had advanced considerably[88] beyond the level represented by his *Analyse* [225] of April 1830. He now tries to associate to each of the primitive equations that form the subject of this paper a "group" of permutations. While this attempt of June 1830 would have to be changed and made more precise, it nevertheless outlined the correct approach.

Encouraged by his insight into the structure of the solutions of an algebraic congruence, Galois tries to link that result with the solution of primitive algebraic equations:

* It is above all in the theory of permutations, in which there is constant need to change the form of the indices, that consideration of the imaginary roots of congruences seems indispensable. As I shall try to indicate very briefly, it gives a simple and easy means of deciding in which case a primitive equation is solvable by radicals. [221, p. 21]

The idea, which he only sketches here, echoes an idea of Lagrange. It is an attempt to decide the solvability of an equation through a condition on the invariance of the numerical value of a rational function of its roots.

Galois starts with an algebraic equation $f(x) = 0$ of degree p^ν. Its p^ν roots are denoted by x_k, where the p^ν values of the index k satisfy the congruence $k^{p^\nu} \equiv k \pmod{p}$. Now let V be a rational function of the roots x_k. As Galois puts it, the function V is "transformed" when the root x_k is replaced by the root with

* C'est surtout dans la théorie des permutations, où l'on a sans cesse besoin de varier la forme des indices, que la considération des racines imaginaires des congruences paraît indispensable. Elle donne un moyen simple et facile de reconnaître dans quel cas une équation primitive est soluble par radicaux, comme je vais essayer d'en donner en deux mots une idée.

index $(ak + b)^{p^r}$; here a and b are arbitrary constants satisfying the conditions

$$a^{p^\nu - 1} \equiv 1, \qquad b^{p^\nu} \equiv b \pmod{p},$$

and r is an integer. Under these conditions, there are $p^\nu \cdot (p^\varphi - 1) \cdot \nu$ ways of interchanging the roots x_k by means of the permutations

$$x_k / x_{(ak+b)p^r}.$$

As a result V will, in general, take on $p^\nu \cdot (p^\nu - 1) \cdot \nu$ different forms. If every function V of the roots of $f(x) = 0$ is invariant under all these permutations, then the equation is solvable by radicals. "Persons acquainted with the theory of equations will see this with ease" ("Les personnes habituées à la theorie des équations le verront sans peine") [221, p. 22]. Conversely, says Galois, again without proof, a primitive equation that fails to satisfy this condition is not solvable by radicals. (Exceptions to this rule are, according to Galois, the degrees 9 and 25.)

Thus, already in June 1830, Galois associated to a given solvable equation of prime power degree a group in the mathematical sense of the word:

* Thus, for each number of the form p^ν one will be able to form a group of permutations such that, when an equation of degree p^ν is primitive and solvable by radicals, then every function of the roots invariant under these permutations will have to admit a rational value. [221, p. 22]

This shows Galois's deep comprehension of the close connection between the solvability conditions of algebraic equations and permutation theory. The point here is to associate to *every* equation a group and to infer from its structure the solvability of the equation by radicals.[89]

This is how far the 18-year-old Galois had advanced when setting down the *Discours préliminaire* in the late summer of 1830. He realized that his deep insights would revolutionize mathematical terminology and thought. He spoke of a new language, "un nouveau langage," in the making . This made him demand intensive intellectual effort of the reader:

** Our goal is to determine the conditions for the solvability of equations by radicals. We can vouch that no subject in pure analysis is more abstruse and perhaps more cut off from the rest. The novelty of the subject has required the use of new terminology and new criteria. [222, p. 21]

---

* Ainsi, pour chaque nombre de la forme p^ν, on pourra former un groupe de permutations tel, que toute fonction des racines invariable par ces permutations devra admettre une valeur rationnelle quand l'équation de degré p^ν sera primitive et soluble par radicaux.

** Le but que l'on s'est proposé est de déterminer des caractères pour la résolubilité des équations par radicaux. Nous pouvons affirmer qu'il n'existe pas dans l'analyse pure de matière plus obscure et peut-être plus isolée de tout le reste. La nouveauté de cette matière a exigé l'emploi de nouvelles dénominations, de nouveaux caractères.

The *Discours préliminaire* occupies just two pages of print. The paper that was to follow it was never written. Nevertheless, in spite of extremely difficult circumstances, Galois made energetic efforts to solve the stated problem. His manuscript "Mémoire sur les conditions de résolubilité des équations par radicaux" [229], dated 16 January 1831 and published posthumously by Liouville in 1846, begins with statements of "principles," that is, explanations of concepts, definitions, and lemmas. In terms of length of presentation, this "Mémoire" represents the greatest concession Galois had ever made to the reader. In the introduction he states that

* this paper is taken[90] from a work that I had the honor of presenting to the Academy a year ago. Since this work has not been understood and the propositions it comprises have been put in doubt, I had to content myself with the presentation, in condensed form, of the general principles and a *single* application of my theory. I beg of my judges at least to pay attention while reading these few pages. [221, p. 33]

Galois speaks of a "single application" that he wants to present; he restricts himself to algebraic equations of prime degree. This restriction explains why the property of normality of subgroups failed to attract attention. At the end of the introduction Galois points to possible additional applications of his theory, such as the unsolvability by radicals of the modular equations (in the theory of elliptic functions).

The content of the main theorem of the "Mémoire" is given in the introduction. The theorem sheds light on Lagrange's analysis of solvability algorithms and satisfies the requirement of an *a priori* criterion for solvability by radicals. The text proper contains the following formulation of that theorem:

** *For an irreducible equation of prime degree to be solvable by radicals, it is necessary and sufficient that, given any two of the roots, the others can be rationally deduced.* [221, p. 49]

As pointed out in the introduction, this fits in purposely with the traditional theory of equations in the sense, say, of the results of Abel. To the extent to which he was familiar with these results, Galois referred to them frequently.

Given the aim of this book, what is of greatest interest about this result is that, while Abel obtained it by computation, Galois obtained it by a group-theoretic argument.

---

* Le Mémoire ci-joint est extrait d'un Ouvrage que j'ai eu l'honneur de présenter à l'Académie il y a un an. Cet Ouvrage n'ayant pas été compris, les propositions qu'il renferme ayant été révoquées en doute, j'ai dû me contenter de donner, sous forme synthétique, les principes généraux et une *seule* application de ma théorie. Je supplie mes juges de lire du moins avec attention ce peu de pages.

** *Pour qu'une équation irréductible de degré premier soit soluble par radicaux, il faut et il suffit que deux quelconques des racines étant connues, les autres s'en déduisent rationnellement.*

As mentioned earlier, the main part of Galois's "Mémoire sur les conditions de résolubilité des équations par radicaux" starts with a section entitled "Principes." The opening sentence is "I begin by stating certain definitions and a number of lemmas all of which are known" ("Je commencerai par établir quelques définitions et une suite de lemmes qui sont tous connus") [222, p. 34]. Galois defines what is meant by a reducible equation and then what is meant by the assertion that it has rational divisors. He makes precise the term "rational divisor" (diviseur rationnel), which is what we now call a divisor, over the domain of rationality (field) of the coefficients. He goes on to explain what it means to adjoin (adjoindre) a quantity to an equation. Then come the definitions of substitution and permutation. Galois draws a sharp distinction: "Substitutions are transitions from one permutation to another" ("Les substitutions sont le passage d'une permutation à l'autre") [221, p. 35]. Oddly enough, the word "group" appears twice among the "known definitions," but is not actually defined. It seems that the term "le groupe" was not yet used consistently as a mathematical term and, therefore, that Galois did not intend to define it. The first time it occurs, it is used in the everyday sense; the second time, closure under multiplication of a set of permutations is spelled out but not always ensured, so that the "group" need not be a group at all. In other words, "groups" are so far just sets of permutations. But in the rest of the text "groups" are true groups closed under multiplication. At this point, Galois's explanations bear on the meaning of permutations and substitutions and show how to compute with them:

* When dealing with functions, the initial permutation for specifying substitutions is quite arbitrary, since in the case of a function of several letters there is no reason why a particular letter should occupy one position rather than another.

However, as one can hardly form for oneself the idea of a substitution without that of a permutation, we shall often, in the text, make use of permutations and shall only consider the substitutions as the passage from one permutation to another.

When grouping substitutions we shall make them all issue from the same permutation.

As one always considers questions for which the initial disposition of the letters influences nothing in the groups that we shall be considering, we shall have to have the same substitutions regardless of the initial permutation. Thus, if one has in the same group the substitutions S and T one is certain to have the substitution ST. [221, pp. 35–36]

---

* La permutation d'où l'on part pour indiquer les substitutions est toute arbitraire, quand il s'agit de fonctions; car il n'y a aucune raison pour-que, dans une fonction de plusieurs lettres, une lettre occupe un rang plutôt qu'un autre.

Cependant, comme on ne peut guère se former l'idée d'une substitution sans se former celle d'une permutation, nous ferons, dans le langage, un emploi fréquent des permutations, et nous ne considérerons les substitutions que comme le passage d'une permutation à une autre.

(continuation see p. 112)

It is interesting to see how the term "le groupe" became more and more precise in the course of Galois's presentation; through spontaneous use an everyday word changes into a technical term. The vagueness is present in the preliminary "definition" at the beginning of the text but is absent from the body of the text; the "group of the equation" introduced in the body of the text is a group in the true sense of the word. Its closure under multiplication, while not proved, is asserted and used throughout the text. This gap is found throughout Galois's theory. It was first closed in 1852 by E. Betti (see II.4).

Following the definitions, Galois states four lemmas essential for the subsequent development. Lemma II asserts the existence of a function V of the roots $a, b, c, \ldots$ of a given equation with simple roots that does not take on two equal values under different permutations of the roots. An example of such a function is

$$V = Aa + Bb + Cc + \cdots$$

for a suitable choice of the integer coefficients $A, B, C, \ldots$ Galois leaves his statement unproved. Such a function V has the property that each root of the given equation can be expressed as a rational function of V. Here Galois gives the sketch of a proof and refers to the fact that this very theorem, without proof, can be found in Abel's posthumous papers.[91] In lemma IV, Galois constructs the entire rational equation satisfied by V. To obtain it, he says, one must multiply all the factors $[V - \varphi(a, b, c, \ldots)]$ obtained as a result of all the permutations of the roots. He then considers the irreducible factor of the equation that contains V. Let $V, V', V'', \ldots$ be its roots. Then lemma IV asserts that if $a = f(V)$ is one of the roots of the given equation, then $b = f(V')$ is another.

Now Galois can prove "proposition I" dealing with the existence and properties of the group associated to a given equation:

* *Given an equation with roots a, b, c, ... there will always be a group of permutations of the letters a, b, c, ... with the following property: 1. All functions of the roots invariant under the substitutions of the group are rationally determinable, and 2. Conversely, all rationally determinable functions of the roots are invariant under the substitutions.* [221, p. 38]

(* continuation from p. 111)

Quand nous voudrons grouper des substitutions, nous les ferons toutes provenir d'une même permutation.

Comme il s'agit toujours de questions où la disposition primitive des lettres n'influe en rien dans les groupes que nous considérerons, on devra avoir les mêmes substitutions, quelle que soit la permutation d'où l'on sera parti. Donc, si dans un pareil groupe on a les substitutions S et T, on est sûr d'avoir la substitution ST.

* *Soit une équation donnée, dont a, b, c, ... sont les m racines. Il y aura toujours un groupe de permutations des lettres a, b, c, ... qui jouira de la propriété suivante:* 1° *Que toute fonction des*

(continuation see p. 113)

This is the Galois group of an equation just as we define it nowadays. In the existence part of the proof Galois presents the group in the form of an array: Let $V, V', V'', \ldots, V^{(n-1)}$ be the roots of the irreducible equation (lemma IV) satisfied by the suitably constructed function V of the roots of the given equation. By lemma III, the m roots of the given equation have the form

$$\varphi(V), \varphi_1(V), \varphi_2(V), \ldots, \varphi_{m-1}(V).$$

The group of the equation consists of the following n arrangements:

$$\begin{array}{l}\varphi(V), \varphi_1(V), \varphi_2(V), \ldots, \varphi_{m-1}(V),\\ \varphi(V'), \varphi_1(V'), \varphi_2(V'), \ldots, \varphi_{m-1}(V'),\\ \varphi(V''), \varphi_1(V''), \varphi_2(V''), \ldots, \varphi_{m-1}(V''),\\ \vdots\\ \varphi(V^{(n-1)}), \varphi_1(V^{(n-1)}), \varphi_2(V^{(n-1)}), \ldots, \varphi_{m-1}(V^{(n-1)}).\end{array}$$

Thus the group of the equation is given first as a group of permutations, or, in accordance with Galois's usage, as a group of arrangements of the roots. But the proof of the properties of that group employs not the permutations but the substitutions that yield the transitions from one permutation to another. This is justified by two subsequent remarks to the effect that the new group, consisting of the corresponding substitutions, has the same properties as the group of permutations. In Galois's words, "The group thus formed will obviously have the same properties as the former" ("Le nouveau groupe ainsi formé jouira évidemment des mêmes propriétés que le premier") [221, p. 39].

Galois bases this solution theory on his concept of the group of an equation and applies it to equations of prime degree. Upon adjunction of a root, the group of the equation remains unchanged or it splits into p groups, which Galois refers to as "partial groups" (des groupes partiels). Suitable adjunctions produce "group diminution" (diminuer le groupe) and finally result in a group consisting of a single substitution [221, p. 46]. When the given equation is of prime degree, this happens after adjunction to the equation of two of its roots. This is how Galois proves the main theorem of this paper *by group-theoretic means.*

To illustrate the use of his method Galois gives the group of the cyclotomic equation $(x^n - 1)/(x - 1) = 0$ and deduces the familiar solution of the quartic

(* continued from p. 112)
racines , invariable[92] *par les substitutions de ce groupe, soit rationnellement connue: 2° Réciproquement, que toute fonction des racines, déterminable rationnellement, soit invariable par les substitutions.*

from the diminution of the symmetric group of order four resulting from adjunction of radicals.

Thus at the beginning of 1831 Galois had associated to every equation (with simple roots) a group of permutations, the group of the equation. He is aware of its fundamental property, namely, closure under multiplication; he spells it out and uses it implicitly, but he does not prove it. The use of the term "le groupe" invites confusion since the word turns up in connection with groups as well as mere complexes of permutations not closed under multiplication; of course, in the absence of closure Galois neither asserts it nor relies on it.

The same confusion surrounds the use of "le groupe" in the paper *Des équations primitives qui sont solubles par radicaux* [231] in which Galois, now possessing a deeper knowledge of his theory, returns to the problem of solvability of a primitive equation. Compared with the "Mémoire" of 1831, this paper shows no deeper insight into the group concept in spite of its preoccupation with Cauchy's study of permutations. This being so, I turn to a discussion of Galois's last work, his famous *Lettre à Auguste Chevalier* [232] of 29 May 1832, the evening before the fateful duel. This document is intended as a scientific testament. Its use of "le groupe" oscillates between "group" and "complex," but it contains the fundamental insight into the role of the special subgroups now called normal. It begins as follows:

* My dear friend,

I have done several new things in analysis.

Some concern the theory of equations; others, integral functions.

In the theory of equations I have found out in which cases the equations are solvable by radicals. This has given me the opportunity to deepen the theory and to describe all the transformations admitted by an equation, even when it is not solvable by radicals. [221, p. 25]

This farewell letter reflects Galois's newly gained insights, allowing him to deal with the case, earlier viewed as devoid of interest, of an equation not solvable by radicals. The property of normality of a subgroup, "la décomposition propre," as Galois calls it, plays a fundamental role.

He begins by summarizing his previous papers. He asserts that there is a great difference between adjoining to an equation one root of an auxiliary equation and adjoining all such roots:

---

* Mon cher ami.

J'ai fait en Analyse plusieurs choses nouvelles.

Les unes concernent la théorie des équations; les autres, les fonctions intégrales.

Dans la théorie des équations, j'ai recherché dans quels cas les équations étaient résolubles par des radicaux, ce qui m'a donné occasion d'approfondir cette théorie et de décrire toutes les transformations possibles sur une équation, lors même qu'elle n'est pas soluble par radicaux.

* In both cases the group of the equation is partitioned by the adjunction into groups such that one passes from one to the other by means of the same substitution; but it is only in the second case that it is certain that these groups have the same substitutions. This is called a "*proper decomposition*".

In other words, when a group G contains another group H, then the group G can be divided into groups that are obtained by performing the same substitution on the permutations of H, so that

$$G = H + HS + HS' + \cdots.$$

It can also be divided into groups with the same substitutions so that

$$G = H + TH + T'H + \cdots.$$

These two kinds of decomposition do not ordinarily coincide. When they do, the decomposition is said to be *proper*. [221, pp. 25–26]

Galois goes on to sketch in a few sentences the development of his solvability theory of algebraic equations involving his insight into the fundamental role of "proper decompositions", that is, of normal subgroups.

** It is easy to see that when the group of an equation does not admit any proper decomposition then one can transform it all one wants; the groups of the transformed equations will always have the same number of permutations.

If, on the contrary, the new group admits a proper decomposition, so that it is divided into M groups of N permutations, then one can solve the given equation by means of two equations, one having a group of M permutations, and the other one of N permutations.

Thus, when one has exhausted in the group of an equation all the possibilities of proper decomposition, then one has groups that one can transform but that always have the same number of permutations.

If each of these groups has a prime number of permutations, then the equation will be solvable by radicals; otherwise, not. [221, p. 26]

---

* Dans les deux cas, le groupe de l'équation se partage par l'adjonction en groupes tels, que l'on passe de l'un à l'autre par une même substitution; mais la condition que ces groupes aient les mêmes substitutions n'a lieu certainement que dans le second cas. Cela s'appelle la *décomposition propre*.

En d'autres termes, quand un groupe G en contient un autre H, le groupe G peut se partager en groupes, que l'on obtient chacun en opérant sur les permutations de H une même substitution; en sorte que

$$G = H + HS + HS' + \cdots.$$

Et aussi il peut se diviser en groupes qui ont tous les mêmes substitutions, en sorte que

$$G = H + TH + T'H + \cdots.$$

Ces deux genres de décompositions ne coïncident pas ordinairement. Quand ils coïncident, la décomposition est dite *propre*.

** Il est aisé de voir que, quand le groupe d'une équation n'est susceptible d'aucune décomposition propre, on aura beau transformer cette équation, les groupes des équations transformées auront toujours le même nombre de permutations.

(** continued on p. 116)

This is Galois's contribution to the development of a general theory of solvability of algebraic equations. The subsequent text contains unproved results implied by his theory and dealing with particular equations. The listing begins with the theorem that the simple group (one without proper decomposition) of least order has order 60, deals next with (Gaussian) imprimitive equations and equations called primitive by Galois, takes up the implications for modular equations (in the theory of elliptic functions), and closes with a few inconclusive ideas on integral transformations.

Toward the end of his letter Galois, quite unlike Gauss whose motto was "Pauca sed matura," writes

* I have often in my life dared to advance propositions that I was not sure of .... [221, p. 32]

The letter concludes with a plea to his friend:

** You will publicly beg Jacobi or Gauss to give their opinion not of the truth but of the importance of the theorems.

After this, there will, I hope, be people who will find it to their advantage to decipher all this mess.

I embrace you warmly. [221, p. 32]

Thus ends Galois's last scientific document. It is dated 29 May 1832. Two days later, on 31 May, he died of his wounds at the age of 20.

Our analysis of Galois's writings has shown how the basic methods, concepts and theorems resulted from his growing insight into the structures he investigated. In the few years of his scientific activity he had reached the fundamental concept of the normality of certain subgroups and was to conceive, in principle, the theory of equations now bearing his name. Its development required the buildup and refinement of group-theoretic resources—a task accomplished not by Galois's

---

(** continued from p. 115)

Au contraire, quand le groupe d'une équation est susceptible d'une décomposition propre, en sorte qu'il se partage en M groupes de N permutations, on pourra résoudre l'équation donnée au moyen de deux équations: l'une aura un groupe de M permutations, l'autre un de N permutations.

Lors donc qu'on aura épuisé sur le groupe d'une équation tout ce qu'il y a de décompositions propres possibles sur ce groupe, on arrivera à des groupes qu'on pourra transformer, mais dont les permutations seront toujours en même nombre.

Si ces groupes ont chacun un nombre premier de permutations, l'équation sera soluble par radicaux; sinon, non.

* Je me suis souvent hasardé dans ma vie à avancer des propositions dont je n'étais pas sûr ....

** Tu prieras publiquement Jacobi ou Gauss de donner leur avis, non sur la vérité, mais sur l'importance des théorèmes.

Après cela, il y aura, j'espère, des gens qui trouveront leur profit à déchiffrer tout ce gâchis.

Je t'embrasse avec effusion.

mathematical contemporaries but by a later generation of mathematicians (see II.4).[93]

Galois forced later developments substantially and in two ways. On the one hand, since he discovered theorems for which he could not give proofs based on clearly defined concepts and calculations, it was inevitable that his successors would find it necessary to fill the gaps. On the other hand, it would not be enough merely to prove the correctness of these theorems; their substance, their group-theoretic core, must be extracted.

Appreciation of the significance of Galois's work grew with the growth of insight into the conceptual underpinning of his theory and the simultaneous creation of a calculus that made it logically and computationally accessible and controllable.

These internal, as well as some external, circumstances caused Galois's work to be understood and appreciated in stages, which differed significantly in their assessment of its core. The first stage, which can well be called the negative stage, lasted until 1846, the year in which Liouville published significant writings from Galois's *Nachlass*. We possess no written document from that period that indicates that Galois had any influence on the development of mathematics. It is true that during this time, immediately after his death, two of his papers appeared; but these brought no more improvement in his scientific standing than had the publication of new papers between the summer of 1830 (when he set down the *Discours préliminaire*) and his death, nor the publication of his letter to Chevalier in the September issue[94] of the *Revue encyclopédique*.

The brevity and fragmentary style of the letter to Chevalier must have rendered it incomprehensible. It employed concepts such as "group of an equation" and methods that Galois had neither defined nor explained in the letter itself or in any of his papers then in print.

The two later stages of the assimilation of Galois's ideas are of interest to us since they represent a phase in the final evolution of the permutation-theoretic group concept. They will be discussed in the next two sections.

1 Liouville's 1846 Edition of the Principal Writings of Galois

As a result of a decision by Galois's family, his *Nachlass* was at first given into the care of his friend Chevalier and then entrusted to the distinguished mathematician J. Liouville.[95] Liouville published Galois's essential writings in the *Journal des mathématiques*, of which he was editor. They appeared in tome XI (1846), pp. 381–444,[96] and consisted of the five papers that appeared in Galois's lifetime and of three items from the *Nachlass*, namely, the letter to Chevalier, the "Mémoire" containing the definition of the group of an equation, and the fragment on the solution of primitive equations.[97]

In his introductory remarks, Liouville praises Galois as one of the few scientists worthy of the name of an inventor. At the same time he rightly chides Galois for excessive brevity of presentation. Given the intrinsic difficulty of the problems with which Galois had concerned himself, this penchant for excessive brevity was an unforgivable flaw.

The year 1846 marks the beginning of the second phase—the first constructive phase—of efforts to give conclusive form to Galois's ideas. The objective significance of Galois's results was recognized and they were incorporated in the edifice of algebra in the middle of the nineteenth century. Nevertheless even this second stage was transitional, just one part of the road to the final distillation of the group-theoretic essence of Galois's solution of algebraic equations.

Liouville's edition of Galois's most important writings restored the then (1846) temporarily tenuous connection between the growth of permutation theory and the group-theoretic formulation of the solvability theory of algebraic equations, and once more set in motion the interaction, decisively important for the eventual emergence of the independent permutation-theoretic group concept, of these seemingly unrelated theories.

Liouville's main contribution in this area was that he saw the connection between the theory of permutations, which Cauchy had managed decisively to advance between 1844 and 1846 (see II.2), and the basic conception of Galois, and that he recognized the potential usefulness of the latter mode of thought though only its bare outline was clear at the time. For all that, Liouville's final evaluation of Galois's results, contained in his prefatory remarks, missed completely the group-theoretic core of Galois's method. In spite of references to some ill-defined "new method" of Galois, Liouville regarded as Galois's central achievement the theorem that an irreducible equation of prime degree is solvable by radicals if and only if its roots are rational functions of any two of them.

Of course it would be completely wrong to blame Liouville for his failure to view Galois's towering achievement in terms other then the current traditional view of the problem of solvability of algebraic equations. The identification of the group-theoretic core of Galois theory and the development of an independent permutation-theoretic group theory were tasks for future mathematicians. In the years just after 1846, all agreed that there were essential gaps in Galois's work, and efforts centered not on clarifying the essence of his method but on closing those gaps by traditional means. The concern with the essence of the method grew gradually. The many works of the "commentary on Galois" genre (see below), which continued to appear well into the eighties, can be conveniently classified as reflecting one or the other of these two concerns.

As for Liouville himself, he had announced in his prefatory remarks in 1846 that he would go beyond Chevalier's footnotes and his own minor remarks and write a commentary on Galois that would complete certain issues and clarify some knotty points in the latter's papers [4, p. 88]. Such a commentary was never written.

2 *The Influence of Galois in Germany, Italy, and England*

One of the earliest efforts to close the gaps in Galois's work is represented by the paper "Über die Beziehungen, welche zwischen den Wurzeln irreductibeler Gleichungen stattfinden, insbesondere wenn der Grad derselben eine Primzahl ist" [651], by Th. Schönemann (1812–68), a *Gymnasium* teacher from Brandenburg, who had published in the thirties and early forties a few papers [647, 648, 649] on the solvability of congruences. According to Schönemann [651, p. 143], C. G. Jacobi (1804–51), whom Alfred Galois had asked to evaluate his brother's writings, called his attention to the papers of Galois that had in the meantime appeared in print and urged him to fill their gaps. Jacobi turned to Schönemann because in 1846 Schönemann, completely unaware of Galois's papers, had published a paper entitled "Grundzüge einer allgemeinen Theorie der höheren Congruenzen ..." [650], which significantly overlapped Galois's *Sur la théorie des nombres*. The result of Jacobi's initiative was Schönemann's "Über die Beziehungen" [651], published in 1853.

In the introduction Schönemann explained the role of Jacobi and went on to say that in proving an algebraic theorem, his primary objective had been

* to put in clear light the aim and underlying principles of Galois in his famous but not yet clarified "*Mémoire sur les conditions de résolubilité des équations par radicaux*".

(* see p. 120)

Schönemann did fill one of Galois's gaps but he could hardly have grasped the deeper content of Galois's papers, in particular, their group-theoretic essence. True, he used permutations of the roots of the given equation and, like Abel and Galois, set up systems of permutations; but he failed to think of them as a new entity. The word "group" or its synonym does not turn up, and closure under multiplication of the group of permutations is not used, let alone proved. Rather Schönemann seems to have wanted to give a rigorous computational proof of the following result:

* The only first degree equation with rational coefficients connecting the roots of an irreducible equation of prime degree is the familiar equation which asserts that the sum of the roots equals the negative of the first coefficient of the irreducible equation. [651, p. 143]

Between 1848 and 1853 there was a further advance in the theory of permutations on a group-theoretic basis in France, Italy, and, to some extent, also in England. Schönemann's paper of 1852–53 was in no way influenced by this development—a state of affairs all too characteristic of the attitude toward Galois in German-speaking lands during the fifties and sixties. In this connection the position of L. Kronecker (1823–91) is especially illuminating.

H. Weber, in his obituary of Kronecker [719, p. 6], reports that, in all likelihood, Kronecker first learned of Galois's theory during his stay in Paris in 1853, at the time, when he had close contacts with Hermite, Bertrand, and other leading French mathematicians who, since 1846, had made significant progress in absorbing Galois's ideas. Nevertheless, Kronecker did not go in the direction indicated by Galois, namely, that of the group-theoretic formulation of the solvability problem and the related study of permutation groups. Instead he concerned himself with the problem suggested by Abel, and indirectly also by Galois, of constructing all equations solvable by radicals. To the extent that they were available at the time, his important papers in this area became part of J. A. Serret's *Cours d'Algèbre supérieure*[98] and of C. Jordan's *Traité des substitutions.*

In his first paper on algebraically solvable equations [424], communicated to the Academy on 20 June 1853 by Lejeune-Dirichlet, Kronecker develops his

(* see p. 119)
die Principien, von welchen Galois in seinem berühmten, aber bis jetzt noch nicht aufgeklärten '*Mémoire sur les conditions de résolubilité des équations par radicaux*' ... ausging, ohne sie vollständig auszusprechen, in ein klares Licht zu stellen.

* Zwischen den Wurzeln einer irreductibelen Gleichung, deren Grad eine Primzahl ist, kann keine Gleichung des ersten Grades mit rationalen Coëfficienten stattfinden, ausser der bekannten, dass die Summe der Wurzeln gleich dem negativen ersten Coëfficienten der irreductibelen Gleichung ist.

program, which is of special value to us because of its assessment of the road taken by Galois.[99] We quote from that paper.

* So far, the investigations of the solvability of equations of prime degree—namely, those of *Abel* and *Galois*, which have served as the basis of all further work in this area —have yielded, in essence, two criteria for deciding the solvability or unsolvability of equations. These criteria, however have shed no light whatsoever on the nature of solvable equations. In fact, we do not really know whether (apart from the equations discussed by *Abel* in Vol. IV of Crelle's Journal[100] and the simplest equations connected with binomial equations) there exist equations satisfying these solvability criteria. We have no idea how to *form* such equations and have not come across them in the course of any other mathematical investigations. We must add that, as it happens, the two above-mentioned and very well known properties of solvable equations found by Abel and Galois tend to obscure rather than to clarify the true nature of these equations. Below, I shall show this to be the case of one of these criteria [namely, the one due to Galois (Wussing)]. Thus, the solvable equations *themselves* have remained in a kind of darkness, mitigated ever so slightly by the virtually ignored and rather special Note of *Abel* (Oeuvres, T. II, p. 266) on the roots of quintic equations with integer coefficients. This darkness can be fully dispelled only by solving the problem: "To find all solvable equations." For not only would we then have infinitely many new solvable equations, but we would, in a sense, have all such equations *before our eyes*, and could find and prove all their properties from the explicit form of their roots. [422, pp. 3–4]

Kronecker refers to E. E. Kummer's paper [441] of 1847 and formulates the main result of his first paper on algebraically solvable equations as follows:

---

* Die bisherigen Untersuchungen über die Auflösbarkeit von Gleichungen, deren Grad eine Primzahl ist – namentlich die *Abel*schen und *Galois*schen, welche die Grundlage aller weiteren Forschungen auf diesem Gebiete bilden – haben im Wesentlichen als Resultat zwei Kriterien ergeben, vermittelst deren man beurtheilen könnte, ob eine gegebene Gleichung auflösbar sei oder nicht. Indessen gaben diese Kriterien über die Natur der auflösbaren Gleichungen *selbst* eigentlich nicht das geringste Licht. Ja, man konnte eigentlich gar nicht wissen, ob (außer den von *Abel* im IV. Bande des Crelleschen Journals behandelten und den einfachsten mit den binomischen Gleichungen zusammenhängenden) überhaupt noch irgend welche Gleichungen existiren, welche die gegebenen Auflösbarkeits-Bedingungen erfüllen. Noch weniger konnte man solche Gleichungen *bilden*, und man ist auch durch sonstige mathematische Untersuchungen nirgends auf solche Gleichungen geführt worden. Dazu kommt noch, daß jene beiden erwähnten und wohl allgemein bekannten, von *Abel* und *Galois* gegebenen Eigenschaften der auflösbaren Gleichungen zufälliger Weise solche waren, die die wahre Natur dieser Gleichungen eher zu verdecken als aufzuklären geeignet sein dürften, wie ich das namentlich von dem einen jener beiden Kriterien späterhin zeigen werde. Und so blieben die auflösbaren Gleichungen *selbst* in einem gewissen Dunkel, welches nur durch die übrigens, wie es scheint, wenig beachtete und ganz spezielle Notiz *Abels* (Oeuvres, T. II, S. 266) über die Wurzeln ganzzahliger Gleichungen fünften Grades ein wenig erhellt wurde und welches nur durch Auflösung des Problems, alle auflösbaren Gleichungen zu finden, vollständig aufgeklärt werden konnte. Denn alsdann hat man nicht bloß unendlich viele neue auflösbare Gleichungen, sondern eben alle möglichen gewissermaßen *vor Augen* und kann an der entwickelten Form ihrer Wurzeln alle ihre Eigenschaften auffinden und erweisen.

* The roots of *every Abelian* equation with integer coefficients can be represented as rational functions of the roots of unity. [422, p. 10]

We add that, following Abel's paper [11] of 1828, Kronecker defines an "Abelian equation" as an equation whose n roots $x_1, x_2, ..., x_n$ satisfy the equations $x_2 = \theta(x_1)$, $x_3 = \theta(x_2)$, ..., $x_n = \theta(x_{n-1})$, $x_1 = \theta(x_n)$ with $\theta(x)$ an entire rational function of x whose coefficients are rational functions of the initial equation. Thus, says Kronecker, "these general *abelian* equations" are "essentially ... just cyclotomic equations" ("diese allgemeinen *Abelschen* Gleichungen" sind "im Wesentlichen nichts Anderes ..., als Kreisteilungs-Gleichungen") [422, p. 11].[101]

Kronecker's second paper on algebraically solvable equations [426] appeared in 1856, that is, in the period—the fifties and sixties—of decisive significance for the evolution of the permutation-theoretic concept. In the introduction to that paper [422, p. 28] Kronecker again speaks of a lack of "precise formulation" ("genaue Fassung"), not in connection with the evolution of the group-theoretic component but with respect to the determination of the concept of "rational dependence" (rationale Abhängigkeit) (later subsumed in the field concept) that Kronecker was in the process of formulating and that he referred to as "domain of rationality" (Rationalitätsbereich). Incidentally, it was the field concept that, toward the end of the century, gave a decisive impuls to the development of the *abstract* group concept (see III.4).

To clarify Kronecker's position vis-à-vis Galois, I quote one of his many relevant statements. It is contained in a letter Kronecker wrote to Lejeune-Dirichlet on 3 March 1856:

** I have ... found a method[102] for the derivation of all the properties of solvable equations of prime degree, whose simplicity and rigor satisfy all reasonable requirements. For the method requires no higher level of mathematical maturity than the problem itself, which is thus disposed of. I shall spend the summer carefully working out these matters. The only reason this paper will be rather lengthy is that there are so many special results, and I must not save space in the precise description of certain new viewpoints [103] missing from the *Galois* and *Abel* fragments, viewpoints without which the requisite rigor cannot be attained [423, pp. 411–412]

---

* daß die Wurzel *jeder Abel*schen Gleichung mit ganzzahligen Coefficienten als rationale Function von Wurzeln der Einheit dargestellt werden kann.

** Ich habe ... eine Methode gefunden zur Herleitung aller Eigenschaften der auflösbaren Gleichungen von Primzahlgraden, deren Einfachheit und Strenge allen gerechten Anforderungen entsprechen dürfte. Denn die Methode verlangt keinen irgend höheren Standpunkt mathematischen Fassungsvermögens als das Problem selbst, welches dadurch erledigt wird. Ich werde den Sommer zur sorgfältigen Ausarbeitung dieser Sachen verwenden, und diese wird nur *deßhalb* ziemlich umfangreich werden weil so sehr viel einzelne Resultate herauskommen, und weil ich keinen Raum sparen darf bei der genauen Bezeichnung gewisser ganz neuer Gesichtspunkte, die nament-
(** continued on p. 123)

We see that, having concerned himself with the problem of constructing all algebraically solvable algebraic equations (that is, algebraic equations solvable by radicals), Kronecker, notwithstanding the depth of his results, made no direct contribution to the development of the permutation-theoretic group concept during the fifties and sixties.[104] This being the case, his papers devoted to the construction of all algebraic equations solvable by radicals are not of direct interest to us. (On the other hand, we must not forget the great significance of Kronecker's number-theoretic investigations, discussed in I.3, for the development of the abstract group concept.)

During that period Kronecker's attitude toward Galois was typical of the attitude of many German mathematicians of the time who, while aiming at sharp formulations of the concepts and methods of Galois, made no decisive contribution to the development of the theory of permutation groups. But the early adoption of the abstract approach did have a strong influence on the development of the concept of an abstract group (see III.4).[105]

Galois's approach met a vastly different response in England. As a result of the concern with the abstract foundations of mathematics, characteristic of the circles around Boole, Cayley, and Sylvester as early as the forties, there developed in England the abstract group concept in the sense of an arbitrary system of elements determined only by defining relations. At first this approach was not generally accepted. But some 25 years later, under different conditions, it was to prove vital to the final evolution of the abstract group concept (see III.4).

During the fifties and sixties, the interaction between the theory of solvability of algebraic equations and the theory of permutations, so important for the development of the permutation-theoretic group concept, was strongest in France and Italy.

The beginnings of this phenomenon of mutual influence are found in the mathematical circles around Liouville. The first important post-Galois publications linking the theory of equations and permutation theory are due to E. Betti (1823–92), one of the leaders of the Italian mathematical school, whose members were closely identified with the Italian fight for national liberation. In a very short time, the Italian school gained international renown in algebraic geometry and the theory of invariants.

Betti's first paper in the series[106] dealing with the two theories appeared in 1851. In 1855, after a number of specialized inquiries into the conceptual and com-

---

(** continued from p. 122)
lich in den *Galois*'schen und *Abel*'schen Fragmenten vermißt werden, und ohne welche die erforderliche Strenge nicht beobachtet werden kann.

putational clarification of various aspects of the complex of problems originating with Galois and the related development of the group idea, Betti embarked on an attempt [37] to present the principles of permutation theory.

Already in his earliest paper of 1851 Betti made it clear that he intended to make more precise the Galois theory now generally known by Liouville's 1846 edition of Galois's essential writings.

* In a Memoir found in vol. XI of Liouville's Journal, Evariste Galois established the foundations of a general theory of solvability by radicals of irreducible algebraic equations and applied it, in particular, to equations of prime degree. In it, Abel's theorem is reestablished, supplemented by means of its inverse and transformed as follows: "For an irreducible equation of prime degree to be solvable by radicals it is *necessary and sufficient* that all the roots be rational functions of any two among them." The profound theory and the deduction from it of the stated theorem are presented in an almost unintelligible manner. This is the result of excessive conciseness. In a note in his Treatise on higher algebra, however, Serret has announced Liouville's intention to publish some day certain developments that will clarify and complete Galois's work. Nevertheless, since Malmsten has already made known to science Abel's theorem, I believe that it will be both proper and useful to show, using theories already developed in Algebra, how it can be transformed into Galois's. In calling attention to this achievement, which the young geometer [that is, mathematician (Translator)] had the time to obtain almost entirely on the basis of his own profound views, I wish that the great Liouville would not deprive the public any longer of the results of the study that he made on the same, for they will undoubtedly provide the stimulus for remarkable progress in Algebra. [31, p. 18]

It is noteworthy that Betti describes Galois's work as the foundation of a general theory. While the content of Betti's paper is of no relevance in the context of this book, it should be noted that he speaks freely of groups of permutations (gruppi di permutazioni). Thus, unlike Serret somewhat later (see below), Betti makes direct use of Galois's terminology.

---

* Evariste Galois, in una Memoria che si trova nel vol. XI del giornale di Liouville, ha posto le basi di una teoria generale sopra la risolubilità per radicali delle equazioni algebriche irriduttibili, e l'ha applicata in particolare a quelle di grado primo. Il teorema di Abel si trova in essa nuovamente stabilito, completato coll'inverso e trasformato in quest'altro: ‚affinchè una equazione irriduttibile de grado primo sia risolubile per radicali, *è necessario e sufficiente* che tutte le radici siano funzioni razionali di due qualunque tra loro'. La profonda teoria, e la deduzione dalla medesima del teorema enunciato sono esposte però in un modo, per concisione esagerata, quasi inintelligibile. Ma il ch. sig. Serret, in una Nota del suo Trattato di Algebra superiore, ha annunziato, che il ch. sig. Liouville ha intenzione di pubblicare un giorno alcuni sviluppi, coi quali egli ha potuto render chiaro e completo il lavoro di Galois. Contuttociò, poichè Malmstén ha posto già la scienza in pieno possesso del teorema di Abel, io credo non sarà disconveniente nè inutile il dimostrare, valendosi di teorie già sviluppate nell'Algebra, come questo si può trasformare in quello di Galois; e richiamando l'attenzione sopra questo frutto, che il giovane geometra ebbe il tempo di ottenere, quasi unicamente, dalle sue profonde vedute, far voti che il celebre Liouville non privi più a lungo il pubblico dei risultati dello studio che ha fatto sulle medesime, e che le renderanno, non può dubitarsi, feconde di molto notevoli progressi nell'Algebra.

Betti's related papers [34–40] show the same unbiased attitude toward Cauchy's theory of permutations. Thus Betti's approach to the solvability theory of algebraic equations can be accurately described as group-theoretic. The most important papers in the series just mentioned are the relatively large "Sulla risoluzione dell'equazioni algebriche" [34] of 1852 and "Sopra la teorica delle sostituzioni" [37] of 1855, intended as a monograph on group theory.

In the introduction to the 1852 paper, Betti glorifies the genius of Galois, brought to light by Liouville's edition of Galois's works, and states that, just as in the earlier paper of 1851, his aim is to systematize Galois's theory and "to add something new to complete the solution of the problem" ("aggiungerne alcuni nuovi per completare la soluzione del problema") [of the solvability of algebraic equations by radicals (Wussing)] [31, p. 32].

The structure of the 1852 paper shows how firmly the permutation-theoretic group concept had become established. The first part ("Parte prima") prepares the group-theoretic auxiliary tools used in the second part, which is devoted to the solution theory. The sections of the first chapter ("Delle sostituzioni"), devoted to permutations, are I. "Principi del calculo delle sostituzioni" and II. "Delle sostituzioni derivate." The sections of the second chapter ("Dei gruppi di permutazioni") are I. "Delle sostituzioni di un gruppo," II. "Dei gruppi derivati simili," and III. "Dei gruppi derivati eguali." The sections of the third chapter ("Della determinazione dei moltiplicatori e divisori di un gruppo") are I. "Equazioni del massimo moltiplicatore e dei divisori di un gruppo," II. "Massimo moltiplicatore dei gruppi di un numero primo di permutazioni sopra un numero eguale di lettere," III. "Massimo moltiplicatore dei gruppi di un numero primo di permutazioni sopra un numero di lettere che ha dei fattori primi differenti tra loro," IV. "Massimo moltiplicatore di un gruppo di un numero primo di permutazioni sopra un numero di lettere che è potenza di un numero primo," and V. "Decomposizione di un gruppo nei suoi divisori primi." Only after this preparation of the group-theoretic tools does Betti deal, in the second part, with the solution theory. The Galois group of an equation is defined in the first paragraph of the first chapter ("Della risolvente de Galois").

I mention a few more elements of the Betti paper of 1852 because of its relevance to the main theme of this book.

Betti defines a "sostituzione" as the operation of transition from permutation—that is, arrangement of n elements x_i—to permutation and employs the notation

$$\binom{x_i}{x_{\varphi(i)}}.$$

Here the function $\varphi(i)$ takes on all n indices i. Betti denotes permutations by

θ, ψ, and so on, and defines the derived permutation $\theta_1 = \psi^{-1}\theta\psi$ of θ by ψ ("la derivata di θ per mezzo di ψ") [31, p. 36]. A group of permutations is defined as follows:

* We call a series of permutations a *group of permutations* if there exists a substitution that enables us to go from one of the permutations to any other and that, when applied to all, merely permutes them among themselves without producing new ones not already belonging to the group. Substitutions that take us from one permutation to all the other permutations of the group are called the substitutions of the group. If the group consists of $\psi_0, \psi_1, \ldots, \psi_{p-1}$, then

$$\psi_m\psi_n = \psi_r,$$

where r is an integer depending on n and m and smaller than p. A group containing n permutations will be said to be *of degree n*. [31, p. 39]

Of special interest is Betti's proof (the first of its kind by anyone) that the Galois group of an equation is actually a group of permutations. Betti's proof closes one of the most critical gaps in the group-theoretic component of the Galois theory. Like Galois, Betti starts with an irreducible algebraic equation of degree μ. Like Galois, he denotes by $V = F(x_0, x_1, x_2, \ldots, x_{\mu-1})$ an always existing function of the roots of the given equation that takes on $\mu! = M$ different values $V_0, V_1, V_2, \ldots, V_{M-1}$ as a result of the application of the M permutations. Betti calls the equation $\theta(V) = 0$ of degree M whose roots are the V_i, $i = 0, 1, \ldots, M-1$, the Galois resolvent—a term not used today:

** Some of the properties of this equation, discovered by Galois, form the foundations, of the present theory. In honor of the great geometer, we shall call it *the Galois resolvent*. [31, p. 59]

Previous theorems enable Betti to write the μ roots of the given equation $f(x) = 0$ in the form

$$\varphi_0(V),\ \varphi_1(V),\ \varphi_2(V),\ \ldots,\ \varphi_{\mu-1}(V),$$

where V is an arbitrary V_i. Following Galois, Betti denotes by $\psi(V) = 0$ the irreducible factor of the equation $\theta(V) = 0$ with root V. Let $\psi(V) = 0$ be of degree ν

* Dicesi *gruppo di permutazioni* una serie di permutazioni tali che una sostituzione mediante la quale si passa da una ad un'altra qualunque di esse, eseguita su tutte, non faccia che permutarle tra loro, senza produrne nessuna nuova che non appartenga già al gruppo. Si dicono sostituzioni del gruppo quelle colle quali si passa da una a tutte le altre permutazioni del medesimo. Se questo siano $\psi_0, \psi_1, \psi_2, \ldots, \psi_{p-1}$ dovrà aversi

$$\psi_m\psi_n = \psi_r,$$

essendo r un valore intero dipendente da m e da n, e minore di p. Un gruppo che contenga n permutazioni lo diremo di *grado* n^{esimo}.

** Questa equazione la quale ha alcune proprietà scoperte da Galois, che fanno il fondamento della presente Teoria, la chiameremo ad onore di questo profondo geometra la *risolvente di Galois*.

with roots $V_0, V_1, \ldots, V_{\nu-1}$. Again following Galois, Betti [31, p. 60] arranges the roots of $f(x) = 0$ in horizontal rows:

* We shall be able to take as roots of (1) [that is, of $f(x) = 0$ (Wussing)] those given by any of the following horizontal lines that form ν different permutations of the same:

$$
(A)\begin{cases}
\varphi_0(V_0),\ \varphi_1(V_0),\ \varphi_2(V_0),\ \ldots,\ \varphi_{\mu-1}(V_0),\\
\varphi_0(V_1),\ \varphi_1(V_1),\ \varphi_2(V_1),\ \ldots,\ \varphi_{\mu-1}(V_1),\\
\vdots\\
\varphi_0(V_{\nu-1}),\ \varphi_1(V_{\nu-1}),\ \varphi_2(V_{\nu-1}),\ \ldots,\ \varphi_{\mu-1}(V_{\nu-1}).
\end{cases}
$$

[31, p. 60]

Betti next wishes to prove that these ν permutations form a group. Briefly, the proof is as follows: The mth permutation goes over into the nth permutation exactly when V_m is replaced by V_n. In view of the rational dependence of the V_i, we have

$$V_m = \lambda(V_n)$$

for a suitable rational function λ. Since $\psi(V) = 0$ is irreducible, the powers of $\lambda(V_n)$ belong to the set $V_0, V_1, \ldots, V_{\nu-1}$, and there exists a least natural number p such that $\lambda^p(V_n) = V_n$. If the values $V_n, \lambda(V_n), \lambda^2(V_n), \ldots, \lambda^{p-1}(V_n)$ do not exhaust all the values $V_0, V_1, \ldots, V_{\nu-1}$, then there are further rational functions $\mu, \mu_1, \mu_2, \ldots$ such that $\mu(V_n), \mu_1(V_n), \mu_2(V_n), \ldots$ represent roots of $\psi(V) = 0$, and the additional roots can be written in the form

$$
(*)\begin{cases}
\mu(V_n),\ \mu\lambda(V_n),\ \mu\lambda^2(V_n),\ \ldots,\ \mu\lambda^{p-1}(V_n),\\
\mu_1(V_n),\ \mu_1\lambda(V_n),\ \mu_1\lambda^2(V_n),\ \ldots,\ \mu_1\lambda^{p-1}(V_n),\\
\vdots
\end{cases}
$$

Betti then argues as follows:

** Now changing V_n to V_m, that is, V_n to $\lambda(V_n)$, we permute cyclically all of the terms of each of the horizontal lines of values $V_n, \lambda(V_n), \lambda^2(V_n), \ldots, \lambda^{p-1}(V_n)$ and (*), that is, all the values $V_0, V_1, V_2, \ldots, V_{\nu-1}$ are permuted among themselves: thus, if we apply any substitution to all [the permutations (Translator)] of (A) that makes them pass from one to another, then they are all permuted among themselves without giving rise to any new permutations; therefore *the ν permutations in (A) constitute a group.* [31, p. 61]

---

* Potremo prendere per radici della (1) (d. i. von $f(x) = 0$, Wg.) quelle date da una qualunque delle seguenti linee orizzontali, che formano ν differenti permutazioni delle medesime:

$$
(A)\begin{cases}
\varphi_0(V_0),\ \varphi_1(V_0),\ \varphi_2(V_0),\ \ldots,\ \varphi_{\mu-1}(V_0),\\
\varphi_0(V_1),\ \varphi_1(V_1),\ \varphi_2(V_1),\ \ldots,\ \varphi_{\mu-1}(V_1),\\
\vdots\\
\varphi_0(V_{\nu-1}),\ \varphi_1(V_{\nu-1}),\ \varphi_2(V_{\nu-1}),\ \ldots,\ \varphi_{\mu-1}(V_{\nu-1}).
\end{cases}
$$

** Ora cangiando V_n in V_m, ossia V_n in $\lambda(V_n)$, si permutano circolarmente in tutti i loro termini ciascuna delle linee orizzontali di valori $V_n, \lambda(V_n), \lambda^2(V_n), \ldots, \lambda^{p-1}(V_n)$ e (*), ossia si permutano (** continued on p. 128)

Betti's 1855 paper "Sopra la teorica delle sostituzioni" [37] is distinguished by strong reliance on symbolism and is concerned with the linear group, a subject studied extensively by C. Jordan in his *Traité des substitutions* of 1870. This work includes the relevant results of Betti as well as of other predecessors of Jordan (see II.4–5).

In sum, one may say that in essence Betti belongs among the commentators on Galois. These were of two types—concerned with the mere closing of the gaps in Galois's work and with the extraction of its group-theoretic core. Betti's place is near the boundary between the two—though he belongs in the second category to the extent that his works deal with the elaboration of the group concept.

Betti's example illustrates the inner dynamic that fostered the development of the permutation-theoretic group concept. This dynamic was the result of a direct link with the line of investigation inspired by Galois. A similar remark applies to T. P. Kirkman's "Theory of Groups and Many-Valued Functions" [344] of 1862.

3 The Development in France

The development in France was more complicated. Liouville's 1846 edition of the most important writings of Galois restored the interaction between the theory of solvability of algebraic equations and group theory based on the theory of permutations. Unlike mathematics in Germany, Italy, and England, however, French mathematics was encumbered by the mistaken idea that Cauchy had given definitive form to the theory of permutations. That is why, in the beginning, there was less progress in the working out of the group-theoretic core of Galois theory in France than elsewhere. But as soon as the lines of development due to Cauchy and to Galois coalesced, there was a deep and rapid advance that disclosed the wide range of the permutation-theoretic group concept throughout mathematics.

The commentary on Galois promised by Liouville in his 1846 edition of Galois's works was never written. On the other hand, we know a few (possibly anecdotal) details of the reception and spread of Galois's ideas among Paris mathematicians immediately after 1846. For example, in a speech in memory of Galois's biographer P. Dupuy, J. Bertrand (1822–1900) quotes Liouville as having said that Galois's

---

(** continued from p. 127)
tra loro tutti i valori $V_0, V_1, V_2, \ldots, V_{\nu-1}$: dunque eseguendo su tutte le (A) una sostituzione qualunque che faccia passare da una ad un'altra delle medesime, esse si permutano tutte quante tra loro, senza che nasca nessuna nuova permutazione; dunque *le ν permutazioni* (A) *costituiscono un gruppo.*

proofs were easy to understand and as having added, after noting his, that is, Bertrand's, amazement, that to reach this conclusion all one had to do was to busy oneself exclusively with them for one or two months and to think of nothing else.

In connection with his promise to write a commentary on Galois, and especially on Galois's difficult "Mémoire," Liouville actually formed a kind of circle for the study of Galois's writings. One of the members of that circle was J.-A. Serret (1819–85). The first edition of Serret's *Cours d'Algèbre supérieure* of 1849, an excellent and innovative textbook of higher algebra, said nothing of Galois's discoveries. We are told by Bertrand that Liouville's failure to write his promised commentary on Galois induced Serret to write a chapter on Galois theory for a planned second edition of his *Cours d'Algèbre supérieure*, which appeared in 1854. Bertrand himself helped correct the galleys and was surprised at the absence of Liouville's name. Serret replied that while he had taken part in the discussion at Liouville's, he had understood nothing. In the end, Serret, out of consideration for Liouville, did not include his lengthy section on Galois theory in the second edition of his book, but filled the 61 pages with completely different material. Nevertheless, he made it clear in the introduction that he had written a commentary on Galois and obliquely reserved the right to its future publication:

> * In changing the object of the 25th lesson that contained elementary theorems on numbers, and in devoting it instead to a complete and detailed exposition of the theory of the new imaginary quantities[107] considered by Galois, I had set for myself the task of facilitating the understanding of a difficult part of the writings of the great geometer. The beautiful Mémoire, entitled *Sur les conditions de résolubilité des équations par radicaux*, contains certain difficulties that I would have strongly wished to clear up by making known, in one of my Notes, the result of my studies on this theory. But the considerations that had held me back at the time of the publication of the first edition of this work continue to impose on me the same constraints today. [616, p. V–VI]

Serret's commentary on Galois was finally included in the third edition of his *Cours d'Algèbre supérieure* published in 1866. This is the first textbook-like presentation of Galois theory and, in particular, the earliest algebra-oriented presentation of group theory.

The remarks that follow are not meant merely as a description of Serret's position in 1866; rather they aim to emphasize the changes in that position from

---

* En changeant l'objet de la vingt-cinquième leçon qui contenait des théorèmes élémentaires sur les nombres, et en la consacrant à l'exposition complète et détaillée de la théorie des nouvelles quantités imaginaires considérées par Galois, je me suis proposé de faciliter l'intelligence d'une partie difficile des écrits de ce grand géomètre. Le beau Mémoire intitulé: *Sur les conditions de résolubilité des équations par radicaux*, ne laisse pas de présenter aussi quelques difficultés que j'aurais vivement désiré éclaircir en faisant connaître, dans l'une de mes Notes, le résultat de mes études sur cette théorie. Mais les considérations qui m'ont retenu lors de la première publication de cet ouvrage m'imposent encore aujourd'hui la même réserve.

1854 to 1866, as well as the inner dynamic of the rapid development leading to the independence of the theory of permutation groups and the gradual rise of the concept of permutation group to a central position in mathematics.

Comparison of the second (1854) and third (1856) editions of Serret's *Cours d'Algèbre supérieure* shows interesting differences[108] in the position of group theory. The two editions differ fundamentally in their content. Nevertheless, the general, and partly historical, remarks contained between the table of contents and the actual text are much the same in both.[109] In both editions algebra is defined as "the Analysis of equations" ("l'Analyse des équations"), and the same words are used to describe the great contributions of Pacioli, Cardano, Ferrari, Lagrange, Gauss, and Abel. Serret states that Abel's studies on the classes of equations solvable by radicals suggested the possibility that these were perhaps the only such classes and hence the question "When, in a word, was an equation solvable by radicals?" ("Dans quel cas, en un mot, une équation peut-elle être résolue algébriquement?") [616, p. 4]. At this point, that is, already[110] in 1854, Serret stresses the merits of Galois:

> * This difficult question was completely answered, at least for irreducible equations of prime degree, by Evariste Gallois [sic! (Wussing)], former student of the Ecole Normale and one of the deepest geometers [that is, mathematicians (Translator)] that France has produced. In a paper presented to the Academy of Sciences in 1831 and published in 1846 thanks to the efforts of Liouville, Gallois [sic! (Wussing)] in effect proved the following beautiful theorem: *For an irreducible equation of prime degree to be solvable by radicals it is necessary and sufficient that, given any two of the roots, the others are rationally deducible.* [616, p. 4]

This is how Serret presents matters in 1854, in the second edition of his textbook on algebra. The third edition contains the same theorems (with Galois's name spelled correctly). All Serret does to acknowledge results found since 1854 is to say that

> ** this important result was the starting point of investigations related to this topic, to which Hermite, Kronecker,[111] Betti, and several other eminent geometers have been dedicated since. [618, Vol. 1, p. 4]

---

* Cette question difficile a été résolue complètement, au moins pour les équations irréductibles de degré premier, par Evariste Gallois, ancien élève de l'Ecole Normale, et l'un des géomètres les plus profonds que la France ait produits. Dans un Mémoire présenté à l'Académie des Sciences en 1831, et publié en 1846 par les soins de M. Liouville, Gallois a, en effet, démontré ce beau théorème: *Pour qu'une équation irréductible de degré premier soit soluble par radicaux, il faut et il suffit que, deux quelconques des racines étant, données, les autres s'en déduisent rationnellement.*

** ce résultat important a été le point de départ des recherches auxquelles se sont livrés depuis sur cette matière MM. Hermite, Kronecker, Betti, et plusieurs autres géomètres éminents.

The previous passage, which had appeared in 1854 and in 1866, indicates a certain precariousness in the evaluation of Galois's contribution by members of the French school in the early sixties. Galois's work is regarded as being of greatest significance in the theory of solution of algebraic equations, but the fundamental role of the group concept in that theory and elsewhere in mathematics is not yet realized. Hence Serret's failure to stress the discovery of the concepts of group and normal subgroup, and his failure to stress the new method. What is singled out is the strongest and deepest theorem in the framework and language of the traditional, Abel-inspired, solvability theory. On the other hand, in view of the Galois-inspired recognition of the significance of permutation theory for solvability theory, Serret discusses permutation theory and adopts Cauchy's terminology and mode of reasoning.

In the second edition of 1854, permutation theory is used exclusively to investigate the number of values of entire rational functions of a certain number of variables and thus is used infrequently. In the third edition, permutation theory takes up about a fifth of the text and is used to prepare the tools of the group-theoretic treatment of the theory of equations in the sense outlined by Galois. The progress achieved between 1854 and 1866 is thus quite considerable.

The third edition of the *Cours d'Algèbre supérieure* came out in two volumes. Compared to the second edition, it is far tighter and more transparent. It is a new book. Instead of 30 lessons with 15 notes at the end of some, the third edition is made up of five sections subdivided into chapters consisting in turn of various numbers of paragraphs. Even in the first sections there are topics, such as continued fractions, absent from the first edition. For the history of group theory, however, by far the most important parts are the new fourth and fifth sections. Section IV, entitled "Les substitutions," consists of five chapters: "(I) Propriétés générales des substitutions," "(II) Propriétés des systèmes de substitutions conjuguées," "(III) Des indices des systèmes conjugués," "(IV) Sur quelques cas particuliers de la théorie des substitutions," and "(V) Applications de la théorie des substitutions." Section V has the heading "La résolution algébrique des équations" and is subdivided into the chapters "(I) Des équations du troisième et du quatrième degré. Considérations générales sur la résolution algébrique des équations," "(II) De l'impossibilité de la résolution algébrique des équations générales au dela du quatrième degré," "(III) Des équations abéliennes," "(IV) Sur un classe d'équations du neuvième degré résolubles algébriquement," and "(V) Sur les équations résolubles algébriquement."

Chapter II of the fourth section is particularly useful in analyzing the interpretation of the word "group." In the first paragraph of that chapter, entitled "Des systèmes conjugués," Serret takes over literally Cauchy's definition of a system of conjugate substitutions[112] corresponding to the permutation-theoretic group concept:

* Given several substitutions formed with n letters. If, when multiplying them one or several times by each other or by themselves, in any order, one always obtains substitutions belonging to the sequence of given substitutions, then the latter form what Cauchy called a *system of conjugate substitutions*, or simply a *conjugate system*. It is clear that any conjugate system contains the substitution equal to unity. [618, Vol. 2, p. 251]

On the basis of this definition, in which closure under multiplication is the sole necessary and sufficient condition, Serret deduces theorems on "systems of conjugate substitutions," that is, permutation groups, and prepares the ground for later applications.

So far Serret follows completely in Cauchy's footsteps. But the Galois conception also appears in this chapter. In the fourth section, "Des groupes de permutations," Serret defines a permutation group.[113] According to him, a "permutation" is a result, an arrangement, whereas a "substitution" is the operation of passing from one "permutation" to another. Given the background of the confused use of the terms "substitution" and "permutation," we must conclude that Serret had in mind the following definition of a permutation group, which adopts Galois's term "le groupe" and the interpretation of a substitution as a transition from permutation to permutation. Serret starts with a system Γ of conjugate substitutions comprising μ substitutions

$$1,\ S_1,\ S_2,\ \ldots,\ S_{\mu-1}$$

on n digits. If we multiply any "permutation" A_0 of these n digits by these μ "substitutions," that is, if, in modern terms, we apply them to A_0, then we obtain μ products

$$A_0,\ S_1A_0,\ S_2A_0,\ \ldots,\ S_{\mu-1}A_0$$

or

$$A_0,\ A_1,\ A_2,\ \ldots,\ A_{\mu-1},$$

"which," says Serret, "form what is called a *group of permutations*" ("qui constituent ce qu'on nomme un *groupe de permutations*") [618, Vol. 1, p. 282]. This definition of "permutation group" by means of the concept of a "system of

* Etant données plusieurs substitutions formées avec n lettres, si, en les multipliant une ou plusieurs fois les unes par les autres ou par elles-mêmes, dans un ordre quelconque, on n'obtient jamais que des substitutions comprises dans la suite des substitutions données, celles-ci constituent ce que Cauchy a nommé un *système de substitutions conjuguées*, ou simplement un *système conjugué*. Il est évident que tout systeme conjugué comprend la substitution égale à l'unité.

conjugate substitutions" establishes a connection between the approaches of Cauchy and of Galois.

Shortly thereafter, the amalgamation of the two approaches is completed in Serret's definition of the substitutions of the permutation group:

* The substitutions of the system Γ, by means of which one passes from permutation to permutation, are called the *substitutions of the group*. [618, Vol. 2, p. 283]

So much for definitions. The amalgamation and interpenetration go so far that, beginning with the end of section IV, that is, 140 pages later, we find in the rest of the text theorems of which the following is a typical example.

** *The substitutions* $1, S_1, S_2, \ldots, S_{\nu-1}$ ... *form a conjugate system. In other words, the* ν *permutations ... form a group*. [618, Vol. 2, p. 420]

We see that the only obstacle to complete amalgamation of the concepts of "system of conjugate substitutions" and "permutation group," an amalgamation that, in terms of content, had been completed long ago, is Serret's insistence—one is inclined to say excessive insistence—on the logically correct but hampering distinction[114] between "permutation" and "substitution." That is why we come across the concept of "substitutions du groupe" (de permutations) but not "le groupe de substitutions," a term belonging to the next period, that of Camille Jordan.

This logical distinction between "substitution" and "permutation," and again, in the use of "system of conjugate substitutions" and "permutation group," can also be found in the work of Betti. But the background of this distinction in Serret's work is not quite the same as in the case of Betti, who relied directly on Galois. At this point, Serret's presentation of permutation theory in Gauchy's terms went objectively beyond the mere preparation of tools for the treatment of equations. What we witness in Serret in 1866 is the transition from the permutation-theoretic to the group-theoretic treatment of equations, and beyond that to the independence of the theory of permutation groups.

There are essentially two places in Serret's book where he presents Galois's accomplishments and results in the solution of algebraic equations.

In section IV, at the end of chapter V, we find the paragraph entitled "Galois's investigations bearing on the preceding theory" ("Recherches de Galois relatives à la théorie précédente"). In that paragraph [618, Vol. 2, pp. 413–420], in connec-

---

* Les substitutions du système Γ, par lesquelles on passe d'une permutation à une autre, sont dites les *substitutions du groupe*.

** *Les substitutions* $1, S_1, S_2, \ldots, S_{\nu-1}$ ... *forment un système conjugué. En d'autres termes, les* ν *permutations ... constituent un groupe*.

tion with Lagrange's results of 1770–71 devoted to the calculation of a function of the roots of an equation given another such function, Serret invokes a theorem of Galois, according to which it is possible to form a function V of the roots of an equation with distinct roots which takes on different values under all $n!$ permutations of the roots, such that each of the roots can be represented as a rational function of V. Serret's proof is a direct repetition of Galois's proof as published by Liouville in 1846.

Serret presents Galois theory proper in the last chapter, "Sur les équations résolubles algébriquement," of section V. He refers explicitly to Galois's paper "Sur les conditions de résolubilité des équations par radicaux" first published in 1846.

From the point of view of content little can be added to Serret's presentation of Galois' accomplishments; Serret follows him very closely. What is of interest are a few of Serret's remarks indicating his own opinion of Galois and of Galois's terminology:

* I have followed the order of the propositions adopted by Galois, but, very frequently, I had to close the gaps in the proofs. [618, Vol. 2, p. 607]

** Galois's investigations rely on the theorems proved under numbers 490 and 492 [existence of the function V referred to above (Wussing)], which thus acquire considerable importance, and on the properties of systems of conjugate substitutions. Galois employs groups of permutations [we discussed this issue above (Wussing)] of which we have spoken ..., but it appeared to us preferable to stick to substitutions. After all, it is just a simple change in the form of the statements of the theorems, dictated by the fact that there is no reason to consider permutations except from he point of view of substitutions by which one passes from one to the other. [618, Vol. 2, p. 608–609]

Nothing need be added to these words. There is probably no better way to describe the coexistence and the now (1866) completed amalgamation of the approaches of Cauchy and Galois.

After these and other preparations, Serret proves that it is possible to associate a group to every equation with simple roots. Here he retains Cauchy's terminology:

---

* J'ai suivi l'ordre des propositions que Galois avait adopté, mais j'ai dû le plus souvent suppléer à l'insuffisance des démonstrations.

** Les recherches de Galois reposent sur les propositions démontrées aux n°s 490 et 492, qui acquièrent ainsi une importance considérable, et sur les propriétés des systèmes de substitutions conjuguées. Galois emploie la considération des groupes de permutations dont nous avons parlé ..., mais il nous a paru préférable de nous en tenir aux substitutions. Au reste, ce n'est là qu'un simple changement dans la forme des énoncés des théorèmes, car il n'y a lieu de considérer les permutations qu'au point de vue des substitutions par lesquelles on passe des unes aux autres.

* *Let $f(x) = 0$ be an equation with n distinct roots $x_0, x_1, x_2, \ldots, x_{n-1}$. There always exists a system of conjugate substitutions G* .... [618, Vol. 2, p. 609]

Otherwise Serret adheres to Galois's terminology and, above all, to Galois's train of thought. His definition of "le *système conjugué propre à l'équation*," which corresponds to Galois's group of the equation, is the same as that of Galois. In a footnote to his concept of "the conjugate system associated with the equation" Serret remarks that

** Galois brings in a group of permutations corresponding to the conjugate system considered here, and calls it the "*group of the equation.*" [618, Vol. 2, p. 613]

The proof itself relies on the same specification of permutations as that of Galois, and is thus, strictly speaking, inconsistent in Serret's sense; the group property is proved by assuming the closure under multiplication of the system of substitutions corresponding to these permutations. Nevertheless, the closure under multiplication is used here immediately; Serret had proved it, following Betti, at the end of section IV [618, Vol. 2, p. 418].

Towards the end, Serret's report on the general theorems found by Galois is reduced to a quotation from Galois in which "group" is replaced by "system of conjugate substitutions." In particular, this is true of the passage in which Galois described the process by which continued adjunction reduces the degree of the equation, which process ends when the group of the equation is reduced to unity.

Close to the end of section V—that is, close to the end of the book—Serret describes new results in Galois theory. At the time (in 1866) the theory continued "to present many difficulties" [618, Vol. 2, p. 634]. Serret refers to the new, but as yet unpublished, investigations on this subject recently submitted to the Paris Academy by C. Jordan [306] (see below). He also refers to a proof of a theorem of Galois communicated to him by Hermite and to Kronecker's paper presented to the Berlin Academy in 1853 by Lejeune-Dirichlet (see above).

4 *Jordan's Commentaries on Galois*

C. Jordan's (1838–1922) two commentaries on Galois illustrate clearly the inner forces set free by the amalgamation of the ideas of Cauchy and Galois. The first is "Commentaire sur le Mémoire de Galois" [306] of 1865 and the second is

---

* *Soit $f(x) = 0$ une équation de degré n dont les n racines $x_0, x_1, x_2, \ldots, x_{n-1}$ sont inégales. Il existe toujours un système de substitutions conjuguées G* ....

** Galois fait intervenir un groupe de permutations correspondant au système conjugué dont il est ici question, et il l'appelle le *groupe de l'équation.*

"Commentaire sur Galois" [316] of 1869. The similarity of the titles notwithstanding, the first commentary is essentially of the gap-filling variety, while the second is unmistakably concerned with making precise the group-theoretic component of Galois theory.

In the 1865 commentary Jordan tries to make Galois's results transparent and verifiable and to fit them into contemporary mathematics. With no group-theoretic preliminaries, Jordan presents Galois's procedure of stepwise adjunction of the roots of a given equation and its effect on the stepwise reduction of the order of the group of the equation. The result is a precise formulation of Galois's theorems on solvability by radicals, some of the statements being direct quotations from Galois. Jordan gives no proofs. The essential difference between this commentary and the third edition of Serret's *Cours d'Algèbre supérieure* is that Jordan immediately uses the term "le groupe de substitutions." The term "transformation of a group," which goes back to Cauchy, is used to describe the normality property of a subgroup:

* If all the roots of an equation $\psi(z) = 0$ are rational functions of any r of them, then the different groups

$$1, a, a_1, \ldots; \; 1, b^{-1}ab, \; b^{-1}a_1b, \ldots; \; 1, c^{-1}ac, \; c^{-1}a_1c, \ldots$$

are all formed by the same substitutions. The unique group is transformed into itself by all the substitutions of G. [306, p. 772][115]

In terms of content, the main contribution of this paper of Jordan is the elaboration of the fundamental role of normal subgroups for the solvability of equations by radicals.

** Let the group G of an equation $F(x) = 0$ contain a group H such that all the substitutions of G transform H into itself. Let N be the number of substitutions in H and $M = N \cdot \nu$ the number of substitutions in G. It is possible to reduce the group of the equation to the substitutions H by solving an equation of degree ν whose group contains just ν substitutions .... Conversely, if the group G contains no group H with the above property [of normality (Wussing)], then it is impossible to reduce the solution of the initial equation to that of other equations whose groups contain a smaller number of substitutions. [306, p. 772]

---

* Si l'équation $\psi(z) = 0$ est telle que toutes ses racines soient des fonctions rationnelles de l'une d'elles r, les divers groupes

$$1, a, a_1, \ldots; \; 1, b^{-1}ab, \; b^{-1}a_1b, \ldots; \; 1, c^{-1}ac, c^{-1}a_1c, \ldots$$

seront tous formés des mêmes substitutions. Ce groupe unique H sera transformé en lui-même par toutes les substitutions de G.

** Si le groupe G d'une équation $F(x) = 0$ est tel, qu'on puisse déterminer un groupe H contenu dans G, et que toutes les substitutions de G transforment en lui-même, soit N le nombre des substitutions de H, $M = N \cdot \nu$ celui des substitutions de G: on pourra réduire le groupe de l'équation aux substitutions H par la résolution d'une équation de degré ν, dont le groupe contient seulement (** continued on p. 137)

The second "Commentaire sur Galois," however, goes beyond Galois in the principal group-theoretic content of Galois theory. This paper of Jordan, comprising just a few pages, and his splendid *Traité des substitutions* of 1870 are the two documents that attest the conclusive development within the French school of the group concept in the permutation-theoretic sense.

Jordan begins his "Commentaire sur Galois" [316] with preliminary remarks ("Préliminaires") in which he presents the basic concepts of group theory required for the development of Galois theory, namely, substitution, unity, the product of two substitutions, group, "the derived group" (le groupe dérivé), the order of a group, the degree of a group [which is the same as the degree of each of its permutations (Translator)], the transitive group, the transform of a group, and simple and composite groups. Jordan defines a substitution as the operation of interchanging the elements of a (finite) set; the unity (unité) substitution leaves all the elements fixed. The product (produit) ab of two substitutions is the substitution corresponding to the successive application of a and b; a^{-1} is defined by the requirement that when multiplied by a it yields unity. After that Jordan defines a group:

* A system of substitutions forms a group if the product of any two substitutions of the system is again a member of the system. [316, p. 141]

Jordan uses this definition to define "le groupe dérivé de a, b, c, ...," where a, b, c, ... are permutations. In modern terms, this is the group generated by a, b, c, ...

** The different substitutions obtained by applying certain given substitutions a, b, c, ... as many times as one wishes and in any order, form a group derived from a, b, c, ... [316, p. 141]

Closure under multiplication is the sole property required of a group. This is the case both in the definition of a group and in the presentation of Galois theory in the text proper; this is all that is proved in establishing the existence of a group. A typical example is given below.

(** continued from p. 136)
ν substitutions .... Au contraire, si le groupe G est tel, qu'il ne contienne aucun autre groupe H jouissant de la propriété précédente, il sera impossible de ramener la résolution de l'équation proposée à celle d'autres équations dont le groupe contienne un moindre nombre de substitutions.

* Un système de substitutions forme un groupe, si le produit de deux substitutions quelconques du système appartient lui-même au systême.

** Les diverses substitutions obtenues en opérant successivement tant qu'on voudra, et dans un ordre quelconque certaines substitutions données a, b, c, ... forment un groupe dérivé de a, b, c, ...

Only in the "Préliminaires" does Jordan prove that every group contains a unit element and, for each substitution a, another substitution a^{-1}, that is, the inverse of a. Terms like "inverse," "reciprocal," or their equivalents, however, are not used:

* The successive powers of a substitution a are distinct up to the power a^{λ} that reduces to unity. Beyond that point they repeat periodically. All groups contain the unit substitution. If a group contains a it also contains

$$a^{\lambda-1} = a^{-1}. \qquad [316, p. 141]$$

The definition of a normal subgroup—the term itself is not used—is based on the concept of a transformation of a permutation a by a permutation b; $b^{-1}ab$ is called the "transform of a by b." If $(a, a_1, \ldots)$ is the group generated by $a, a_1, \ldots$ (in Jordan's terminology, the derived group), then $(b^{-1}ab, b^{-1}a_1b, \ldots)$ is called the "transform of the group $(a, a_1, \ldots)$ by b." If it coincides ("se confond") with $(a, a_1, \ldots)$, then it is said to commute with b (to be "permutable" with b). Apart from terminology, this coincides with Galois's "proper decomposition." There follows the definition of a simple group:

** A group is simple if it does not contain another group whose substitutions are permutable (except the one formed by the single substitution 1), and composite otherwise. [316, p. 142]

The above use of the digit 1 for the unit ("unité") is also found in Cauchy. At the end of the "Préliminaires" we find a proof of the theorem that the order of a subgroup divides the order of the group. The term "subgroup" (or its synonym) is not used; Jordan uses only the verb "contenir" (to contain).[116]

*** A group contains another if it contains all of its substitutions ... *Theorem:* If a group H is contained in another group G, then its order n divides the order N of G. [316, p. 142]

The group-theoretic preliminaires are used in the text proper, which is devoted to the solution of algebraic equations. To illustrate Jordan's group-theoretic arguments, I reproduce the decisive part of the text devoted to a vital aspect of Galois theory, namely, the association of a permutation group to an algebraic equation with simple roots.

---

* Les puissances successives d'une substitution quelconque a sont toutes distinctes, jusqu'à l'une d'elles a^{λ} qui se réduit à l'unité; au delà elles se reproduisent périodiquement. – Donc tout groupe contient la substitution unité; et tout groupe qui contient a contiendra

$$a^{\lambda-1} = a^{-1}.$$

** Un groupe est simple, s'il ne contient aucun autre groupe auquel ses substitutions soient permutables (sauf celui qui est formé de la seul substitution 1): composé dans le cas contraire.
*** Un groupe en contient un autre, s'il contient toutes ses substitutions ... *Théorème:* Si un groupe H est contenu dans un autre groupe G, son ordre n divise N, ordre de G.

The existence of the Galois group of an equation is stated in Jordan's "Théorème Ier":

* Let $F(x) = 0$ be an equation with distinct roots $x_1, \ldots, x_m$, to which one may have adjoined certain auxiliary quantities $y, z, \ldots$. There always exists a group of substitutions of the roots $x_1, \ldots, x_m$ such that a function of the roots is rationally expressible if and only if its numerical value is unaltered by the substitutions of that group. [316, p. 145]

The first half of the proof of this fundamental theorem of Galois theory, which establishes the connection between the theory of equations and permutation theory, is the same as Galois's proof. Having earlier proved Galois's lemma II, Jordan assumes the existence of a function V that takes on $m!$ different values under all the $m!$ permutations of the x_i; like Galois, he chooses V_1 as one of the values of $V = M_1x_1 + M_2x_2 + \cdots$ with suitable coefficients M_i. He denotes by $1, a, b, c, \ldots$ all possible permutations and by $V_a, V_b, \ldots$, the result of applying $a, b, \ldots$ to V_1. Then V_1 satisfies the equation $(X - V_1)(X - V_a)(X - V_b)(X - V_c) \cdots = 0$ with rational coefficients. Jordan considers that irreducible factor of that equation (which may coincide with the full equation) of which V_1 is a root and denotes it by

$$(*) \qquad (X - V_1)(X - V_a)(X - V_b) \cdots = 0.$$

Galois had defined the permutations $1, a, b, \ldots$ [on the irreducible factor (Translator)] as the group of the equation. Now Jordan proves the second half of the existence theorem, to the effect that (to use modern terms) a rational function of the roots belongs to the smallest field containing the coefficients of the equation and the quantities adjoined to the equation if and only if it admits the permutations of the group; in Jordan's terms it is "exprimable rationellement." Thus it remains to prove the group property of $1, a, b, \ldots$, that is, closure under multiplication:

** It remains to prove that the substitutions $1, a, b, \ldots$ form a group. This presents no difficulty. [316, p. 146]

The coefficients of the equation (*) in X belong to the field of the original equation. "Since it is a function of the indeterminate X with rational coefficients, it remains unchanged by the substitutions $1, a, b, \ldots$" ("étant une fonction de l'in-

* Soit $F(x) = 0$ une équation dont les racines $x_1, \ldots, x_m$ sont toutes inégales, et à laquelle on peut supposer adjointes certaines quantités auxillaires $y, z, \ldots$ Il existera toujours entre les racines $x_1, \ldots, x_m$ un groupe de substitutions tel, que toute fonction des racines, dont les substitutions de ce groupe n'altérent pas la valeur numérique, soit rationnellement exprimable, et réciproquement.
** Il reste à démontrer que les substitutions $1, a, b, \ldots$ forment un groupe, ce qui n'offre pas de difficulté.

déterminée X, à coefficients rationnels, ne devra varier par aucune des substitutions 1, a, b, ...") [316, p. 146]. By applying one of the permutations, say a, we obtain

* $$(X - V_a)(X - V_{a^2})(X - V_{ba}) \cdots .$$

In order that this new polynomial be identical with

$$(X - V_1)(X - V_a)(X - V_b) \cdots$$

for all X, it is necessary that, apart from order, the quantities

$$V_a, V_{a^2}, V_{ba}, \ldots$$

be the same as the quantities V_1, V_a, V_b, ... But by hypothesis, two different substitutions give for the function V two essentially different values. It follows that, apart from order, a, a^2, ba, ... are the same as 1, a, b, ... Thus, if a and b are any two substitutions of the sequence 1, a, b, ..., then ba belongs to that sequence. Thus the substitutions of that sequence form a group. [316, p. 146].

This proves the multiplicative closure, and thus the group property of this complex of permutations. It is important to note the terminology employed by Jordan. Unlike Galois, who referred to any complex of permutations as "groupe," Jordan calls a set of permutations a "suite" and not a "groupe," and goes on to give a proof that the special "suite" he considers is a "groupe."[117]

The group introduced by this existence theorem depends on the choice of the quantities y, z, ... adjoined to the equation. Jordan denotes this group as "the group of the equation relative to the adjoined quantities y, z, ... ("le groupe de l'équation relatif aux quantités adjointes y, z, ...") [316, p. 146]. Among all the groups obtained in this manner, one group is noteworthy in that it can be designated, "in an absolute sense," as "the group of the equation" ("d'une manière absolue," "le groupe de l'équation"). That is the group which is obtained when no quantities are adjoined to the equation.

Already this "Commentaire" ... contains the theorem about the uniqueness of the orders of the factor groups, a subject Jordan explored in greater detail

---

* $$(X - V_a)(X - V_{a^2})(X - V_{ba}) \ldots .$$

Pour que ce nouveau polynôme soit identique à

$$(X - V_1)(X - V_a)(X - V_b) \ldots,$$

quelque soit X, il faut nécessairement que les quantités

$$V_a, V_{a^2}, V_{ba}, \ldots$$

ne soient autres que les quantités V_1, V_a, V_b, ... à l'ordre près. Mais par hypothèse, deux substitutions distinctes donnent pour la fonction V des valeurs essentiellement différentes. Il faudra donc que a, a^2, ba, ... soient identiques à l'ordre près à 1, a, b, ... Donc a et b étant deux substitutions quelconques de la suite 1, a, b, ..., ba appartiendra à cette suite. Donc les substitutions de cette suite forment un groupe.

one year later in his *Traité*. Theorem IX of Jordan's Commentary asserts the following:

* Let $F(x) = 0$ be an equation whose group G is composite. Let G, I, I_1, ... be a sequence of groups such that 1°. each of these groups is contained in its predecessor and commutes with its substitutions [the normality property (Wussing)], 2°. each group is the most general among those that satisfy this double condition [the maximality property (Wussing)]. Let the respective orders of these groups be N, N/ν, $N/\nu\nu_1$, ... Then the solution of the proposed equation wil' depend on the solution of successive equations whose groups are simple and contain, respectivelyl ν, ν_1, ... substitutions. [316, p. 152]

After proving this theorem, Jordan continues:

** This theorem readily suggests the idea of classifying equations with composite group according to the number and value of the composition factors ν, ν_1, ... [note the terminology! (Wussing)]. But the sequence of groups G, I, I_1, ... can be determined in several ways. This being so, it is absolutely necessary to justify this classification; to prove that no matter which way one chooses this sequence one will always find the same factors, apart from order. [316, p. 152–153]

For the express purpose of saving space, Jordan refrained from supplying, at this point, a proof of the last statement.

5 *Jordan's Traité des Substitutions et des Équations Algébriques*

Jordan's *Traité des substitutions et des équations algébriques* [317] shows that the development of the permutation-theoretic group concept was now at last complete. Though this work presents no aspects of the general theory that are not already found in the "Commentaire," it does contain some details that seem notable in view of the developments that would occur after 1870.

Our first comment concerns, the *Traité* as a whole. The title of this comprehensive work of 667 quarto pages is excessively modest and therefore misleading. The work represents not only a definitive solution of the problem formulated by Galois, but also a review of the whole of contemporary mathematics from the

---

* Soient $F(X) = 0$ une équation dont le groupe G soit composé: G, I, I_1 ... une suite de groupes tels 1°. que chacun d'eux soit contenu dans le précédent et permutable à ses substitutions, 2°. qu'il soit aussi général que possible parmi ceux qui satisfont à cette double propriété; N, N/ν, $N/\nu\nu_1$, ... les ordres respectifs de ces groupes: La résolution de l'équation proposée dépendra de celle d'équations successives, dont les groupes seront simples, et contiendront repectivement ν, ν_1, ... substitutions.

** Ce théorème suggère naturellement l'idée de classer les équations à groupe composé d'après le nombre et la valeur des facteurs de composition ν, ν_1, ...: mais la suite des groupes G, I, I_1, ... peut souvent être déterminée de plusieurs manières: il est donc absolument nécessaire, pour justifier cette classification, de prouver que, de quelque manière que l'on choisisse cette suite, on trouvera toujours, à l'ordre près, les mêmes facteurs.

standpoint of the occurrence of group-theoretic thinking in permutation-theoretic form. We shall return to this point in II.5.

As already noted, the permutation-theoretic group concept is fully formulated in the *Traité* . This comes through clearly in the terminology. The term "system of conjugate substitutions" does not appear at all. Jordan uses Galois's "group" (le groupe) and, strangely enough, the term "bundle" (faisceau). In Jordan's definition, the one condition that a complex of permutations must satisfy to be a group is closure under multiplication:

* One says that a system of substitutions forms a *group* (or a *bundle*) if the product of any two of the substitutions of the system is again a member of that system. [317, p. 22]

This is the same definition as that in his "Commentaire sur Galois."

The independent role of the group concept finds expression in the whole organization of the *Traité*. While book I discusses congruences, book II is devoted to a discussion of permutations from a group-theoretic point of view. The section headings refer to primitive and imprimitive groups, simple and composite groups, and the alternating group. The terminology is the same as that of the "Commentaire sur Galois" except for the change from "sequence" (suite) in "Commentaire" to "system" (système) in the *Traité*.

In book II, §IV [317, p. 42–48], Jordan proves the uniqueness of the "composition factors" ("facteurs de composition") of a group G; this proof had been omitted for lack of space in the "Commentaire." It relies entirely on group-theoretic considerations. In distinction to its treatment in the "Commentaire," this theorem (later named for Jordan) is formulated in the *Traité* as a purely group-theoretic result; the connection with the theory of solvability of equations is not even mentioned.

Jordan also treats the historical starting point of the permutation-theoretic group concept, that is, the problem of the number of values taken on by a function of n variables as a result of their permutation, from a strictly group-theoretic point of view (book II, §V). The correspondence between functions and groups is elaborated with particular lucidity. Jordan relies on material which had been thoroughly researched and comprehensively presented by a number of mathematicians from Lagrange to Cauchy and Kirkman.

This part of the work also discusses the isomorphism of permutation groups—a model for a variety of future investigations. The definition of the isomorphism of groups is entirely modern, except for its historically unavoidable but logically

* On dira qu'un système de substitutions forme un *groupe* (ou un *faisceau*) si le produit de deux substitutions quelconques du système appartient lui-même au système.

pointless permutation-theoretic disguise of the abstract group concept. In Jordan's words,

* A group Γ is said to be *isomorphic* to a group G if it is possible to establish between their substitutions a correspondence such that: 1° to each substitution of G there corresponds a unique substitution of Γ and to each substitution of Γ one or more substitutions of G; 2° to the product of any two substitutions of G there corresponds the product of their respective corresponding substitutions [this is, of course, our definition of a homomorphism from G to Γ (Translator)]. [317, p. 56]

This is followed shortly by a distinction:

** An isomorphism is said to be *meriedric* if many substitutions of G correspond to the same substitution of Γ, and *holoedric* in the opposite case. [317, p. 56]

Here we have an example of the interesting phenomenon of the incomplete exploitation of the idea inherent in a concept—in this case the concept of "isof morphism of permutation groups." The absence of the abstract formulation o- the group concept, the clinging to the view that the elements of a group must be permutations, barred Jordan from discovering the theorem that asserts the isomorphism of the composition factors of different composition series of the same group—a theorem proved by O. Hölder in 1889 precisely on the basis of an abstract view of the nature of group elements.

Nevertheless, the part of the Jordan-Hölder theorem established by Jordan entirely sufficed for the complete mastery of the problem put forward by Abel and Galois of finding all equations solvable by radicals. This problem runs through the whole of the *Traité*. Book III, chapter II is concerned with the groups of abelian and galois equations solvable by radicals, and books IV and V give the final solution. The presentation of Jordan's solution, however, lies far outside the aims of the present work.

Jordan's "Commentaire sur Galois" of 1869 and his *Traité* of 1870 mark the end[118] of the process of development of the permutation-theoretic group concept, a development that, through preoccupation with the Cauchy type of permutation theory, had lasted five to ten years longer in France than in Italy and in England. Meanwhile, around 1870, the way was opened for a more abstract view of groups in Germany (I.3) and in England (III.4). But this view was not immediately accept-

---

* Un groupe Γ est dit *isomorphe* à un autre groupe G, si l'on peut établir entre leurs substitutions une correspondance telle: 1° que chaque substitution de G corresponde à une seule substitution de Γ, et chaque substitution de Γ à une ou plusieurs substitutions de G; 2° que le produit de deux substitutions quelconque de G corresponde au produit de leurs correspondantes respectives.

** L'isomorphisme sera dit *mériédrique*, si plusieurs substitutions de G correspondent à une même substitution de Γ, *holoédrique* dans le cas contraire.

ed by all. Jordan himself did not follow the road of abstraction and subsequent axiomatization that completely deprived the group elements of their concrete character. In the introduction to his *Traité*, Jordan dismissed such approaches in general, and Kronecker's approach in particular, in these words:

> * We would have liked to present a larger part than we have of the work on equations of this illustrious author. Various causes have prevented us from so doing: the totally arithmetic nature of his methods, so different from our own; the difficulty of having completely to reconstitute a number of proofs most often barely indicated; and finally, the hope of seeing one day these beautiful theorems, which are now the envy and despair of geometers, grouped in one body of coherent and complete doctrine. [317, p. viii]

This passage illustrates the misunderstanding that Jordan had failed to resolve. The "despair" of mathematicians in the face of difficulties connected with the application of group theory could be surmounted not by turning away from, but rather by turning toward, a more abstract concept of a group. It was precisely the latter tendency that would prove the essence of the progress achieved in the period following Jordan's *Traité* (see III.2–4).

---

* Nous aurions désiré tirer un plus grand parti que nous ne l'avons fait des travaux de cet illustre auteur sur les équations. Diverses causes nous en ont empêché: la nature tout arithmétique de ses méthodes, si différentes de la nôtre; la difficulté de reconstituer intégralement une suite de démonstrations le plus souvent à peine indiquées; enfin l'espérance de voir grouper un jour en un corps de doctrine suivi et complet ces beaux théorèmes qui font maintenant l'envie et le désespoir des géomètres.

5 The Theory of Permutation Groups as an Independent and Far-Reaching Area of Investigation

The development of the concept of a permutation group marked the first stage in the evolution of the abstract group concept. In the next chapter we shall consider the factors that forced the broadening of the concept of a permutation group and that liberated group theory by dispensing with the need to represent group elements by means of permutations.

This far-reaching process of abstraction was paralleled in time, for a period of decades, by a process that had begun in the 1840s and consolidated the theory of permutation groups as an independent area of investigation. This consolidation revealed a multiplicity of possible applications that propelled the permutation- theoretic group concept to the position of a fundamental concept of mathematics. In turn, that insight gave direct support to the shift in the direction of the abstract group concept. We shall look from the latter point of view at the rich historical material represented by papers on the theory of permutation groups.

Since the stages of a historical development do not end on a specific day and at a specific hour, it is not surprising that the development of the permutation-theoretic group concept went hand in hand with the growing independence of the theory of permutation groups and the application of that theory to broad areas of mathematics. This phenomenon can be seen most clearly in Serret's *Cours d'Algèbre supérieure* of 1866 and Jordan's *Traité* of 1870, which must therefore be regarded as being crucial not only in terms of content but also in terms of the historical development of group theory; this all the more so because Serret and Jordan, especially the latter, had also elaborated earlier attempts to apply the permutation-theoretic group concept to analysis, geometry, and other areas such as theoretical mechanics. In what follows, I shall occasionally disregard strict chronology and take into consideration earlier cases of applied permutation theory and of the application of the permutation-theoretic group concept.

1 Group Theory as an Independent Area of Investigation in the Work of Serret

The level of independence achieved by permutation-based group theory in 1866 can be gauged with all desirable clarity by a study of section IV, chapter II of Serret's *Cours d'Algèbre supérieure*. One paragraph is titled "Du problème général qui fait l'objet principal de la théorie des substitutions." The main problem of the theory of *permutations* is already formulated as a group-theoretic problem:

* The general problem considered in the theory of substitutions can be stated as follows: *What are the systems of conjugate substitutions that can be formed with n given letters?* [618, Vol. 2, p. 256]

Serret adds that the problem remains unsolved in spite of its tremendous importance for algebra and in spite of the great efforts of many scientists. (This is still true today.) Serret concludes with these words:

** The science of today has just a few general propositions on this subject that we shall establish here. [618, Vol. 2, p. 256]

To clarify the level of development and independence of permutation-theoretic group theory, I shall state the main theorems selected by Serret. For the sake of brevity, I shall use modern terminology.

Theorem I: The $n!$ permutations of n numbers form a group containing the alternating group, whose order is $n!/2$.

Theorem II: If a group contains all cyclic permutations whose order is a given number $p \leqslant n$, then the order of the group is $n!$ or $n!/2$. If p is even, then that order is always $n!$.

Theorem III: All the permutations on n numbers that commute with a permutation T on n numbers form a group.

Theorem IV: Let T be a permutation on n numbers. All the permutations S on n numbers such that STS^{-1} is a power of T form a group.

Corollary I to Theorem V: For n numbers, we can form groups of order $n \cdot \varphi(n)$ and $n \cdot \varphi(n)/t$, where $\varphi(n)$ is the Euler function and t is a divisor of n. (This theorem follows from a theorem discovered by Jordan [305], which is an extension of Serret's Theorem III.[119])

Corollary II to Theorem V: For $n - 1$ numbers, we can form groups of order $\varphi(n)$ and $\varphi(n)/t$.

Theorem VI: Let

$$1,\ S_1,\ S_2,\ \ldots,\ S_{\mu-1} \quad \text{and} \quad 1,\ T_1,\ T_2,\ \ldots,\ T_{\nu-1}$$

be two groups of permutations on n numbers of respective orders μ and ν. If the two groups commute and are disjoint, then by multiplying all the S_i on the right, or on the left, by all the T_j we obtain a group of order $\mu \cdot \nu$.

---

* Le problème général que l'on a en vue dans la théorie des substitutions peut être énoncé dans es termes suivants: *Quels sont les systèmes de substitutions conjuguées que l'on peut former avec n lettres données?*

** La science ne possède aujourd'hui sur ce sujet qu'un petit nombre de propositions générales que nous allons établir ici.

Theorem VII: Given n numbers. Suppose $n_1 \leqslant n$, n_1 an integer $\geqslant 1$, let $\nu = m_1 n_1$, and let ν be a divisor of n. Further, suppose $n_2 \leqslant m_1$, n_2 an integer $\geqslant 1$, and $m_2 n_2 \mid m_1$. Again, suppose $n_3 \leqslant m_2$, $m_3 n_3 \mid m_2$, ... For any ν numbers chosen from among the given n numbers it is always possible to form a group of order

$$n_1^{m_1} n_2^{m_2} n_3^{m_3} \cdots .$$

(The content of this theorem is due to Cauchy. Serret quotes Cauchy and refers to [108].)

Theorem VIII: Given a group of permutations on m numbers of order μ and a group of permutations on p numbers of order ω, one can construct a group on mp numbers of order $\mu^p \omega$.

Theorem IX: If a group on n numbers contains all third-order cycles that can be formed with $n - 1$ numbers as well as additional permutations, then it contains all third-order cycles on n digits.

Theorem X: If a group of permutations on n numbers, $n > 4$, does not contain all third-order cycles on $n - 1$ numbers but contains all such cycles on $n - 2$ numbers, then the permutations of the group that permute the two remaining numbers can permute them only among themselves.

These are the main theorems given by Serret that deal with the construction of possible permutation groups.

In the next chapter (chapter III), "Des indices des systèmes conjugués", Serret follows Cauchy (see II.2) in defining the index of a permutation group relative to the symmetric group on n letters rather than the index of a subgroup of an arbitrary permutation group on the same number of letters.

In this same chapter Serret summarizes the results of Cauchy and then those of Bertrand, Mathieu, and himself. As regards the solution of algebraic equations, Galois's work made it obvious that the decisive question is that of a lower bound greater than 2 on the group indices (relative to the symmetric group on the same number of letters). Here too Serret summarizes the relevant results from the work of Cauchy.[120] As for his own contribution, Serret, after bridging a gap in Bertrand's paper [28], gives a complete proof [612] of the fact that no permutation group on a set of n letters can have index greater than 2 and less than n. This theorem is formulated as follows:

* *For n different from* 4, *the index of a system of conjugate substitutions on n letters cannot at the same time be greater than* 2 *and less than n.* [618, Vol. 2, p. 290]

---

* *L'indice d'un système de substitutions conjuguées, formées avec n lettres, ne peut être en même temps supérieur à* 2 *et inférieur à n, à moins que n ne soit égal à* 4.

Further paragraphs in that chapter deal with theorems exploring the connection between multiple transitivity of groups and group indices. Here Serret again follows Cauchy but also takes into consideration the results of Mathieu.[121]

2 *New Aspects of the Conceptual Content of Permutation Groups in the Work of Serret and Jordan*

In addition to the phenomenon of independence of permutation theory, we witness already in the third edition of Serret's *Cours d'Algèbre supérieure* another extremely important phenomenon. As a result of the application of the concept of a permutation group outside the theory of equations, the conceptual content of permutation theory was forced into the open. To be sure, in Serret's case, and as early as 1870 in Jordan's case in his *Traité,* we see this only as a tendency; in other words, in the work of neither of these authors is it possible to discern an actual, conscious, that is, definition-based, extension of the concept of a permutation group. All the relevant additions—loosely connected in the work of Serret but systematically ordered by Jordan—are based conceptually on the definition of a permutation group. Looked at abstractly, however, they coincide with the concept of a finite group of linear substitutions (see III.3), and thus foreshadow, in content, if not conceptually, the representation theory developed by Molien, Frobenius, and Schur at the end of the nineteenth century (III.3).

Already in 1859 [617], following Galois [122] and Hermite [276], Serret envisioned applications of permutation theory going beyond the theory of algebraic equations, such as the application of group-theoretic thinking to a set of fractional linear functions of one variable. These extensions appear in Serret's third edition (1866) under the rather neutral heading "Sur quelques cas particuliers de la théorie des substitutions" in section IV of chapter IV. Of particular interest is the section "Des fonctions analytiques propres à représenter les substitutions" [618, Vol. 2, pp. 356–363 and subsequent applications].

Serret starts with substitutions determined by functions of the form $(ax + b)/(a'x + b')$, which he calls "linear functions." The coefficients can be arbitrary numbers. He defines $\theta x = ax + b/a'x + b'$ and then by recursion.

$$\theta^2 x = \theta\theta x, \ \ldots, \ \theta^m x = \theta\theta^{m-1} x.$$

Then the cofficients of

$$\theta^m x = \frac{a_m x + b_m}{a'_m x + b'_m}$$

can be expressed in terms of a, b, a', b'.

Now Serret asks for necessary and sufficient conditions for the μth power of θx to "reproduce x" (reproduire x) [618, Vol. 2, p. 133]. After elementary calculations he obtains the relation

$$(a + b')^2 - 4(ab' - ba') \cos^2 \frac{\lambda\pi}{\mu} = 0,$$

where λ is relatively prime to μ. By assuming that the coefficients of θx are real, by noting that it is no restriction of generality to suppose that $ab' - ba' = 1$, and by a suitable convention on signs, Serret obtains

$$a + b' = 2 \cos \frac{\lambda\pi}{\mu}, \qquad (\lambda, \mu) = 1,$$

as the required necessary and sufficient condition for $\theta^\mu x = x$.

By means of this result he can solve the following problem:

* Given the linear function $(a_m x + b_m)/(a'_m x + b'_m)$, find a linear function $\theta x = (ax + b)/(a'x + b')$ such that one has, identically,

$$\theta^m x = \frac{a_m x + b_m}{a'_m x + b'_m} \quad \text{and} \quad \theta^\mu x = x. \qquad [618, Vol. 2, p. 335]$$

In the next section, "Des fonctions rationnelles linéaires prises suivant un module premier," Serret presents results established, for the most part, by himself [617]. They involve an amalgamation of analysis, number theory, and permutation theory and help, in a roundabout way, to advance group theory.

** The developments that will follow have consequences that are of interest for number theory and that are, above all, useful in the theory of substitutions. [618, Vol. 2, p. 336]

Serret considers the functions

$$\theta z = \frac{az + b}{a'z + b'}, \qquad a, b, a', b' \text{ integers.}$$

All numbers are taken modulo p, where p is an odd prime. Serret eliminates the case when θz reduces to a constant. He calls the expression $\Delta = ab' - ba'$, of interest in what follows, the determinant (le determinant) of the linear function θz, and divides the functions θz into two classes: θz belongs to the

* Etant donnée une fonction linéaire $(a_m x + b_m)/(a'_m x + b'_m)$, trouver une fonction linéaire $\theta x = (ax + b)/(a'x + b')$ telle, que l'on ait identiquement $\theta^m x = (a_m x + b_m)/(a'_m x + b'_m)$ et $\theta^\mu x = x$.

** Les développements qui vont suivre conduisent à de conséquences interessantes pour la théorie des nombres et qui sont surtout utiles dans la théorie des substitutions ....

first class if its determinant is a quadratic residue mod p, and to the second class otherwise.

Serret shows[123] that there are altogether $N = (p + 1)\,p(p - 1)$ linear functions. He imposes on this finite set of functions a composition rule $\theta\theta_1 z$ (a kind of multiplication), which amounts to successive application of the operations designated by θ_1 and θ, respectively. Using the rule of multiplication of determinants, Serret concludes that

> * it is clear that the set of all linear functions forms a group such that the product of a number of functions of the group is again part of that group. One sees from the preceding that the same is true of just the set of linear functions of the first class but not of just the set of linear functions of the second class. [618, Vol. 2, p. 338]

This represents an advance of group-theoretic thinking and of the group concept itself into a new area of mathematics lying outside the domain of application of the concept of "permutation group," though this extension by Serret is, so to speak, spontaneous and not definitive.

The air of transition in the content and scope of the term "le groupe" is also evident in the part of the *Cours d'Algèbre supérieure* that contains results obtained by Serret after 1859.

Using Euler's reasoning (see I.3), Serret defines the order ("l'ordre") of a linear function modulo p abstractly, that is, independently of the nature of the elements involved. Since there are only finitely many functions θz, the sequence $z, \theta z, \theta^2 z, \ldots$ must be periodic. The smallest positive number n such that[124] $\theta^n z \equiv z \pmod p$ is called the order of the function. Serret sets $\theta^0 z \equiv z \pmod p$ and $\theta^{-1} z \equiv \theta^{n-1} z \pmod p$ and (in modern terms) proves that θz, and every power $\theta^e z$ with e relatively prime to p, all generate modulo p cyclic groups of order n. $\theta^{-1} z$ is called the inverse of θz:

> ** The function $\theta^{-1} z \ldots$ is called the inverse of θz. One can immediately obtain its value. [618, Vol. 2, p. 340]

Serret shows that $\theta^{-1} z = (-b'z + b)/(a'z - a)$.

In the following seven numbers of this section Serret considers the problem of constructing all linear functions of order n and of counting the number of different kinds of such functions. We shall say just enough about this material

---

* Il est évident que l'ensemble de toutes les fonctions linéaires forme un groupe tel, que le produit de plusieurs fonctions du groupe fait aussi partie de ce groupe. On voit par ce qui précède qu'il en est de même de l'ensemble des seules fonctions linéaires du premier genre (lineare Funktionen der 1. Klasse, WG.), mais non pas de l'ensemble des seules fonctions du deuxième genre.

** La fonction $\theta^{-1} z \ldots$ sera dite l'inverse de θz; on peut obtenir immédiatement sa valeur.

to make understandable what is typical in Serret's thinking and reasoning in connection with the development of group theory.

For the congruence $\theta^n z \equiv z \pmod{p}$ to hold it is necessary and sufficient that

$$\theta^n = \frac{(a + b' + 2t)^n - (a + b' - 2t)^n}{2t} \equiv 0 \pmod{p}$$

with

$$2t = \sqrt{(a + b')^2 - 4\Delta}.$$

The possibility that θz is of the first order, that is, that θz reduces to z, is ruled out. This must be taken into consideration in the final account. The discussion of this condition requires consideration of the following cases:

Case 1. $t \equiv 0 \pmod{p}$.
Case 2. t^2 is a quadratic residue mod p.
Case 3. t^2 is a quadratic nonresidue mod p.

The case $t \equiv 0 \pmod{p}$ is simplest. Together, $\theta^n \equiv 0 \pmod{p}$ and $t \equiv 0 \pmod{p}$ imply that the order of the linear function is p and that the function θz belongs to the first class. To obtain all the linear functions of order p we must take the different solutions of the system of congruences

$$a + b' \equiv 2\sqrt{\Delta} \pmod{p},$$

$$ab' - ba' \equiv \Delta \pmod{p}.$$

The values $a' = 0$, $b' = 1$ yield the integral linear functions and the value $a' = 0$ yields the fractional linear functions. The total number of linear functions of order p is $N_p = (p + 1)(p - 1)$. Since a number less than p is relatively prime to p, the order of a power of a linear function of order p is also p. In formulating these results Serret uses the word "group." Since the possibility $\theta z = z$ has been ruled out beforehand, none of the "groups" contains a unit element. Serret writes,

* The result is that the N_p functions whose existence we have just established form $p + 1$ groups each of which contains $p - 1$ functions that are powers of any one of them. [618, Vol. 2, pp. 343 to 344]

In the second case, when t^2 is a quadratic residue mod p, the order n of a linear function mod p is $p - 1$ or a divisor of $p - 1$, and the number N_{p-1} of such functions is $N_{p-1} = \frac{1}{2}(p + 1)p(p - 2)$. In the third case, when t^2 is a quadratic

---

* Il en résulte que les N_p fonctions dont nous venons d'établir l'existence forment $p + 1$ groupes renferment chacun $p - 1$ fonctions qui sont les puissances de l'une quelconque d'entre elles.

nonresidue mod p, the order n is $p + 1$ or a divisor of $p + 1$, and the number N_{p+1} of such functions is $N_{p+1} = \frac{1}{2}p^2(p - 1)$.

Serret formulates these results as follows:

* These N_{p-1} linear functions can be divided into $\frac{1}{2}(p + 1)p$ groups each containing $p - 2$ functions .... [618, Vol. 2, p. 348]

** These N_{p+1} linear functions can be divided into $\frac{1}{2}p(p - 1)$ groups each containing p functions .... [618, Vol. 2, p. 352]

We see that "le groupe" is used here in the everyday sense of a collection of elements.

To construct the linear functions, we need only know a primitive root of each of the congruences

$$i^{p-1} \equiv 1 \pmod{p}, \qquad i^{p+1} \equiv 1 \pmod{p}.$$

Also, $N_p + N_{p-1} + N_{p+1} = N - 1$, the number of linear functions exclusive of the identity function $\theta z = z$.

In the next section, "Des fonctions analytique propres à représenter les substitutions," Serret explains how a number-theoretic function can "represent" a substitution. Suppose we wish to permute n values. We assign these n values $0, 1, 2, \ldots, n - 1$ to a variable z as indices. To determine the permutation we assign to z the values to be permuted in their natural order and apply a number-theoretic function $f(z)$ to the indices. Then, says Serret, "the substitution is represented" by $f(z)$. As an illustration, Serret gives the results, due to Hermite [276], on the conditions that must be satisfied by polynomials so that they represent permutations of p digits incongruent modulo p.[125] This leads to the following main result: Let $f(z)$ be a polynomial of degree $p - 2$ with integral coefficients. Let $f^m(z)$ be the mth power of $f(z)$ whose degree has been reduced to less than p by means of the relation $z^p \equiv z \pmod{p}$. The polynomial $f(z)$ represents a permutation of p digits incongruent modulo p if and only if the coefficient of z^{p-1} in all the functions $f^2(z), f^3(z), \ldots, f^{p-2}(z)$ is congruent to zero modulo p. If the polynomial $f(z)$ represents a permutation of p digits incongruent modulo p, then, clearly, so does the polynomial

$$\theta z = \alpha f(z + \beta) + \gamma.$$

We can choose the values of α, β, and γ so that the coefficient of the highest power

---

* Ces N_{p-1} fonctions linéaires peuvent être partagées en $\frac{1}{2}(p + 1)p$ groupes contenant chacun $p - 2$ fonctions ....

** Ces N_{p+1} fonctions linéaires peuvent être partagées en $\frac{1}{2}p(p - 1)$ groupes contenant chacun p fonctions ....

of z is 1 and the coefficient of the power one lower as well as the constant term vanish. Then θz has the form

$$\theta z = a_1 z + a_2 z^2 + \cdots + a_{\nu-2} z^{\nu-2} + z^{\nu}, \qquad \nu \leqslant 2.$$

When they represented permutations, Hermite called these functions "reduced substitutions" (substitutions réduites). Serret speaks of "reduced functions." A reduced function θz represents a permutation if and only if, after reduction by means of $z^{p-1} \equiv 1 \pmod p$, the constant terms in the second, third, ..., $(p-2)$th powers are congruent to zero modulo p.

In the section "Des substitutions rationnelles et linéaires," Serret—again following Hermite—extends these arguments to linear functions.

The considerations sketched above imply that the entire linear functions $az + b$ represent permutations of the p digits $0, 1, 2, ..., p-1$ (which play the rôle of indices). Since their number is $p(p-1)$, the number of permutations is $p(p-1)$, and—says Serret, using the language of Cauchy—"It is obvious that they form a conjugate system" ("Il est évident qu'elles forment un système conjugué") [618, Vol. 2, p. 363]. Similarly, the rational functions $\theta z = (az + b)/(a'z + b')$ modulo p can be used[126] to represent permutations of $p + 1$ indices $0, 1, 2, ..., p-1, \infty$. For the *totality* of linear functions, Serret obtained the following result:

* *The set of all linear substitutions relative to a prime modulus p forms a triply transitive conjugate system of order* $(p+1)p(p-1)$. *Moreover, the linear substitutions whose determinant is a quadratic residue form a doubly transitive conjugate system of order* $\frac{1}{2}(p+1)p(p-1)$. [618, Vol. 2, p. 366]

In the last two sections of the chapter, entitled respectively, "De quelques propriétés des substitutions linéaires" and "Sur les substitutions de cinq et de sept lettres," Serret investigates—to use modern terminology—the structure of the permutation groups represented in this manner. A striking feature of the notation is the consistent use of a letter to denote a set of functions as well as an operator. One result of this is the merger of the terms "functions that represent permutations" and "permutations." For example, Serret denotes by Ez the totality of the $p(p-1)$ entire linear functions. If φz is a particular linear function, then "the substitutions of the form $\varphi^{-1}E\varphi z$ form a conjugate system of order $p(p-1)$. The substitutions of that system fix the index z such that z is infinite." Since there

* *L'ensemble des toutes les substitutions linéaires, relatives à un module premier p, constitue un système conjugué trois fois transitif d'ordre* $(p+1)p(p-1)$. *En outre, les substitutions linéaires dont le déterminant est résidu quadratique forment un système conjugué deux fois transitif d'ordre* $\frac{1}{2}(p+1)p(p-1)$.

are in all $(p + 1)p(p - 1)$ linear functions modulo p, it follows that there are $p + 1$ conjugate systems of order $p(p - 1)$. To obtain the complete system of $(p + 1)p(p - 1)$ "substitutions," we multiply a system of order $p(p - 1)$ whose permutations fix one index by the system of powers of a linear "substitution" of order $p + 1$. This theorem of Serret (compare [618, Vol. 2, p. 368]) (together with his results—omitted here—on the generation, by number-theoretic means and through Hermite's results on "reduced systems," of the required permutations that fix certain indices) makes possible the construction of permutation groups on a prescribed number of letters. One example of the use of such considerations is furnished by Kronecker's proof [428] of the existence of a group of permutations of order 168 on 7 digits—a result included by Serret in this chapter.

Whereas the extensions of the theory of permutations presented by Serret take up little space in his *Cours d'Algèbre supérieure* of 1866, it is just such applications that are one of the two main components of Jordan's *Traité* of 1870. Here the major theme is the survey of all of mathematics by areas in which the theory of permutations had been applied or seemed likely to be applicable, largely on the basis of Jordan's own results;[127] this dominates even the presentation of group theory as an independent branch of investigations. In Serret's *Cours*, the stress on the presentation of group theory is definitely greater.

The *Traité des substitutions et des équations algébriques* [317] begins with an introduction of extraordinary scope, which demonstrates the scientific starting point, as well as the greatness and limitations, of Jordan's conception. In a certain sense the *Traité* contains formulations of problems belonging to a development yet to come; these would be solved a few years later, with the aid of a new conception of groups (III.2).

Jordan begins his introduction by stressing the fundamental difference between Galois's papers on the theory of equations and the relevant papers of Galois's predecessors from Lagrange to Abel. By associating to every equation a (permutation) group whose structure mirrors its essential properties, including its possible solution by radicals, Galois had supplied the ultimate basis of the theory of equations. This assigned a special role to the theory of permutations. As the foundation of all questions bearing on the theory of equations, permutation theory became an independent area of investigation. Jordan adds that

> * the theory of substitutions that thus becomes the foundation of all the questions concerning equations is nevertheless but little advanced. [317, p. VII]

---

* la théorie des substitutions, qui devient ainsi le fondement de toutes les questions relatives aux équations, n'est encore que peu avancée.

In view of this development and of algebraic questions that had just arisen, Jordan concludes that the theory of equations has acquired a new and very different character. The old quest for solutions of given equations in the form of transparent solution formulas has become the study of the structure of algebraic number fields. Jordan describes this change without using the field concept, in these words:

* From this advanced point of view [that is, that of the Galois theory (Wussing)] the problem of solution by radicals that not long ago had seemed to be the only object of the theory of equations appeared to be no more than the first link in a long chain of questions concerning transformations of irrationals and their classification. [317, p. VI]

This passage makes it abundantly clear that Jordan had reached a view of algebra that was far more modern and future-oriented than the view Serret had reached just four years before. This explains Jordan's remark that, besides studying Galois's works, he had been reading Serret. He continues:

** It is the assiduous reading of this Book that has initiated us into Algebra and has inspired in us the desire to contribute to its progress. [317, p. VIII]

In Jordan's view, all applications of the theory of permutations follow from its role as "the foundation of all questions about equations." He designates in the introduction two large groups of problems to be studied in the *Traité* with the aid of permutation theory:

1. The study of the "division" (Teilung) of transcendental functions [expressing $f(x/n)$ in terms of f when f is the exponential function or an elliptic function (Translator)]. This study had already given Galois the opportunity to make a new and brilliant application of his method. It had been advanced by Hermite who, using Galois's methods, built on the relevant work of Abel and Gauss. In the sixties, Brioschi in Italy and Kronecker in Germany had obtained decisive results in this area.
2. The use of permutation theory in studying the new direction of development of analytic geometry. This new development had been initiated by O. Hesse, largely in the fifties, in papers on the number of inflection points of cubic curves. Through these papers, algebra, with its many then very modern tools (determinants in the case of Hesse), began to appear as the direct representation

---

* De ce point de vue élevé, le problème de la résolution par radicaux, qui naguère encore semblait former l'unique objet de la théorie des équations, n'apparaît plus que comme le premier anneau d'une longue chaîne de questions relatives aux transformations des irrationnelles et à leur classification.

** C'est la lecture assidue de ce Livre qui nous a initié à l'Algèbre et nous a inspiré le désir de contribuer à ses progrès.

of geometric propositions. Cayley, Clebsch, Kummer, and other followers of Hesse used other algebraic means (III.1) to elaborate this approach.

It thus appears that Jordan had grasped an objective tendency that pointed to the use of group theory in geometry and that was to gain general acceptance two years after the *Traité* as a result of the Erlangen Program. But whereas the Erlangen Program developed an entirely new view of groups and used the new group concept directly for geometric purposes (see III.2) Jordan remained in the realm of applications of permutation groups; geometric problems were "translated" into algebraic equations and this, says Jordan [317, p. vii], made possible, without further difficulties, the application of Galois's methods.

For effective handling of the second issue, Jordan had to fashion a calculus of permutation groups that would allow him to use the group-theoretic approach corresponding to a given geometric problem. He was then able to classify earlier, isolated results by group-theoretic means; more specifically he could describe such results in terms of statements about subgroups of the "linear group" (le groupe linéaire), a group that he had investigated in great detail. To illustrate the extent of this recasting of geometric problems in group-theoretic form, we mention the fact that Jordan speaks of "Steiner" groups, in spite of the fact that Steiner made neither explicit nor implicit use of group-theoretic methods. What Jordan had in mind was, of course, the group his method associates to Steiner's genuinely geometric problem.[128]

Jordan's conception was related to his own earlier investigations and to those of other mathematicians—for example, to T. P. Kirkman's work of the sixties, which involved the representation of isometries of polyhedra effected by means of permutation groups. It must be stressed, however, that, whatever their common subject matter, these approaches were based exclusively on the permutation-theoretic view of groups, and that it was F. Klein who first viewed the isometries themselves as elements of a group (III.3).

It is obviously unnecessary to go into details. Nevertheless, I shall use Jordan's "linear group" to clarify his principal conception. This group is the central idea of the whole *Traité* in the sense that Jordan regards his applications of group theory as a "structure problem" of the "linear group."

In chapter II of the second book of the *Traité* [317, pp. 88–249], entitled "Des substitutions linéaires," Jordan refers to investigations of analytic representations of permutations included by Serret in his *Cours d'Algèbre supérieure*. Thus, let S be an arbitrary permutation (Jordan writes, of course, "substitution") on k quantities denoted by l_x, $x = 0, 1, \ldots, k-1$. S takes each l_x to an $l_{\varphi(x)}$, where $\varphi(x)$ is a suitable polynomial in x. If k is a prime p, then, as Hermite had shown, $\varphi(x)$ represents a permutation if and only if $\varphi(x)$ and its first $p - 2$ powers can be reduced to polynomials of degree $p - 2$ by means of the relation $x^p \equiv x \pmod p$. Jordan denotes the permutation represented in this manner by

$$S = |x\varphi(x)|.$$

In particular, $|x\alpha x + \beta|$ with $\alpha \neq 0$ is a permutation. Jordan claims that the symbol $|x\varphi(x)|$ can represent every permutation on p quantities: "Toute substitution pourra donc être représentée par le symbole $|x\varphi(x)|$" ("Every substitution can therefore be represented by the symbol $|x\varphi(x)|$") [317, p. 89].

It is possible, using Hermite's approach as a model, to decide when $\varphi(x)$ represents a permutation if k is a prime power. However, Jordan prefers to use a method works for all composite natural numbers $k = m$.

Let l be a quantity with n indices, each of which takes the m values $0, \ldots, m - 1$, with which we calculate modulo m. Thus the quantities $l_{0,0}, \ldots, l_{x,x'}, \ldots$ form a finite set of m^n elements. The permutations

$$A_{\alpha,\alpha'}, \ldots = |x, x', \ldots, x + \alpha, x' + \alpha', \ldots|$$

change the indices $x, x', \ldots$ of $l_{x,x'}, \ldots$ to $x + \alpha, x' + \alpha', \ldots$ modulo m and form a transitive group F of order m^n.

Now Jordan tries to find the general form of the permutations that commute with the group F; that is, he tries to embed F in a group of which F is a normal subgroup. He finds that all such permutations are of the form $S = TA_{\delta,\delta'}, \ldots,$ where T is a permutation of the form

$$|x, x', \ldots, ax + bx' + \cdots, a'x + b'x' + \cdots, \ldots|,$$

which he also writes in the form

$$\begin{vmatrix} x & ax + bx' + \cdots \\ x' & a'x + b'x' + \cdots \\ \vdots & \vdots \end{vmatrix}.$$

Such an expression represents a permutation if and only if the determinant

$$\Delta = \begin{vmatrix} a & b & \ldots \\ a' & b' & \ldots \\ \vdots & \vdots & \end{vmatrix}$$

is relatively prime to m. These permutations form a group that Jordan calls the linear group of degree m^n ("le groupe linéaire du degré m^n") and whose properties—such as order, systems of generators, composition factors, subgroups, canonical form of the elements, etc.—he proceeds to investigate.

It is clear that what is involved in the case of the elements of the linear group of degree m^n is interchanges of a finite number of quantities that are represented by an indexed letter to make them amenable to calculations. But the linear substitutions corresponding to these interchanges are not viewed as elements of a group. Throughout the *Traité*, Jordan fails to go beyond thinking of a group as a group of permutations. On the other hand, from a historical viewpoint, the conception associated with the term "analytic representation of permutations" is the transitional phase between the concept of a permutation group and the concept of a (finite) group of linear substitutions in which the linear substitutions are themselves group elements (III.3).

It remains to clarify Jordan's methodological starting point in connection with his application of permutation groups to geometry [317, book III, chapter III, pp. 301–333].

By way of a preliminary clarification of his ideas [317, p. 301ff], Jordan notes that one of the commonest problems of geometry is to determine points, curves, and surfaces satisfying given conditions. When the number of solutions is finite, then the coordinates of the required points, or the parameters that determine the given curves or surfaces, are determined by a system of equations whose number is equal to the number of the unknowns. If we eliminate all but one unknown, say x, then the degree of the final equation X gives the number of solutions of the problem. If all the roots of X are distinct and x_0 is one of them, then substitution of x_0 in the initial equation gives the corresponding values of y, ... as functions of x_0.

Thus the required points, curves, and surfaces are determined as soon as the equation X is solved. The required geometric entities correspond to the different roots $x_0, x_1, \ldots$ of X. Later Jordan denotes by $x_0, x_1, \ldots$ the required points, curves and surfaces as well as the roots of X.

In general, there are many solutions of the problem, but they are linked by geometric relations. The latter imply the existence of a polynomial function $\varphi(x_0, x_1, \ldots)$ whose group contains the group of X. Conversely, if we could be certain that we knew all the *geometric* relations that solve the posed problem (or at least those relations the rest depend on), then the group of the equation X would contain all the permutations of the group associated to the equation $\varphi(x_0, x_1, \ldots)$. Jordan continues as follows:

* But this kind of certainty is difficult to obtain in spite of the care brought by able geometers to the study of these problems. Thus it would not be impossible that the equations to which these problems give rise have sometimes a form even more particular than that which we shall find in leaning on the results of our predecessors. [317, pp. 301–302]

(* see p. 159)

In the *Traité*, group theory is applied to geometry only in the sense that the group of the algebraic equation to which the geometric problem gives rise is applied to the study of that equation. This is both cause and consequence of Jordan's exclusive reliance in the *Traité* on the concept of a permutation group. The decisive application of the group idea to geometry, above all as an ordering principle of geometry, required the transition to a more comprehensive group concept, the concept of a group of transformations.

Similarly, Jordan's treatment of the monodromy group did not go beyond the conceptual content of a permutation group [317, pp. 277–279]. The term "le groupe de monodromie" is due to Jordan, but the related content and the use of the word "monodrom" go back to V. Puiseux (1820–83) and Ch. Hermite.

In 1850 Puiseux [577] had investigated equations whose coefficients are rational functions of a complex variable z; the underlying domain of rationality consists of all rational functions of z with arbitrary coefficients, so that the roots of the equation depend on z. Puiseux investigated the interchanges of the roots resulting from analytic continuation of these functions of z along all possible closed paths issuing from, and returning to, a fixed z_0, and found that these interchanges of the roots form a group. One year later Hermite proved [274] that under a certain assumption on the underlying domain of rationality this permutation group is the Galois group of the equation.

Jordan followed up these results in his *Traité*. By taking as the underlying domain of rationality the field of rational functions of z with rational coefficients, Jordan obtained in place of the monodromy group the so-called "algebraic group" (le groupe algébrique) and proved that the monodromy group was a normal subgroup of the algebraic group.

Finally, we observe that Jordan used the results involving his "linear group" in the study of the group of the modular equation[129] and of the theory of elliptic functions; thus he remained within the limits of the concept of a permutation group. While Jordan's approach enabled him to give a unified presentation of results due to Galois, Betti, Hermite, Mathieu, Jacobi, Kronecker, Krikman, and others, it must nevertheless be pointed out that even in this area of mathematical investigation, and in particular in the application of group theory to the theory of functions, the realization of the tremendous fruitfulness of group-theoretic thinking came only with the transition to the concept of a group of transformations (III.3).

(* see p. 158)

Mais une semblable certitude est difficile à obtenir, malgré le soin apporté par d'habiles géomètres à l'étude de ces problèmes. Il ne serait donc pas impossible que les équations auxquelles ces problèmes donnent naissance eussent parfois une forme plus particulière encore que celle que nous allons trouver, en nous appuyant sur les résultats obtenus par nos prédécesseurs.

In sum, Jordan's 1870 *Traité des substitutions* marks the high point of the range of applications, and of the general importance, of the concept of a permutation group. At about this time, and owing in part to Jordan's own 1869 paper on groups of motions (III.2), developments had begun that would lead to the eventual broadening of the group concept from its restricted sense of a group of permutations.

The *Traité* marks a break in the evolution and application of the permutation-theoretic group concept. It was an expression of Jordan's deep desire to bring about a conceptual synthesis of the mathematics of his time. That he tried to achieve such a synthesis by relying on the concept of a permutation group, which the very next phase of mathematical development would show to have been unduly restricted, makes for both the glory and the limitations of his book.

This striving for a mathematical synthesis based on key ideas is a striking characteristic of Jordan's scientific activity, as appears from the affectionate biography written by H. Lebesgue [451]. Jordan's career was marked, in the main, by a consistently high productivity that extended into his old age and embraced all areas of mathematics, including applications. In spite of radically different political and social views—he was a conservative and a strict Catholic—Jordan resembled Evariste Galois in a certain spiritual sense. Both men grappled with the question of how it is possible to make decisive advances in mathematics.[130] Both strove to replace computation with synthetic considerations and to uncover structural frameworks. In fact, in the introduction to the *Traité*, Jordan emphasizes the essential role of three fundamental concepts—transitivity, primitivity, and normal subgroup—that point the way to insights into structure. In his words:

* Three fundamental concepts begin, however, to make themselves clear; that of primitivity, already indicated in the works of Gauss and Abel; that of transitivity, which belongs to Cauchy; and finally the distinction between simple and composite groups. This last concept, the most important of the three, is again due to Galois. [317, p. VII]

This makes it clear why Lebesgue, in his eulogy of Jordan delivered a year after the latter's death, states that Jordan had followed Galois's example and direction and was one of those mathematicians who replace the mathematics of quantities with the mathematics of qualities.

Seen in this light, the *Traité* of 1870 corresponded exactly to Jordan's scientific objective, which led to his attempt to fuse and synthesize arithmetics and geometry by means of the permutation-theoretic concept of a group. Jordan's interest in

---

* Trois notions fondamentales commencent cependant à se dégager; celle de la primitivité, qui se trouvait déjà indiquée dans les Ouvrages de Gauss et d'Abel; celle de la transitivité qui appartient à Cauchy; enfin la distinction des groupes simples et composés. C'est encore à Galois qu'est due cette dernière notion, la plus importante des trois.

analysis situs (that is, topology), an interest that dates from 1867, represents yet another attempt at a synthesis, a recasting of the known in new forms. It led him in the direction that B. Riemann had followed before him and that H. Poincaré was later to follow in his turn. It is noteworthy that when G. Cantor's papers on the "theory of manifolds" made their appearance and at first met virtually unanimous rejection by his colleagues, Jordan was one of the earliest to adopt Cantor's set-theoretic thinking and to apply it fruitfully to specific problems; one need only cite his definition of a curve and the Jordan curve theorem.

Jordan's ready acceptance of the new stemmed also from his desire to subsume existing results under some system of powerful new ideas; in algebra such a system, in his opinion, had to be sought in the direction of Galois's conceptions. The adaptability of the term "order" (l'ordre) and the multiplicity of associations generated by its use made him think it admirably suited for a central role in the unified mathematics of which he dreamed. He thought that mathematics could be derived in bold strokes from a yet-to-be-developed "théorie de l'ordre." Such was the guiding thought of his work in topology and in his application of set theory to geometry and analysis. In Jordan's words, these investigations

* *had almost constantly as a goal the deepening of the theory of order from the point of view of analysis as well as of pure geometry.* [451, p. 58]

Jordan shared with the most far-seeing and creative mathematicians of the end of the nineteenth century the need to view the known from a higher—that is, a unifying —perspective. The striving for a unity to be achieved by the formulation of sufficiently deep ideas yielded a significant number of factors helpful for the development of the concept of an abstract group. In this situation the *Traité*, although based on a group concept soon to become obsolete, had much impact on the development of abstract group theory; this was spurred by the strong influence Jordan—the author of the impressive and popular *Cours d'analyse* of 1882 and the publisher, since 1885, of the *Journal des mathématiques pures et appliquées* founded by Liouville—exerted on the character of the mathematical sciences in France. In spite of the fact that Jordan had, so to say, written the *Traité* too soon, it was precisely this book (see III.2) that promoted the extension of the permutation-theoretic group concept undertaken by F. Klein and S. Lie. At the same time, the *conception* of the *Traité* had been superseded[131] in the early seventies in spite of its remarkable individual achievements and the overview it attained of the mathematics of the previous decade, an overview based on its leading idea of a group of permutations.

---

* *ont eu presque constamment pour but d'approfondir la théorie de l'ordre au double point de vue de la géométrie pure et de l'analyse.*

In this sense, J. Houël's detailed review of 1871, though describing the influence of the *Traité* in almost prophetic terms, and giving a remarkably accurate account of Jordan's achievements, fails to sense the impending historical termination. The review begins,

* The work that we announce will, we are sure, exert a considerable influence on the progress of the most important theory of algebra. Also, until such time as we are able to speak in more detail of the beautiful discoveries of the author, we believe that we must make for our readers an analysis of the different questions treated and solved by M. Jordan. The theory of substitutions has always been cultivated in France, and we are glad that it is to a French geometer that the theory owes a fresh advance and some important additions. [68, p. 161]

The review ends with these words:

** Such is the analysis of the main questions solved by M. Jordan. This analysis was made easy for us by the developments that the author has given in different compendia. The preceding works come at a favorable moment, since the progress of analytic geometry has enabled M. Jordan, as we have seen, to give applications that add great interest and a new attraction to the so difficult theory of substitutions. [68, p. 169]

It is precisely Houël's last remark that is an extremely interesting document of the times. Geometry is correctly designated "as a new attraction to the theory of permutations." When Houël wrote his review, F. Klein was at work on the Erlangen Program, and this launched a far deeper penetration of the group-theoretic viewpoint into geometry than Houël had anticipated. The permutation-theoretic viewpoint was, in Hegel's multiple sense of the word, "repealed" (aufgehoben) by this development. The decisive moments for the post-1870 evolution of the abstract group concept are those at which the permutation-theoretic group concept invaded geometry and permutation theory as an independent discipline was left behind. From now on, the theory of permutation groups lost its role as the pacemaker of the group concept. We may therefore dispense with the detailed study of the rapid development of knowledge in the area of permutation groups. I shall merely state that the turn toward abstract group theory was complete, and

---

* L'Ouvrage que nous annonçons exercera, nous en sommes sûr, une influence considérable sur les progrès de la théorie la plus importante de l'Algèbre: aussi, en attendant que nous puissons parler d'une manière plus détaillée des belles découvertes de l'auteur, nous croyons devoir faire à nos lecteurs une analyse des différentes questions traitées et résolues par M. Jordan. La théorie des substitutions a toujours été cultivée en France, et nous sommes heureux de reconnaître que c'est à un géomètre de notre pays qu'elle doit un nouveau progrès et d'importantes additions.

** Telle est l'analyse des principales questions résolues par M. Jordan. Cette analyse nous a été rendue facile par les développements qu'a donnés l'auteur dans différentes recueils. Les travaux qui précédent viennent au moment favorable; car les progrès de la Géométrie analytique ont permis, comme on l'a vu, à M. Jordan, de donner des applications qui ajoutent un grand intérêt et un nouvel attrait à la théorie si difficile des substitutions.

that it was accompanied until 1900 by intensive research in the area of permutation groups. Also, during the transition period there were just as many studies of permutation groups that used a mixture of abstract and permutation-based methods and concepts as there were summaries of group theory relying on more or less explicit use of the abstract group concept (III.4).

III Transition of the Concept to a Transformation Group and the Development of the Abstract Group Concept

1 The Theory of Invariants as a Classification Tool in Geometry

1 The Conceptual Background of Algebraic Invariant Theory

In contrast to the obvious unifying tendencies of form and content in number theory, in the solvability theory of algebraic equations and in analysis in the first half of the nineteenth century, the question of the inner connection among various "geometries" remained unanswered until the fifties. The efforts to remove this discrepancy gave rise to the study of geometric relations. This first unifying tool in geometry was implicitly group-theoretic. Under the circumstances, the analytic conception was bound to be more fruitful than the synthetic one (I.2). This was convincingly demonstrated by the large number of results in (algebraic) invariant theory, a theory that appears in retrospect to have been a transitional stage on the way to the later, explicitly group-theoretic classification of the whole edifice of geometry.

The British geometric school centered around A. Cayley (1821–95) and J. J. Sylvester (1814–97). From its inception in the forties, it was directly influenced by the analytic approaches to projective geometry, in particular by the French school around M. Chasles (1793–1880) and by the ideas of J. Plücker.[132] Cayley would often refer to himself as the direct successor of Plücker; for example, he regarded Plücker's discovery of the inflection points of plane algebraic curves as the greatest discovery in the whole history of geometry, and valued the problem itself as the stimulus for his own invariant-theoretic contributions.[133]

Given our interest in the history of the group concept, we must note certain essential characteristics of the extensive invariant-theoretic contributions of Cayley and Sylvester, such as the use of deep algebraic-number-theoretic tools and the immediate adoption of an abstract viewpoint in which space is regarded as a number manifold. These factors helped the British school achieve a leading position in the fifties and sixties.

It is reasonable to designate 1841 as the year of the adoption of algebraic-number-theoretic tools in geometry. In that year G. Boole (1815–64) took up the invariant-theoretic approaches that originated with Lagrange. Boole, a most independent-minded, nonacademic individualist, has nowadays the reputation of being the most undervalued mathematician of the nineteenth century. By far the greatest part of this retrospective recognition goes to his contribution to mathematical logic. But in his paper "Researches on the Theory of Analytic Transformations" of 1841 [48], Boole pointed out a shortcoming of the idea of "plain" invariance, derived from Lagrange (I.3) and exemplified by unimodular substitutions. Such a concept of invariance allowed only unimodular transformations of the variables. Boole posed the more general problem of finding expressions in the coefficients of a binary form that remain fixed, or are at most multiplied by a power of the determinant of the form when the variables of the form undergo an arbitrary linear transformation. Boole also gave the first substantial definition of the term "covariant." This occurs in his extensive paper "Exposition of a General Theory of Linear Transformations" [49], in which Boole used partial differential equations to obtain from given invariants and covariants further invariants and covariants.

It is here that Cayley enters the picture. In the paper "On the Theory of Linear Transformations" [135] of 1845, in which he referred specifically to Boole's "Exposition" [49],[134] Cayley began the search for covariants under linear transformations.[135] At first, under the influence of C. G. J. Jacobi, Cayley used the calculus of determinants in his search for decisive methods for geometric purposes. His use of "hyperdeterminant" for "invariant" points to his initial intellectual sources. These sources also supplied the stimulus for Cayley's development of the abstract matrix calculus.

It would take us too far afield to assess all of Cayley's work. Nor is it our aim to give a detailed account of the now completely forgotten symbolism and terminology of algebraic invariant theory. In view of my group-theoretic and geometric concerns, I wish above all to present Cayley's contribution to invariant theory and to emphasize the decidedly abstract nature of Cayley's considerations that gave rise (III.4) to the abstract conception of a group as a system of defining relations. The theory of invariants yielded the long-sought tool with which to bring to light the deeper connections between metric and projective geometry (I.1).

This is the sense in which the British school of geometry went beyond the French school around Chasles; while E. Laguerre (1834–86) never went beyond a discussion of the axiomatics of the distance definition of Euclidean geometry, Cayley used what is now known as the "Cayley metric" to embed (euclidean) metric geometry in the general scheme of projective geometry. In 1854 the theory of invariants or, as Cayley called it, the theory of quantics, became the preliminary stage of the explicitly group-theoretic ordering scheme in geometry, which was given final form by Klein in his Erlangen Program of 1872. The Erlangen Program dispensed with the invariant-theoretic terminology and approach.

2 *Cayley and the Theory of Quantics*

From the very beginning, Cayley, in his *geometric* objectives, went beyond the stimuli provided by Boole's ideas. Already in 1845 he adopted completely the standpoint of n-dimensional geometry and may very well have had in mind the classification of geometry as an aim now within reach after the progress in the investigation of forms. Cayley's work in this area defied the historical pattern, in the sense that each of the papers that came in a steady stream from his pen was a success, and apparent deviations and peripheral problems turned out to be indispensable preliminaries. Already in 1846, one year after taking over Boole's problem of what Sylvester would later (1851) call the invariants and covariants of a form, Cayley formulated a comprehensive program of future investigations of quantics and their complete fundamental systems,[136] a program that represented the essence of investigations on linear transformations [129, p. 95].

An intercession by Hamilton made it necessary for Cayley to define the boundary between his investigations and the calculus of quaternions. After that, Cayley published, among other things, special papers on cubic ternary forms [137] and a whole series of papers on higher-order curves and surfaces. In a paper on hyperdeterminants of binary forms [139] of 1851, Cayley adopted Sylvester's terminology: Instead of "hyperdeterminant derivative" or briefly "hyperdeterminant," Cayley now writes "invariant" and "covariant."

* In making use of Sylvester's new terms, I call a *Covariant* of a given function any function which does not change form as a result of a linear transformation of the variables, and an *Invariant* any function of the coefficients alone with that property. [129, p. 577]

The elaboration of the problem that Cayley took over from Boole in 1846 required the adoption and perfection of a suitable terminology and symbolism. (Given the prolixity of that long-dead "language," I shall all but refrain from its presentation.) Then came increased precision: What was investigated was not the invariants and covariants of "functions" but of forms, or as Cayley put it, of "quantics."

In 1854 there appeared the first of Cayley's ten innovative papers on quantics under the programmatic title "An Introductory Memoir upon Quantics" [142]; the titles of the remaining nine differed only in the identifying number. At the beginning of the first of these papers Cayley defines a quantic as a homogeneous form and explains that

the term Quantic is used to denote the entire subject of rational and integral functions, and of the equations and loci to which these give rise. ... [142, p. 221]

---

* En me servant des nouveaux termes de M. Sylvester, je nomme *Covariant* d'une fonction donnée, toute fonction qui ne change pas de forme en faisant subir aux variables des transformations linéaires quelconques, et *Invariant* toute fonction des seuls coefficients qui a la propriété mentionnée.

The term "degree" of a quantic coincides with the degree of a form as presently defined. Cayley calls the indeterminates of a form "facients" and chooses the following symbolism:

A quantic of the degrees m, m', ... in the sets $(x, y, \ldots)$, $(x', y', \ldots)$ &c. will for the most part be represented by a notation such as

$$(* \between x, y, \ldots \overset{m}{\between} x', y', \ldots \overset{m'}{\between} \ldots),$$

where the mark* may be considered as indicative of the absolute generality of the quantic .... [130, pp. 222–223]

In other words, this symbol is equivalent to a sum of terms of the form

$$x^{\alpha} y^{\beta} \cdots x'^{\alpha'} y'^{\beta'} \cdots,$$

multiplied by arbitrary coefficients. If in place of the * we have $(a, b, \ldots)$, then what is intended is the specific quantic with coefficients $a, b, \ldots$.

From the beginning, Cayley stresses the geometric objective of his investigations:

To avoid complexity, it is proper to take the facients themselves as coordinates, or at all events to consider these facients as linear functions of the coordinates ... I consider that there is an ideal space of any number of dimensions, but of course, in the ordinary acceptation of the word, space is of three dimensions... [130, p. 222].

Cayley begins [130, p. 561] his sixth paper [145] on quantics (1859) with the words "I PROPOSE in the present paper to consider the geometrical theory." The paper is actually a clarification of the geometric foundations—more specifically, a clarification of the relation of projective geometry and metric (euclidean) geometry with the aid of invariant theory based on what later became known as the Cayley metric. This clarification involves a discussion of the concept of distance. The treatment becomes abstract in that Cayley dispenses with the customary notion of the "distance" between two points and calls "a notion of distance" every relation that—as he writes—satisfies the condition

$$\text{Dist.}\,(P, P') + \text{Dist.}(P', P'') = \text{Dist.}(P, P'')$$

for arbitrary positions of three points P, P', P''.[137]

In the case of homogeneous coordinates, the connection with the theory of forms is that the theory of binary forms is identical with 1-dimensional geometry, that of ternary forms with plane geometry, and so on. For the sake of clarity, Cayley restricts himself to an explicit treatment of these two cases:

The present memoir relates to the geometry of one dimension and the geometry of two dimensions, corresponding respectively to the analytical theories of binary and ternary quantics [130, p. 561].

3 The Cayley Metric and the Determination of the Relation between Metric Geometry and Projective Geometry

What follows is a brief sketch of the Cayley metric in 1-dimensional geometry. While the ideas are the same as those underlying the usual exposition of the Cayley metric in modern textbooks, the symbolism and the use of algebraic forms make for a considerable difference.

Cayley uses homogeneous coordinates and writes $x, y = a, b$ for $x : y = a : b$. In the case of 1-dimensional geometry, the line is, so to speak, the support of the space. In Cayley's words,

In geometry of one dimension we have the line as a space or *locus in quo*, which is considered as made up of points. [130, p. 563]

The equation $x, y = a, b$ determines a point on the line. This relation is equivalent to the quantic

$$(* \between x, y)^1 = 0.$$

This quantic also determines a point on the line. An equation

$$(* \between x, y)^m = 0$$

represents a system of m points; if U is a quantic of degree m, then

$$U = 0$$

is the equation of a "point-system of the order m"—Cauchy calls it a "quadric" for $m = 2$, a "cubic" for $m = 3$, and so on. If the discriminant of the form vanishes, then the quantic has multiple roots, that is, we obtain pairs of coincident points. For the "quadric" $(a, b, c \between x, y)^2 = 0$, this condition is $ac - b^2 = 0$ or $a, b = b, c$, and for the "cubic" $(a, b, c, d \between x, y)^3 = 0$ it is

$$a^2d^2 - 6abcd + 4ac^3 + 4b^3d - 3b^2c^2 = 0.$$

For point systems of higher order Cayley refers the reader to his earlier paper [144] of 1857.

Now Cayley states the essential issue—in paragraph 152 of [145]—in these words:

Any covariant of the equation

$$(* \between x, y)^m = 0,$$

equated to zero, gives rise to a point-system connected in a definite manner with the original point-system. And as regards the invariants, the evanescence of any invariant implies a certain relation between the points of the system; the identical evanescence of any covariant implies

> relations between the points of the system, such that the derived point-system obtained by equating the covariant to zero is absolutely indeterminate. [130, p. 564]

In the special case, when for two point pairs

$$\begin{aligned}(a, b, c \between x, y)^2 &= 0,\\ (a', b', c' \between x, y)^2 &= 0\end{aligned}$$

the invariant $ac' - 2bb' + ca'$ vanishes, then we have a harmonic set of four points; this is the link to projective geometry on the line. In particular, we can now decide when two point pairs are connected by an involution. This result had been deduced by Cayley in his fifth memoir on quantics, in 1858.

Next Cayley introduces two different lines with two different coordinate systems and obtains the condition for four points on one line to be projectively related to four points on the other line. This makes possible the transition from one basic geometric element on one *locus in quo* to another such element on another *locus in quo*.

Then Cayley brings the two lines into coincidence but retains, at first, their coordinate systems. He starts with a fixed point pair, which he calls "the Absolute" [130, p. 583], and chooses another point pair such that it and the Absolute are related by an involution. Then there are two further inscribed points ("the centre and the axis") that are harmonically related to the Absolute. A point pair thus inscribed in the Absolute is said to be a "point-pair circle," or simply a "circle."

Then Cayley defines the two points of such a "circle" as equidistant from the "centre"; there are the points P, P', and P'' with P' as "centre," then the points P', P'', P''' with P''' a suitable new point and P'' a new center, and so on. In this way the line is subdivided into segments of "equal length." Further, in Cayley's words,

$$\text{Dist.}(P, P') + \text{Dist.}(P', P'') = \text{Dist.}(P, P'').$$

He goes on to present this idea in terms of his calculus. He takes $(a, b, c \between x, y)^2 = 0$ as the equation of the "Absolute," obtains

$$(a, b, c \between x, y)^2\,(a, b, c \between x', y')^2 \cos^2\theta - \{(a, b, c \between x, y \between x', y')\}^2 = 0$$

as the equation of the "circle" with "centre" (x', y'), and shows that the distance is a function of the expression

$$\frac{(a, b, c \between x, y \between x', y')}{\sqrt{(a, b, c \between x, y)^2}\,\sqrt{(a, b, c \between x', y')^2}},$$

nowadays written as

$$\frac{axx' + b(xy' + x'y) + cyy'}{\sqrt{ax^2 + 2bxy + cy^2}\sqrt{ax'^2 + 2bx'y' + cy'^2}},$$

and that it is a covariant of the "Absolute." As the distance function, satisfying the abstract condition

$$\text{Dist.}(P, P') + \text{Dist.}(P', P'') = \text{Dist.}(P, P''),$$

Cayley takes

$$\text{arc cos}\frac{axx' + b(xy' + x'y) + cyy'}{\sqrt{ax^2 + 2bxy + cy^2}\sqrt{ax'^2 + 2bx'y' + cy'^2}}$$

and explains that

we may in general assume that the distance is equal to the arc in question, viz. that the distance is

$$\cos^{-1}\frac{(a, b, c \between x, y, \between x', y')}{\sqrt{(a, b, c \between x, y)^2}\sqrt{(a, b, c \between x', y')^2}}$$

or, what is the same thing,

$$\sin^{-1}\frac{(ac - b^2)(xy' - x'y)}{\sqrt{(a, b, c \between x, y)^2}\sqrt{(a, b, c \between x', y')^2}}.$$ [130, p. 584]

This is the approach to the general distance function. In paragraph 213, Cayley considers the special case of the ordinary (euclidean) metric:

The foregoing is the general case, but it is necessary to consider the particular case where the Absolute is a pair of coincident points. [130, p. 585]

Application of invariant theory to the linear *locus in quo* augmented by this "Absolute" yields the metric relations of euclidean geometry.

Then Cayley treats the case of the plane. Here the "Absolute" is the conic section

$$(a, b, c, f, g, h \between x, y, z)^2 = 0.$$

Application of invariant theory to the plane *locus in quo*, consisting of the points of the plane, or, dually, of the lines of the plane, augmented by the special conic section $x^2 + y^2 + z^2 = 0$, again yields the ordinary euclidean metric in the plane. In this way Cayley has succeeded in embedding metric geometry in projective geometry:

In ordinary plane geometry, the Absolute degenerates into a pair of points, viz. the points of intersection of the line infinity with any evanescent circle, or what is the same thing, the Absolute is the two circular points at infinity. The general theory is consequently modified, viz. there is not, as regards points, a distance such as the quadrant, and the distance of two lines cannot be in any way compared with the distance of two points; the distance of a point from a line can be only represented as a distance of two points.

I remark in conclusion, that *in my own point of view*, the more systematic course in the present introductory memoir on the geometrical part of the subject of quantics, would have been to ignore altogether the notions of distance and metrical geometry; for the theory in effect is, that the metrical properties of a figure are not the properties of a figure considered *per se* apart from everything else, but its properties when considered in connexion with another figure, viz. the conic termed the Absolute. The original figure might comprise a conic; for instance, we might consider the properties of the figure formed by two or more conics, and we are then in the region of pure descriptive [i.e. projective (Wussing)] geometry: we pass out of it into metrical geometry by fixing upon a conic of the figure as a standard of reference and calling it the Absolute. Metrical geometry is thus a part of descriptive geometry, and descriptive geometry is *all* geometry, and reciprocally; and if this be admitted, there is no ground for the consideration, in an introductory memoir, of the special subject of metrical geometry; but as the notions of distance and of metrical geometry could not, without explanation, be thus ignored, it was necessary to refer to them in order to show that they are thus included in descriptive geometry. [130, p. 592]

Given my objectives, I need not concern myself with the application of invariant-theoretic methods to concrete geometric problems, such as points of inflection of higher-order plane curves, bitangents to surfaces, and so on, and thus with the description of the extremely rich literature on the so-called "equation problems of geometry" (to use a later term of F. Klein [342, p. 166]), or with the connection between invariant-theoretic methods and problems of differential geometry. On the other hand, it is important that I give a measure of attention to an aspect of invariant theory that has an abstract orientation.

In the hands of Cayley, Sylvester, and others, the investigations of forms, originally a means to geometric ends, became divorced from geometry. In Sylvester's terminology, the aim of the theory of invariants was the construction of a "complete asyzygetic system," that is, of a finite, complete "fundamental system" of (independent) invariants and covariants, such that every invariant and covariant of a form, or a system of forms, could be expressed as an entire rational function with numerical coefficients in the invariants and covariants of the fundamental system. By 1856 Cayley and Sylvester managed to solve this problem in a few special cases.[138] In 1868 P. Gordan (1837–1912), whom his colleagues later called the "king of invariants," proved the existence of a finite fundamental system for every binary form, and in 1870 he proved an analogous result for every finite system of binary forms. This problem—in Sylvester's terminology, the problem of syzygies—has an obviously close connection with the determination of all independent irreducible algebraic relations between the invariants and covariants of a finite system of forms. It strongly influenced Cayley's formulation of the group concept and, on the other hand, was one of the historical roots of ideal theory. (Of the mathematicians whose work contains similar approaches and who preceded Cayley and Sylvester, I mention Jacobi and Eisenstein.[139]) Finally[140], in 1890–91 young D. Hilbert proved the decisive basis theorem, to the effect that every finite system of forms has a finite fundamental system [277, 278].

Given our concerns at this point, it is important to note—without going into details—that the two problems on forms set forth by Cayley and Sylvester provided an important stimulus for the development of the theory of invariants up to the eighties and, through the work of Salmon, Aronhold, Gordan, Clebsch, Brioschi, and others, had a marked influence on the development of geometry and higher algebra. This influence prepared the ground for the realization and ready acceptance of the Erlangen Program. On the other hand, this same influence, in spite of, or perhaps because of, the formidable symbolic calculus associated with the theory of invariants, delayed the transition to the abstract way of looking at the elements of a system (such as an ideal, a module, a ring or a field) and thus also the thoroughgoing process of abstraction in group theory.[141]

4 The Cayley Metric as the Historical Root of the Erlangen Program

Because of his involvement with the extensive investigations aimed at the determination of complete fundamental systems of invariants and covariants of forms and systems of forms. Cayley did not follow up his initial results on metrics associated with various "Absolutes," and thus he failed to discover the connection between this problem and noneuclidean geometries.[142] As is well known, in 1870–71 Felix Klein embarked on the task of scrutinizing the metrics associated with all types of quadratic curves and quadric surfaces as "Absolutes," and thus discovered plane and solid hyperbolic and elliptic geometries. Klein was fully aware of the divergence of the different directions of development in geometry and, having learned from Clebsch in Göttingen the invariant-theoretic conception of geometry, had the means "to encompass through a unified overall conception" ("durch eine einheitliche Gesamtauffassung zu umspannen") the contrasts between related directions of investigation [346, p. 52]. This is Klein's retrospective description of his impression of the overall state of affairs in geometry between 1866 and 1868, when he was Plücker's assistant.

Actually, Klein's close connection with Cayley's metric conceptions is one of the intellectual roots of the Erlangen Program of 1872 and is thus of the greatest interest for the history of group theory.

In a preliminary announcement[143] of 1871, "Über die sogenannte Nicht-Euklidische Geometrie" [352], Klein still based himself completely on Cayley's ideas. This does not contradict the fact that he obviously took into consideration the conceptions that had been developed in the meantime by Helmholtz and Riemann (III.3) and that he was intent on clarifying the connection between euclidean and noneuclidean geometries. In 1871 Klein wrote,

* The need to make intuitively clear the very abstract speculations that led to the construction of the three geometries, motivated my search for examples of metrics that could be thought of as representations of these geometries and would thus make obvious their individual consistency. [346, pp. 246–247]

** I wish to construct plane and space representations of the three geometries that would afford a complete overview of their characteristic features. In this way I shall demonstrate that these representations are not only interpretations of the geometries in question but that they explain their inner essence. ...

The representations in question utilize the plane and space as the respective objects of the metric that is different from, and is a generalization, in the sense of projective geometry, of the usual metric. This generalized metric was essentially constructed by Cayley. Cayley's point of view, however, is altogether different from the present one. Cayley constructs his metric in order to show how metric (euclidean) geometry can be viewed as a special part of projective geometry. He investigates in detail only the plane. He shows how to construct a plane metric relative to a given conic section as "absolute." If that conic section degenerates into an imaginary point-pair, then we have the usual metric used in euclidean geometry; that metric is obtained if the two imaginary fundamental points coincide with the two circular points.

We shall extend this general Cayley metric to space. Compared with Cayley's exposition, I use a more geometric approach. Let an arbitrary quadric surface be given. The line joining two given points in space intersects the surface in two points. The two given points and the two points on the surface have a certain cross ratio and *the product of an arbitrary constant c and the logarithm of that cross ratio is defined to be the distance between the two points*. Similarly, given two planes, we can pass through their line of intersection two planes tangent to the fundamental surface. Together with the two given planes they determine a certain cross ratio. *We call the product of an arbitrary constant c′ and the logarithm of that cross ratio the angle between the two given planes*. [346, pp. 247–248]

---

* Das Bedürfnis, die sehr abstrakten Spekulationen, welche zur Aufstellung der dreierlei Geometrien geführt haben, zu versinnlichen, hat dahingeführt, Beispiele von Maßbestimmungen aufzusuchen, die als Bilder der genannten Geometrien aufgefaßt werden könnten, und damit zugleich die innere Folgerichtigkeit jeder einzelnen in Evidenz setzten.

** Ich will nun hier zunächst für die dreierlei Geometrien sowohl in der Ebene als im Raume Bilder aufstellen, welche ihre Eigentümlichkeiten vollkommen übersehen lassen. Sodann werde ich zeigen, daß diese Bilder nicht nur Interpretationen der genannten Geometrien sind, sondern daß sie deren inneres Wesen darlegen... .

Die fraglichen Bilder betrachten als Objekt der Maßbestimmung die Ebene resp. den Raum selbst und benutzen nur eine andere Maßbestimmung als die gewöhnliche, welche, im Sinne der projektivischen Geometrie, als eine Verallgemeinerung der gewöhnlichen Maßbestimmung erscheint. Es ist diese verallgemeinerte Maßbestimmung im wesentlichen von Cayley aufgestellt worden; bei ihm sind nur die leitenden Gesichtspunkte ganz anderer Art als die hier vorliegenden. Cayley konstruiert diese Maßbestimmung, um zu zeigen, wie die (Euklidische) Geometrie des

(** continued on p. 177)

Klein summarizes his explanation of the three possible cases:

* 1. *The fundamental surface is imaginary.* This yields elliptic geometry.

2. *The fundamental surface is real, not a ruled surface, and convex.* This is the assumption underlying hyperbolic geometry.

3. (Transition case) *The fundamental surface degenerates into an imaginary curve.* This is the assumption underlying ordinary parabolic geometry. [346, p. 252]

Elsewhere, Klein writes in 1872:

** Metric geometry is just the investigation of the projective properties of space forms relative to a fixed given conic—the imaginary circle at infinity. [346, p. 106]

The preliminary communication was followed in 1871 and 1873 by the two papers "Über die sogenannte Nicht-Euklidische Geometrie" [353, 359], which gave a detailed account of the result obtained in 1871.

What is of interest to us is the gradual moving away from the invariant-theoretic view of geometry toward explicitly group-theoretic classification in geometry.[144]

(** continued from p. 176)
Maßes als ein besonderer Teil der projektivischen Geometrie aufgefaßt werden kann. Er betrachtet dabei des näheren nur die Ebene. Er zeigt, wie man in der Ebene auf Grund der projektivischen Vorstellungen eine Maßbestimmung treffen kann, die sich auf einen beliebig gegebenen Kegelschnitt als ‚absoluten' Kegelschnitt bezieht. Degeneriert dieser Kegelschnitt in ein imaginäres Punktepaar, so hat man eine Maßbestimmung, wie die von uns (in der Euklidischen Geometrie) angewandte ist; man erhält geradezu die gewöhnliche Maßbestimmung, wenn man die beiden imaginären Fundamentalpunkte mit zwei bestimmten Punkten der Ebene, nämlich den beiden Kreispunkten, zusammenfallen läßt.

Diese allgemeine Cayleysche Maßbestimmung soll hier kurz auf den Raum übertragen werden, wobei ich mich, gegenüber der Cayleyschen Auseinandersetzung, einer mehr geometrischen Darstellungsweise bediene. Sei eine beliebig anzunehmende Fläche zweiten Grades als ‚fundamentale' Fläche gegeben. Zwei gegebene Raumpunkte bestimmen durch den Durchschnitt ihrer Verbindungslinie mit der Fläche zwei Punkte der letzteren. Die beiden gegebenen Punkte haben zu diesen beiden ein gewisses Doppelverhältnis, und *der mit einer willkürlichen Konstanten c multiplizierte Logarithmus dieses Doppelverhältnisses soll die Entfernung der beiden Punkte genannt werden.* Analog, wenn zwei Ebenen gegeben sind, so lassen sich durch die Durchschnittslinie derselben zwei Tangentialebenen an die Fundamentalfläche legen. Dieselben bestimmen mit den beiden gegebenen Ebenen ein gewisses Doppelverhältnis. *Der mit einer willkürlich zu wählenden Konstanten c^1 multiplizierte Logarithmus dieses Doppelverhältnisses ist es, den wir als Winkel der beiden gegebenen Ebenen bezeichnen.*

* 1. *Die Fundamentalfläche ist imaginär.* Dies ergibt die elliptische Geometrie.

2. *Die Fundamentalfläche ist reell, nicht geradlinig und umschließt uns.* Die Annahme der hyperbolischen Geometrie.

3. (Übergangsfall) *Die Fundamentalfläche ist in eine imaginäre Kurve ausgeartet.* Die Voraussetzung der gewöhnlichen parabolischen Geometrie.

** Die metrische Geometrie ist ja nichts anderes, als die Untersuchung der projektivischen Eigenschaften der räumlichen Gebilde unter Zugrundelegung eines ein für allemal gegebenen Kegelschnittes, des unendlich fernen imaginären Kreises.

2 Group-Theoretic Classification of Geometry: The Erlangen Program of 1872

From the sixties to the eighties algebra was, in a sense, dominated by the theory of invariants.[145] At present, however, the results of that prodigious mathematical activity, with its intricate symbolic calculus, have been all but absorbed by the abstract disciplines, especially ideal theory, and the "language" and symbolism of invariant theory are dead.

Given the concerns of this book, the invariant-theoretic classification in geometry was and is of interest as a historical transition to the group-theoretic classification in geometry, both in the objective sense of the change inthe mathematicians' overall conception of the problem of geometry as well as in the subjective sense of the progressive change in the geometric views of one mathematician, namely, Felix Klein. His Erlangen Program of 1872 was the consummation of the transition, initially his own, from the invariant-theoretic conception of the geometry problem to implicit group-theoretic thinking and then on to the explicit application and manipulation of thinking in terms of groups.

From the point of view of group theory, what was really new was that the existing group-theoretic thinking inherited from the solution of algebraic equations was imprinted on the whole system of geometric thought and, in turn, in a kind of dialectical interplay, the imprinted permutation-theoretic group concept underwent modification. The progressive evolution of the Erlangen Program brought a change in the manner of picturing a group and in its definition, and contributed to the formulation of the concept of "group" as a group of transformations.

Thus Klein's so-called Erlangen Program of 1872 signified a genuine break in the history of the rise and development of group theory, and it is from this point of view that we shall study its protohistory and history.

1 *The Intellectual Roots of the Erlangen Program*

Felix Klein added autobiographical notes and a few mathematical-historical notes to his *Gesammelte Mathematische Abhandlungen* [346, 347, 348] published in 1921–23. In these he describes how, as a young man, in the late sixties, he gradually acquainted himself with the dominant lines of geometric thought, how he became a researcher, and how his initial invariant-theoretic conception of geometry evolved into the main ideas of the Erlangen Program.

Since Easter 1866 he had occupied the position of assistant to J. Plücker. Largely through Plücker's assigning him to prepare and edit his *Neue Geometrie des Raumes, gegründet auf die Betrachtung der geraden Linie als Raumelement* [567], Klein learned line geometry. At the time of Plücker's death in 1868, essentially just the first part of that work had been completed. The publication of the second volume was entrusted to A. Clebsch. Though not quite twenty years old, by that time Klein had acquired the reputation of an authority in line geometry. He joined Clebsch in Göttingen and was assigned by him the task of making the necessary additions to the second volume.[146]

A. Cayley and L. Cremona maintained regular contacts with Clebsch, and so Klein became familiar with their investigations in invariant theory. In the summer of 1869, he attended Clebsch's course on invariants of binary quadratic forms. From the end of August 1869 to the middle of March 1870, Klein stayed in Berlin and was one of the most active participants in the mathematical seminar of Kummer and Weierstrass. He was not enthusiastic about the lectures and hardly attended them, though he later regretted this, and he did full justice to Weierstrass in his *Vorlesungen über die Entwicklung der Mathematik im 19. Jahrhundert* [403, p. 284].

All this notwithstanding, the short stay in Berlin brought Klein rich rewards. It marked the beginning of friendships with S. Lie and O. Stolz (1842–1905), which were to yield human and scientific rewards for years to come, and it gave Klein his first opportunity to hear—through Stolz—of noneuclidean geometry. Klein writes of his first contact with noneuclidean geometry as follows:

* The thought occurred to me that there must be an extremely close connection between it [noneuclidean geometry (Translator)] and Cayley's general projective metric. ... In the final session of Weierstrass's seminar, in the middle of March 1870, I lectured on Cayley's metric and concluded with the question of its possible connection with noneuclidean geometry. Weierstrass rejected the idea by declaring that the distance between two points was the necessary foundation of geometry and, as a result, insisted on defining a straight line as the shortest curve joining two points. [346, pp. 50–51]

Klein was disappointed. He left Berlin and stayed in Paris, with Lie, from the end of April until the middle of July 1870 when the Franco-Prussian war broke

---

* Erfasste damals gleich den Gedanken, daß diese mit Cayleys allgemeiner projektiver Maßbestimmung auf das engste zusammenhängen müsse. ... Bei Weierstraß habe ich in dem Schlußseminar, Mitte März 1870, über Cayleysche Maßbestimmung vorgetragen und geradezu mit der Frage geschlossen, ob hier nicht eine Beziehung zur Nicht-Euklidischen Geometrie vorliege. Weierstraß lehnte dies ab, indem er die Entfernung zweier Punkte als notwendigen Ausgangspunkt für die Grundlegung der Geometrie erklärte und dementsprechend die Gerade als kürzeste Verbindungslinie definiert wissen wollte.

out. It is necessary to mention the fact—reported by Lie himself [466, p. XXII]—that already in 1863 Lie had attended a brief lecture by L. Sylow (1832–1918) on Galois theory. In Paris, as a result of their close contact with G. Darboux, Klein and Lie learned of the latest results of the French school of geometry but also, above all, of the theory of permutation groups. In 1921 Klein commented on the time spent in Paris:

* We [Klein and Lie (Wussing)] lived in adjoining rooms and looked for new scientific stimulation largely through personal contacts, especially with younger mathematicians. Camille Jordan impressed me very much. His traité des substitutions et des équations algébriques, just published, was to me a book with seven seals. [346, p. 51]

Since his Paris stay, at the latest, Klein had all the intellectual prerequisites to realize his aim in geometry—an overview that would assign to each particular geometry a rational place in an overall system. He wrote in 1921,

** Since the time in Bonn I had wanted to understand, in the contest of warring mathematical schools, the mutual relations of neighboring, externally dissimilar and yet essentially related, lines of investigation, and to encompass their contrasts by means of a unified overall conception. In this respect there was a great deal for me to do in geometry. [346, p. 52]

During his second stay in Göttingen, from the beginning of 1871 until the end of September 1872, Klein attempted a classification of geometry on an invariant-theoretic foundation (III.1). His main ideas took shape in almost daily discussions with Clebsch, in an extensive correspondence with Lie and in constant contact with Stolz, who was spending the summer term of 1871 in Göttingen:

*** The ideas about the connection between noneuclidean geometry and Cayley's metric, which had developed somewhat since my Berlin days, came to the fore, and I managed not only to convince Stolz of the correctness of my ideas but to develop an independent projective justification of the whole theory based on v. Staudt's approaches. All that time Stolz was not only my strict critic but also my main support in matters of the literature. He had studied in detail Lobachevski, Joh. Bolyai and v. Staudt, something I could never force myself to do, and could answer all my questions. [346, pp. 51–52]

---

* Wir wohnten Zimmer an Zimmer und haben neue wissenschaftliche Anregung wieder wesentlich in persönlichem Verkehr gesucht, insbesondere mit jüngeren Mathematikern. Einen großen Eindruck machte mir Camille Jordan, dessen traité des substitutions et des équations algébriques eben erschienen war und uns ein Buch mit sieben Siegeln erschien.

** Mein Interesse war schon von meiner Bonner Zeit her darauf gerichtet, im Widerstreite der sich befehdenden mathematischen Schulen das gegenseitige Verhältnis der nebeneinander herlaufenden, äußerlich einander unähnlicher und doch ihrem Wesen nach verwandter Arbeitsrichtungen zu verstehen und ihre Gegensätze durch eine einheitliche Gesamtauffassung zu umspannen. Innerhalb der Geometrie gab es in dieser Hinsicht noch viel für mich zu tun.

*** Die Ideen über den Zusammenhang der Nicht-Euklidischen Geometrie mit der Cayleyschen Maßbestimmung, die sich seit meiner Berliner Zeit bereits einigermaßen weitergebildet hatten,
(*** continued on p. 181)

After Klein and Lie's stay in Paris and their introduction to Jordan's *Traité*, but not sooner, the classification of geometries could, and inevitably would, acquire an explicit group-theoretic character. In much the same way, Lie had derived from the *Traité* and the works of the French school on the geometric theory of differential equations decisive impulses for his later papers on contact transformations (III.3). In 1921 Klein wrote,

* Since the fall of 1871 I had attempted to clarify the mutual relations between a consistent projective mode of thought ... and developments of Möbius's barycentric calculus as well as the basic ideas of Hamilton's quaternions. That is how, in November 1871, I arrived at the fundamental idea of my Erlangen Program, worked out in October 1872. I presented it ... for the first time in my second paper on noneuclidean geometry [that is, in [359] (Wussing)], which, due to printing delays, appeared after the Erlangen Program. In that paper, depending on the group that, in this treatment, is the basic element of the geometry, I speak of different "geometric methods," an expression I later dropped on Lie's advice. [346, p. 52]

After his appointment to Erlangen, when he joined the faculty and senate of that university, Klein propounded his so-called Erlangen Program under the characteristic title "Vergleichende Betrachtungen über neuere geometrische Forschungen" [358].[147] But here the geometric content of the Erlangen Program, essential from our point of view, remained in the background. It was not until 1875–76 that Klein embarked on investigations of the theory of permutations (III.3). It is worth mentioning that, from an epistemological point of view, the preparatory writings for the Erlangen Program and the program itself have contributed greatly to the clarification of the grave philosophical difficulties affecting scientists and mathematicians. Following B. Riemann and H. Helmholtz, Klein elaborated the basic difference between objective "physical" space and the space concept in geometry, and frequently, for example, in meetings of the German

(*** continued from p. 180)
traten in den Vordergrund, und es gelang mir, Stolz nicht nur von ihrer Richtigkeit zu überzeugen, sondern auf Grund der Ansätze v. Staudts zu einer unabhängigen projektiven Begründung der Gesamttheorie vorzudringen. Stolz war alle die Zeit nicht nur mein strenger Kritiker, sondern auch mein literarischer Anhalt. Er hatte Lobatschewsky, Joh. Bolyai und v. Staudt genau studiert, wozu ich mich nie habe zwingen können, und stand mir bei allen meinen Fragen Rede und Antwort.

* Ich habe im Herbst 1871 insbesondere daran gearbeitet, die konsequente projektive Denkweise ... mit den Entwicklungen von Möbius' baryzentrischem Kalkul und den Grundanschauungen von Hamiltons Quaternionen in klare gegenseitige Beziehung zu setzen. So ist im November 1871 der Grundgedanke meines im Oktober 1872 ausgearbeiteten Erlanger Programms ... entstanden. Ich habe ihn zuerst in meiner zweiten Abhandlung über Nicht-Euklidische Geometrie ... zur Darstellung gebracht, die infolge äußerer Druckschwierigkeiten erst nach dem Erlanger Programm erschien. Es ist dort noch, je nach der Gruppe, welche man bei der Behandlungsweise der Geometrie zugrunde legen will, von verschiedenen ‚geometrischen Methoden' die Rede, eine Ausdrucksweise, die ich später auf Anraten von Lie fallen gelassen habe.

Scientists and Physicians, took a stand against the widespread conceptual confusion and the resulting attacks on noneuclidean geometry. His position was consistently materialist.

2 *Klein's Gradual Development of the Erlangen Program*

In his biographical notes, Klein singles out the papers in which he had worked out his classification of geometry, which he saw dimly at first and then, in November 1871, with complete clarity. There are six of these papers; I list them not in the order of their publication but in the order in which they were written.

A. "Deux notes sur une certaine famille de courbes et de surfaces" [350], coauthored with S. Lie during their stay in Paris and dated 13 June 1870. It was communicated to the Paris Academy by Chasles.
B. "Über diejenigen ebenen Kurven, welche durch ein geschlossenes System von einfach unendlich vielen vertauschbaren linearen Transformationen in sich übergehen" [351], also coauthored with Lie and dated March 1871.
C. "Über die sogenannte Nicht-Euklidische Geometrie (1. Aufsatz)," dated 19 August 1871.
D. "Über Liniengeometrie und metrische Geometrie" [356], dated October 1871.
E. "Über die sogenannte Nicht-Euklidische Geometrie (2. Aufsatz)" [359], dated 8 June 1872.
F. "Vergleichende Betrachtungen über neuere geometrische Forschungen" [358], dated October 1872. This was the so-called Erlangen Program.

Given our concerns, an examination of these papers—here identified for brevity as A, ..., F, respectively—for group-theoretic viewpoints is very rewarding. Among other things, it shows clearly how their intellectual roots in invariant-theoretic trains of thought and their spontaneous implicit group-theoretic thinking led Klein and Lie to the recognition of the "intimate" intellectual contact of their work with existing group-theoretic forms of thought. We see just as clearly how the invariant-theoretic ideas gave rise to the extensive use of coordinate transformations as the main tool of research, a tool that gradually became itself the object of investigation. Further, there is the unmistakable use of Helmholtz's motion-geometric way of thinking (III.3), obvious from direct quotations.

What Klein does not mention in his autobiographical notes is the obviously fruitful influence that close contact with physics—through his own early interest in that subject, and his work as Plücker's physics assistant—had on the development of his group-theoretic thinking in geometry.[148] In paper B, a kinematic assumption even becomes the principle underlying the mathematical problem. The authors refer explicitly to a then popular textbook of kinematics[149] and stress that

* we shall make consistent use of a well-known geometric argument that we want to state with all clarity at the very beginning of our paper.

That argument is used in the investigation of all geometric objects for which one knows transformations that take them into themselves.

The argument in question can be summarized in the theorem that asserts *that any other geometric object connected to the original object by a relation that is not destroyed by the associated transformations, goes over under these transformations into geometric objects that stand in the same relation to the original object.* [346, p. 424]

In their text, Klein and Lie explain the importance of this principle and show its connection with the question of plane curves that are mapped into themselves by a simply infinite system of commuting transformations.[150] Also—and this is a point to which Lie was to return in a later line of investigation (III.3)—a footnote explains that "simply infinite" refers to a continuous 1-dimensional manifold.

In view of this book's concern with the history of group theory, I shall discuss the papers A–F under the following headings:

1. the gradual enlargement of the class of applied and discussed transformations, as an extension of the method of investigation in invariant theory;
2. the adoption of the group-theoretic terminology in the sense of "group of permutations," with the sense of the term "group" changed to that of "group of transformations";
3. the group-theoretic formulation of the subject matter of geometry.

Paper A makes no reference to the group concept and does not involve thinking in terms of groups inasmuch as the question of the closure of the set of admissible transformations is not considered. Nevertheless, a kind of indirect group-theoretic issue is involved, in that one asks for curves invariant under linear transformations:

** We shall consider curves that are transformed into themselves by an infinity of linear transformations that make it possible, in general, to map each point of the curve to any other. [346, p. 416]

---

* eine geometrische Schlußweise mit Konsequenz angewandt werden, die wir, obwohl sie durchaus nicht neu ist, gleich hier beim Eingange unserer Arbeit mit Bestimmtheit bezeichnen wollen.

Dieselbe findet bei der Untersuchung aller solcher geometrischer Gebilde ihre Stelle, bei welchen man Transformationen kennt, durch die sie in sich selbst übergeführt werden.

Sie läßt sich in dem allgemeinen Satze zusammenfassen: *daß ein beliebiges anderes Gebilde, welches zu dem ursprünglichen in irgendeiner durch die zugehörigen Transformationen unzerstörbaren Beziehung steht, durch diese Transformationen in solche Gebilde übergeht, welche dieselbe Beziehung zu dem ursprünglichen haben.*

(** see p. 184)

A similar problem is stated for surfaces. Such curves and surfaces are denoted by the letter V.

In paper B, the "content" of A, "to the extent to which it deals with *plane curves*," is treated "in greater detail" [346, p. 426]. It is stated in passing that the aim of the paper is to find

* those plane curves that are transformed into themselves by a closed simply infinite system of commuting transformations. [346, p. 427]

The approach is group-theoretic in the sense that, as the title of the paper indicates, the authors actually determine "closed systems of transformations" in the plane.

This change was made deliberately. The authors note in their prefatory remarks [346, p. 427] that consideration of closed systems of commuting transformations is "intimately related" to investigations that turn up in "substitution theory" and thus in the theory of algebraic equations. At this point they refer in a footnote to Serret's *Cours d'Algèbre supérieure* and Jordan's *Traité des substitutions et des équations algébriques* and note that "there is *one* deep difference" between these authors and themselves in that the former work with "discretely variable quantities," whereas they use "continuously variable quantities." Thus, the moment they came across the permutation-theoretic group concept, Klein and Lie realized that their investigations initiated an extension of that concept.

The concept "closed system" (geschlossenes System) is made precise in §1 of paper B, entitled "Einfach unendliche, geschlossene Systeme von vertauschbaren linearen Transformationen. W-Kurven." There are five admissible classes of transformations, of little interest to us. They are systems, depending on a continuously varying parameter, that are closed under successive application of the transformations. We quote:

** *Two arbitrary transformations, applied in succession, in any order, yield the same new transformation.*

This new transformation is itself a transformation of the system.

In view of the first property, the transformations of the system are called *commutative*. In view of the second property, the system is called *closed.* [346, p. 430]

(** see p. 183)

Les courbes que nous allons considérer sont celles qui se transforment en elles-mêmes par une infinité de transformations linéaires, permettant d'amener en général chaque point de la courbe en chaque autre.

* diejenigen ebenen Kurven, welche durch ein geschlossenes System von einfach unendlich vielen unter sich vertauschbaren linearen Transformationen in sich übergehen.

** *Zwei beliebige Transformationen des Systems geben, hintereinander angewandt, unabhängig von ihrer Reihenfolge, dieselbe neue Transformation.*

(** continued on p. 185)

The last sentence has two footnotes that form a bridge to the group theory that had been developing independently. The first footnote—after the word "second"—explains that in the case of a *simply* infinite class of transformations, the first property follows necessarily from the second; as a counterexample, the authors mention the noncommutative triple infinity of linear transformations fixing a conic. In the second footnote—following the word "closed"—the word "group" appears for the first time in the work of Klein and Lie:

> * Thus the expression a "closed system of transformations" corresponds exactly to what is designated in the theory of substitutions as "a group of substitutions." [346, p. 430][151]

Papers C and D belong essentially to the invariant-theoretic mode of classification in geometry and do not treat the group concept as pivotal (III.1). On the other hand, in E—the paper to which Klein referred occasionally as the first, provisional version of the Erlangen Program [346, p. 314]—transformation groups are mentioned routinely.

In the first section of paper E, Klein makes precise his concept of an n-dimensional manifold. It consists of all n-tuples

$$(x_1, x_2, \ldots, x_n)$$

—called elements—of complex numbers. In §2 he defines a transformation of the manifold to be the transition of each element to an associated element (or elements). It is stressed that the nature of the transformations, which may be given by equations, is "irrelevant"—this in spite of the remark that what will occur in most cases are algebraic transformation equations. On the other hand, it is explicitly stated that the equations corresponding to the transformations are invertible and the

> ** inverted equations represent what is to be called the *inverse transformation.* If we denote a transformation by a letter such as A, B, ..., and the composition of two transformations A, B by the symbol (product) AB, then the inverse of A will be denoted by A^{-1} .... Now let there be given a sequence of transformations A, B, C .... *If this sequence has the property that the composite of any two of its transformations yields a transformation that again belongs to the sequence, then the latter will be called a group of transformations.* [346, p. 316]

(** continued from p. 184)

Diese neue Transformation ist selbst eine Transformation des Systems.

Mit Rücksicht auf die erste Eigenschaft heißen die Transformationen des Systems *vertauschbar*, mit Bezug auf die zweite heißt das System *geschlossen.*

* Der Ausdruck ‚ein geschlossenes System von Transformationen' entspricht also ganz dem, was man in der Theorie der Substitutionen als eine ‚Gruppe von Substitutionen' zu bezeichnen pflegt.

** umgekehrten Gleichungen repräsentieren, was die *umgekehrte Transformation* heißen soll. Be-

(** continued on p. 186)

A footnote on the phrase "transformation group" emphasizes that this concept comes from permutation theory and mentions the fact that Klein and Lie had earlier called "that which is here called a transformation group" a "closed system of transformations" [346, p. 316]. It is thus clear that in the summer of 1872 Klein had made a great deal of progress in adapting himself to the permutation-theoretic form of group theory. This will soon become even clearer.

Viewed abstractly, Klein's definition of a group is fully equivalent to Jordan's definition (II.4) in the *Traité*. While assuming the invertibility of the transformation equations and permutations, respectively, both assume the closure of the set with respect to the operation. If we set aside the axioms of associativity and of the existence of an identity, then, to be sure, the closure axiom suffices for finite but not infinite groups. In this sense, Jordan's use of the group concept in the *Traité* fitted the object of his investigations. In Klein's case, however, the definition did not meet the requirements, for the membership of the inverse transformation in the "group" is used implicitly and is not explicitly stipulated in the definition of a group. It is interesting to note that, shortly thereafter, Lie recognized and corrected this deficiency in his investigations (III.3).[152]

Further study of papers E and F confirms our conviction, already described above, that, by the summer of 1871, Klein had adapted the permutation-theoretic group concept. The fact that around this time, that is, in 1871–72, Klein did not adopt Jordan's term "isomorphic" is not significant. Klein called two groups, isomorphic in Jordan's sense, "similar":

* Two transformation groups are said to be *similar* if we can associate the transformations of one group to the transformations of the other group so that composition of corresponding transformations yields corresponding transformations. [346, p. 317]

By way of an example, Klein states that we obtain similar groups if we associate each of the transformations A of one group to the transformations $C^{-1}AC$ of the other group, where C is a different transformation. Without proof, Klein states a theorem (true under certain restriction) to the effect that each of two similar

(** continued from p. 185)
zeichnet man ... eine Transformation durch einen Buchstaben $A, B, \ldots$, die Zusammensetzung zweier Transformationen A, B, durch das Symbol (Produkt) AB, so wird die umgekehrte Transformation von A durch A^{-1} darzustellen sein ... Sei nun eine Reihe von Transformationen A, B, C ... gegeben. *Wenn diese Reihe die Eigenschaft besitzt, daß je zwei ihrer Transformationen zusammengesetzt eine Transformation ergeben, die selbst wieder der Reihe angehört, so soll sie eine Transformationsgruppe* heißen.

* Zwei Transformationsgruppen heißen *ähnlich*, wenn man die Transformationen der einen Gruppe so den Transformationen der anderen Gruppe zuordnen kann, daß die Zusammensetzung entsprechender Transformationen entsprechende Transformationen ergibt.

groups is obtained from the other by the use of an auxiliary transformation. He can then interpret similar groups as groups that "arose *from one another through the application of a transformation C*" ("die *durch Anwendung einer Transformation C aus einander* entstanden sind") [346, p. 317].

In §3, Klein introduces the new concept of the principal group ("Hauptgruppe"):

* Geometry cannot ... concern itself with any properties of geometric objects in space except those that are independent of the position in space occupied by these objects and of their absolute magnitude. Nor can it (always without the assistance of a third body) differentiate between the properties of an object and those of its mirror image. These sentences characterize a group of transformations of space—we shall call it the *principal group* (Hauptgruppe)—whose transformations preserve the totality of geometric properties of an object. This group consists of the 6-ply infinite collection of motions, the simply infinite collection of similarity transformations, and the transformation of reflection in a plane. [346, p. 318]

In paper E, §4 is entitled "Die verschiedenen Methoden der Geometrie sind durch eine zugehörige Transformationsgruppe charakterisiert." Klein proceeds from the fact that the larger the group (the term "subgroup" does not appear), the smaller the number of preserved properties of a geometric object in space. The "difference among the usual methods of geometry" ("Unterschied der in der Geometrie üblichen Methoden") can be accounted for "by the nature of the transformation groups adjoined in the process of study" ("in der Art der bei der Behandlung adjungierten Transformationsgruppen") [346, p. 319].

3 The Group-Theoretic Content of the Erlangen Program

A few months later, in paper F, the Erlangen Program proper, Klein changed "methods of geometry" to "geometries." With this change, the heading of §4 of E—"The different methods of geometry are characterized by the associated transformation group"—expresses Klein's central idea. The group concept turns out to be the magic wand with which to order geometry. As Klein puts it, to each "geometry" there is "adjoined" (adjungiert) a definite transformation group.

---

* Die Geometrie kann sich ... überhaupt nur mit solchen Eigenschaften der räumlichen Gebilde befassen, welche unabhängig sind von der Stelle im Raume, die von den Gebilden eingenommen wird, sowie von der absoluten Größe der Gebilde. Auch kann sie nicht (immer ohne Zuhilfenahme eines dritten Körpers) zwischen den Eigenschaften eines Körpers und denen seines Spiegelbildes unterscheiden. Durch diese Sätze ist eine Gruppe räumlicher Transformationen charakterisiert — sie mag die *Hauptgruppe* genannt werden —, deren Transformationen die Gesamtheit der geometrischen Eigenschaften eines Gebildes unberührt lassen. Es setzt sich diese Gruppe zusammen aus den sechsfach unendlich vielen Bewegungen, aus den einfach unendlich vielen Ähnlichkeitstransformationen und aus der Transformation durch Spiegelung an einer Ebene.

* The essence of the different geometric methods developed in modern times consists in the introduction of such more general groups in place of the principal group. [346, p. 319]

At the time Klein thought that any group of geometric transformations had to contain the principal group. The group he immediately turned to was the group of all collineations—including complex collineations—which (as he put it) is the group "adjoined" (adjungiert) to "projective" (projektivische) geometry.[153]

** It [projective geometry (Wussing)] perceives in objects in space only that which is not changed by collineation transformations, and thus *adjoins to itself ... the group of collineation transformations.* [346, p. 319]

In §5 of paper E, Klein extends this principle to n-dimensional manifolds and the paper culminates in the theorem that asserts

*** that the study of a manifold of constant curvature is contained in the projective study. [346 p. 323]

The aim of the Erlangen Program (paper F) is the classification of geometry. The discrepancies and apparent divergence of the different geometric disciplines, whose dynamics gave rise to implicit group-theoretic thinking, can now be reconciled by the explicit application of group-theoretic methods. In his prefatory remarks, Klein elucidates this objective as follows:[154]

† Strictly speaking, we do not develop an essentially new idea but we do delimit clearly and plainly what has been thought more or less definitely by some. What gives additional justification to the publication of such comprehensive considerations is the fact that, as a result of its rapid development in recent years, geometry, which after all is of uniform content, has been fragmented into virtually distinct disciplines, each developing rather independently of the others. [346, pp. 460 to 461][154]

§1 of the Erlangen Program contains a definition of "group" and a clarification of the concept of "principal group" (Hauptgruppe). There are no differences of

---

* In der Einführung solcher allgemeinerer Gruppen an Stelle der Hauptgruppe besteht das Wesen der verschiedenen geometrischen Methoden, die sich in der Neuzeit entwickelt haben.
** Sie faßt an den räumlichen Dingen nur das auf, was durch kollineare Umformungen nicht geändert wird, *sie adjungiert sich also ... die Gruppe aller kollinearen Umformungen.*
*** daß die Behandlung einer Mannigfaltigkeit von konstantem Krümmungsmaße in der projektivischen Behandlung enthalten ist.
† so entwickeln wir wohl keinen eigentlich neuen Gedanken, sondern umgrenzen nur klar und deutlich, was mehr oder minder bestimmt von manchem gedacht worden ist. Aber es schien um so berechtigter, derartige zusammenfassende Betrachtungen zu publizieren, als die Geometrie, die doch ihrem Stoffe nach einheitlich ist, bei der raschen Entwicklung, die sie in der letzten Zeit genommen hat, nur zu sehr in eine Reihe von beinahe getrennten Disziplinen zerfallen ist, die sich ziemlich unabhängig voneinander weiterbilden.

content between F and E. However, in F the difference between "space" (Raum) and "manifold" (Mannigfaltigkeit) is clearer than in E, and so is the difference between the "physical image" (sinnliches Bild) of geometry and its "abstract form" (abstrakte Form). The latter is immediately dismissed, however, by the view that the study of manifolds is a "generalization of geometry" (Verallgemeinerung der Geometrie):

* We peel off the mathematically inessential physical image and see in space only an extended manifold; more specifically, following the usual representation of a point as space element, a triply extended manifold. Following the analogy with space transformations, we speak of transformations of the manifold; they also form *groups*. In contrast to the group of space transformations, however, no group is distinguished by its significance; they are all of equal value. In this way, the following comprehensive program arises as a generalization of geometry:

Given a manifold and in it a transformation group. To investigate the geometric objects belonging to the manifold with regard to properties that are unchanged by the transformations of the group. [346, p. 464]

Next, as a kind of concession to the then dominant invariant-theoretic terminology, Klein immediately formulates this principle in "invariant-theoretic terms" (invariantentheoretisch). This means, according to the point of view that he adopted at the time, "the theory of the relations of any given objects which are unchanged by the group" ("die Lehre von den bei der Gruppe unverändert bleibenden Beziehungen irgendwelcher vorgelegter Gebilde") [346, p. 464]. Using the mixed terminology characteristic of this transitional period, he states the above problem as follows:

** *Given a manifold and in it a transformation group. To develop the invariant theory relating to that group*. [346, p. 464]

The next section explains the method of investigation to be adopted, and ends in the main theorem:

---

* Streifen wir jetzt das mathematisch unwesentliche sinnliche Bild ab und erblicken im Raume nur eine mehrfach ausgedehnte Mannigfaltigkeit, also, indem wir an der gewohnten Vorstellung des Punktes als Raumelement festhalten, eine dreifach ausgedehnte. Nach Analogie mit den räumlichen Transformationen reden wir von Transformationen der Mannigfaltigkeit; auch sie bilden *Gruppen.* Nur ist nicht mehr, wie im Raume, eine Gruppe vor den übrigen durch ihre Bedeutung ausgezeichnet; jede Gruppe ist mit jeder anderen gleichberechtigt. Als Verallgemeinerung der Geometrie entsteht so das folgende umfassende Problem:

Es ist eine Mannigfaltigkeit und in derselben eine Transformationsgruppe gegeben; man soll die der Mannigfaltigkeit angehörigen Gebilde hinsichtlich solcher Eigenschaften untersuchen, die durch die Transformationen der Gruppe nicht geändert werden.

** *Es ist eine Mannigfaltigkeit und in derselben eine Transformationsgruppe gegeben. Man entwickle die auf die Gruppe bezügliche Invariantentheorie.*

* *If we replace the principal group with a larger group, then only some of the geometric properties are retained. The others are no longer properties of the space objects themselves but rather of the system that results when we add to the latter a distinguished geometric object. This distinguished object (assuming that it is given in a definite way) is defined by the following property: If we fix it, then of the elements of the given group only those remain as transformations of the space that belong to the principal group.*

In this theorem lies the peculiar nature of the new geometric directions that we are about to discuss, and their relation to the elementary method. What characterizes them is the fact that instead of the principal group, they base the study on an extended group of space transformation. Insofar as their groups are included, their mutual relation is determined by a corresponding theorem. The same is true of the different methods of study of multiply extended manifolds, methods yet to be considered. [346, pp. 465–466]

This explains the method of investigation. In subsequent sections Klein analyzes in this way the logical position of projective geometry, line geometry, inversive geometry, and Lie's spherical geometry. Here also he provides links to invariant-theoretic ideas. He asserts, for example, that

** *the theories of binary forms and of plane projective geometry, with a conic section as foundation, are equivalent.* [346, p. 471]

To help clarify Lie's papers written at about the same time (III.3), Klein included a section (§9) on contact transformations and went on to indicate in §10 how all this could be carried over, independently of the physical image, to arbitrarily extended manifolds [346, p. 486] [of *n* dimensions (Wussing)].

The Erlangen Program ends with very interesting "Concluding Remarks" (Schlußbemerkungen) and includes in the appendix seven "Notes" in which Klein enlarges on issues treated too briefly in the text, with a view to eliminating at the earliest opportunity any misunderstandings about the epistemological status of geometry.

At the moment, I am interested in the second topic of the concluding remarks.

---

* *Ersetzt man die Hauptgruppe durch eine umfassendere Gruppe, so bleibt nur ein Teil der geometrischen Eigenschaften erhalten. Die übrigen erscheinen nicht mehr als Eigenschaften der räumlichen Dinge an sich, sondern als Eigenschaften des Systems, welches hervorgeht, wenn man denselben ein ausgezeichnetes Gebilde hinzufügt. Dieses ausgezeichnete Gebilde ist (soweit es überhaupt ein bestimmtes ist) dadurch definiert, daß es, fest gedacht, dem Raume unter den Transformationen der gegebenen Gruppe nur noch die Transformationen der Hauptgruppe gestattet.*

In diesem Satze beruht die Eigenart der hier zu besprechenden neueren geometrischen Richtungen und ihr Verhältnis zur elementaren Methode. Sie sind dadurch eben zu charakterisieren, daß sie an Stelle der Hauptgruppe eine erweiterte Gruppe räumlicher Umformungen der Betrachtung zugrunde legen. Ihr gegenseitiges Verhältnis ist, sofern sich ihre Gruppen einschließen, durch einen entsprechenden Satz bestimmt. Dasselbe gilt von den verschiedenen zu betrachtenden Behandlungsweisen mehrfach ausgedehnter Mannigfaltigkeiten.

** *die Theorie der binären Formen und die projektivische Geometrie der Ebene unter Zugrundelegung eines Kegelschnittes sind gleichbedeutend.*

Klein writes that it "is to identify a number of problems that promise, when set out in terms of the analysis given here, to be both important and profitable" (es "soll einige Probleme kennzeichnen, deren Inangriffnahme nach den hier gegebenen Auseinandersetzungen als wichtig und lohnend erscheint") [346, p. 488]. In this way, by pointing out what was especially worth doing in the future, the Erlangen Program actually became a program, a clue to future investigations—although things did not in fact develop in the manner anticipated by Klein at this time,[155] and Klein himself followed ways different from those he was here proposing.

As regards the contemporary version of the Galois theory of equations, that is, Galois theory as presented in Jordan's *Traité*, Klein abandoned the position he had reached in the Erlangen Program. As a result, he bypassed completely the permutation-theoretic group conception. He maintained that what Galois's theory and his own program have in common is the investigation of "groups of changes" (Gruppen von Änderungen).

* To be sure, the objects the changes apply to are different: there one deals with a finite number of discrete elements, whereas here one deals with an infinite number of elements of a continuous manifold. Nevertheless, given the identity of the group concept, the comparison can be pursued ... [346, p. 489]

For all that, Klein states immediately that the true objects of investigation in Galois theory are the groups, or "substitution theory" itself, and that the theory of equations is merely an *application*(!) of substitution theory. Thus Klein too views group theory as an independent mathematical discipline and not as an ad hoc system of concepts and methods for the mastery of geometry or of equations. Klein says that just as there is a theory of permutation groups,

** we insist on a *theory of transformations*, a study of groups generated by transformations of a given type. An application of transformation theory is that study of manifolds which is a consequence of the use of transformation groups as a foundation. [346, p. 489]

We close the group-theoretic analysis of the Erlangen Program and the papers that prepared it with two observations that touch on its influence.[156] First, when we consider the transformation groups treated by Klein we are struck by the fact

---

* Die Objekte, auf welche sich die Änderungen beziehen, sind allerdings verschieden; man hat es dort mit einer endlichen Zahl diskreter Elemente, hier mit der unendlichen Zahl von Elementen einer stetigen Mannigfaltigkeit zu tun. Aber der Vergleich läßt sich bei der Identität des Gruppenbegriffes doch weiter verfolgen ....

** verlangen wir eine *Transformationstheorie*, eine Lehre von den Gruppen, welche von Transformationen gegebener Beschaffenheit erzeugt werden können ... Als eine Anwendung der Transformationstheorie erscheint die aus der Zugrundelegung der Transformationsgruppen fließende Behandlung der Mannigfaltigkeit.

that he had not studied the affine group—that is, he had put no group between the principal group and the group of collineations. Later (1921) Klein referred to this fact as the "result of a one-sided tradition" ("Ergebnis einseitiger Tradition") [346, p. 320; in a footnote to E], and added that at the time he had not arrived at "a full appreciation of the works of Möbius and Grassmann" (zur "vollen Würdigung der Arbeiten von Möbius und Grassmann") and had not been able to present the full significance of the algebraic invariant theory. It was not until 1895–96 that Klein—in his lectures on number theory—emphasized the affine group. Using nonhomogeneous coordinates, he defined it as the totality of integral linear substitutions.

The word "affine" (affin) had been coined by Euler in 1748[157] and taken up by Möbius in his *Der barycentrische Calcul* of 1827. At the same time, following the advice of his philologist friend Weiske, Möbius also introduced the term "collineation" [523, Introduction, p. XII]. The modern concepts of "group of similarities" (äquiforme Gruppe) and "similarity geometry" (äquiforme Geometrie) for what Klein called "principal group" (Hauptgruppe) and "metric geometry" (metrische Geometrie) came much later. This modern terminology, including also the distinguishing of the group of congruence-preserving transformations, is first introduced and used consistently in *Lehrbuch der analytischen Geometrie* by Köhler and Heffter [267] of 1905. The term "subgroup" (Untergruppe) appears in Klein's work, in connection with the Erlangen Program, for the first time in 1893; a new printing enabled Klein to add a footnote correcting a passage in the text of the Erlangen Program that dealt with the concept of "field relative to the generating group" ("Körper mit Bezug auf die erzeugende Gruppe") [346, p. 473], by which Klein meant the object (or objects) that remained invariant under all transformations of the group.

A second observation pertains to Klein's own assessment of the Erlangen Program. In the second volume of his well-known *Vorlesungen über die Entwicklung der Mathematik im 19. Jahrhundert* [404], Klein devotes just a short paragraph to the Erlangen Program. This is part of a larger subdivision, entitled "Freiere Erfassung der linearen Invariantentheorie, mit Einordnung der Vektoranalyse,"[158] dealing with the inner dynamic of algebraic invariant theory. This assessment of the Erlangen Program is flawed by the fact that it fails to take into consideration the influence of the program on the development of abstract group theory. Also, it is deficient on another score. Klein believes that the Erlangen Program is important for having set the stage for the (special) theory of relativity, and he goes to a great deal of trouble to prove this by means of a kind of play on words.[159] Clearly, at this point Klein is already defending a view that he was to adopt late in life, in connection with his extensive studies of the theory of relativity [399, 400,

401, 402]. Klein failed to appreciate the revolutionary physical content of the theory of relativity, regarding it as a mere physical interpretation of essentially mathematical theories that Riemann, Cayley, Sylvester, Klein himself, and Minkowski had developed much earlier.[160]

3 Groups of Geometric Motions; Classification of Transformation Groups

1 Motion Geometry and Groups of Geometric Motions

Logically and historically, there is a distinction between the use of group-theoretic reasoning in geometry and the use of motions or transformations as group elements. This distinction is crucial in an account of the developments that led to the abstract group concept.

My analysis of Jordan's *Traité* of 1870 showed an admittedly very advanced application of the permutation-theoretic group concept to geometry (among other things). The use of the group concept, in the form of a group of transformations, for the purpose of classification in geometry brought a change in that concept; in this sense, the Erlangen Program represents a true break in its history.

It is necessary to consider the change in the group concept in another, larger, context: motion-geometric thinking, of which clear expressions are encountered already in the sixties.

Since the Industrial Revolution at the end of the eighteenth century there has been a steady development of productive forces in the capitalist countries of Europe and North America. Against this background, the activities of Cauchy, Fresnel, Hamilton, Jacobi, Gauss, Fr. Neumann, Bunyakovski, Kirchhoff, Helmholtz, Stokes, Maxwell, Riemann, Gibbs, and many others have led, since the middle of the nineteenth century, to the development of essentially new aspects of the connection between the natural sciences—in particular, physics—and mathematics. Above and beyond the traditionally close relation between mathematics and mechanics, we witness a kind of amalgamation of these disciplines and a mathematization of optics, hydrodynamics, electrodynamics, thermodynamics, and crystallography. In this mathematical physics, the relation between physical—in particular mechanical—motion and coordinate transformation shifted in favor of the physical view: The coordinate transformation was consciously relegated to the secondary role of a mathematical reflection of actual physical-mechanical motion. This trend gave rise to a branch of research that aimed at studying the geometry of physical space through the description of its motions—more specifically, the description of its motions by means of point transformations in which the points represent material points and are subject to analytic conditions connected with the mathematical treatment of a physical body as a system of points.

In historical and philosophical terms, this development is a reflection of mechanical materialism. It is symptomatic rather than accidental that these viewpoints were emphasized by H. Helmholtz, whose declared aim was to "reduce" the whole of "physics to mechanics," that is, point mechanics and central forces. This train of thought continued in the work of Jordan and Klein and led to the idea that the mathematics associated with the motion-geometric conception should be pursued as the study of groups of motions. This development motivated the extension of the (permutation-theoretic) group concept to the concept of a transformation group.

Riemann's 1854 habilitation lecture "Über die Hypothesen, welche der Geometrie zugrunde liegen," published in 1866, attracted a great deal of scientific attention. Helmholtz's papers of the period between 1866 and 1870 dealing with the problems of space and geometry refer to Riemann. Of these papers, the one most related to the ideas of Riemann, and also best known, is "Über die Thatsachen, welche der Geometrie zum Grunde liegen" [271], submitted to the Göttinger Gesellschaft der Wissenschaften in 1868. This paper was preceded, in 1866, by the paper "Über die thatsächlichen Grundlagen der Geometrie" [270].[161]

The two papers are an attempt to axiomatize geometry. Helmholtz's basic assumption[162] is that geometry is the structure of objective space, and as such it can be described, and should be defined, in terms of the possible motions of physical bodies. In Helmholtz's words:

> * My starting assumption was that all the original measurements in space rest on the verification of congruence, and so the system of space measurement must assume conditions under which the only thing we can talk about is the verification of congruence. [270, p. 614]

The possibility of realizing congruence is tied to the assumption of mobility of bodies and is treated in four groups of axioms that deal with the continuity and dimensionality of space, the existence of mobile rigid bodies, free mobility, and the independence of form of rigid bodies under rotation.

Without analyzing the individual postulates and the group-theoretic objections raised against Helmholtz's system by Lie, de Tilly, Klein, Poincaré, Pasch, Hilbert, and others,[163] I wish to point out that Helmholtz advanced the point of view that aimed at understanding the geometry of objective space through motion. A related technical provision was that the motions that lead to the congruence of two geometric objects play a distinguished role and are realized analytically as isometries by means of point transformations. This view, and the group theoretic insight that motions form groups, opened the way for the formulation of the concept of a transformation group.

Earlier (I.1), I discussed the development that led to the separation of the mathematical and physical conceptions of space. The emergence of motion-theoretic

---

* Mein Ausgangspunkt war der, dass alle ursprüngliche Raummessung auf Constatirung von Congruenz beruht, und dass also das System der Raummessung diejenigen Bedingungen voraussetzen muß, unter denen allein von Constatirung der Congruenz die Rede sein kann.

thinking corresponded to a kind of dialectical counterdevelopment—brought about by an application-oriented intertwining of mathematics and physics, especially mechanics, that had been rapidly advancing since the middle of the nineteenth century—that consisted in the identification of [geometric (Translator)] motion with physical motion. Historically, both motives served as the decisive assumptions of a mode of thought that made possible the perfection of the concept of a transformation group and the classification of such groups, in advance of all pragmatic justifications of the concept itself. While it is true (III.2) that between 1866 and 1872, owing to the joint efforts of Lie and Klein, the concept of a transformation group had been elaborated with much precision, it nevertheless remained in what may be called a raw state—to some extent the result of an ad hoc attempt to render group theory applicable in the form given it by the theory of permutation groups, by extending the scope of the group concept. What followed in historical terms was the differentiation of the extended concept of "transformation group"; its counterpart, in logical terms, was the classification of such groups.

Here the leading position belongs to Jordan's large-scale classification of groups of motions in his "Mémoire sur les groupes de mouvements" [313] of 1868–69, whose results had been made public in 1867 in an announcement in the Paris *Comptes rendus* [310]. This paper was directly influenced by the contemporary development of crystallography, especially by the French school around A. Bravais (1811–63).[164] What is remarkable is that Jordan failed to adopt the group concept, in the form that he had given it in the "Mémoire," in his *Traité* two years later.

In the "Mémoire" Jordan investigated the physically possible motions of a rigid body, that is, motions that can be obtained by composition of translations and rotations; he left out reflections, deformations and, more generally, all coordinate transformations that cannot be mechanically realized in the case of rigid bodies.

As a result of this restriction Jordan could claim that *every* motion of a solid in space can be thought of as a twist ("mouvement hélicoïdal") and is completely determined if we know

1. the position in space of the axis A of the twist,
2. the angle r of rotation about A, and
3. the magnitude of the displacement t along A.

Jordan denotes such a twist by the symbol $A_{r,t}$. The result of two successive motions $A_{r,t}$ and $A'_{r',t'}$ is a third motion $A''_{r'',t''}$. Just as in the case of permutations Jordan writes, $A_{r,t}A'_{r',t'}$ yields $A''_{r'',t''}$.

The aim of the paper is not the actual determination of the resultant of two such motions. Jordan points out that such determination is carried out in traditional textbooks of mechanics, including his own *Cours d'Analyse*. Instead, the content of his paper is group theoretic. The closure of the set of twists under composition permits Jordan to use group-theoretic reasoning. He is concerned with the group of motions generated by $A_{r,t}$, $A'_{r',t'}$, $A''_{r'',t''}$, ...:

* *Given* certain motions $A_{r,t}$, $A'_{r',t'}$, $A''_{r'',t''}$, etc., form the different motions that could result from the combination of the given motions executed successively any number of times and in any order. [313, p. 167]

The term "group" is used without reference to its permutation-theoretic origin. Jordan continues thus:

** The group formed by the required motions has the following characteristic property: If M' and M'' are any two motions belonging to this group, then $M'M''$ will also belong to it. [313, p. 167]

Thus all that Jordan requires of a "group," in abstract terms, is closure of its elements with respect to the underlying operation.

With the help of the group concept taken over without much ado from the theory of permutations, Jordan formulates his problem in completely group-theoretic terms. He states that variation of "motions that serve as a point of departure" ("mouvements, qui servent de point de depart") yields infinitely many groups of motions. However, "We shall show that all of these groups can be derived from a limited number of different types" ("Mais nous démontrerons que ces groupes se ramènent tous à un nombre restreint de types différents") [313, p. 168]. Jordan solved his problem and obtained 174 different types of groups of motions, which he listed at the end of the two-part paper [313, pp. 339–345].

The principal group-theoretic tool of the investigation was a theorem taken (as Jordan points out) from permutation theory, a result concerning the transformation of one motion by means of another:[165]

*** Let $M = A_{r,t}$ and $N = B_{\varrho,t}$ be any two motions; $M^{-1} = A_{-r,t}$ is the motion equal and opposite to $A_{r,t}$. Then $M^{-1}NM = B'_{\varrho,t}$, where B' is the axis into which M transforms B. [313, p. 171][165]

We see that, in Jordan's formulation, the problem posed in the "Mémoire" is the problem of group generators. Without reference to Cauchy and Cayley, Jordan

---

* Etant donné certains mouvements $A_{r,t}$, $A'_{r',t'}$, $A'_{r'',t''}$ etc. former les mouvements divers qui peuvent résulter de la combinaison de ceux-là exécutés successivement un nombre quelconque de fois et dans un ordre quelconque.

** Le groupe formé par les mouvements cherchés jouit de cette propriété caractéristique, que si M' et M'' sont deux mouvements quelconques faisant partie de ce groupe, $M'M''$ en fera partie.

*** *Soient* $M = A_{r,t}$ *et* $N = B_{\varrho,\tau}$ deux mouvements quelconques; $M^{-1} = A_{-r,-t}$ le mouvement égal et contraire à $A_{r,t}$: on aura $M^{-1}NM = B'_{\varrho,\tau}$, B' étant l'axe transformé de B par M.

described the conceptual content of the term "generator" by means of the phrase "motions that serve as a point of departure" ("mouvements, qui servent de point de depart"). It should be noted, however, that, in the "Mémoire," Jordan did not construct systems of generators of groups of motions.

It is this very linking of two advances—the introduction of groups of geometric motions and of the question of their generators—that produced the advance that we encounter in Klein's studies of groups of isometries of regular polyhedra. This advance enabled Klein not only to apply to geometry the fundamental principles of permutation theory but also to work out the concept of a (discrete) group of transformations. The recognition and clarification of that concept's potential to fuse geometry, algebra, and the theory of functions set in motion a far-reaching development.

Klein's interest in the group-theoretic treatment of symmetries of regular polyhedra coincided with the creation of the Erlangen Program. Therefore, my separate description of this aspect of Klein's work served only the methodological purpose of demonstrating the changing nature of the concept of a group in the context of yet another class of problems. Besides Jordan, it was essentially only Klein who had attained, and suitably emphasized, a view of the isomorphism of certain permutation groups and certain groups of symmetries of regular polyhedra—an idea pointing toward the future abstract group concept (III.4).

Already in the Erlangen Program, Klein brought up the question of the connections between groups and the theory of equations and asked for (among other things) a group-theoretic study of regular fields.[166] As a means toward the analytic treatment of this problem, Klein suggested that the complex numbers be represented on what we now call the Riemann sphere.[167] (This admittedly sketchy suggestion is found in the Erlangen Program in §6: Die Geometrie der reziproken Radien. Die Interpretation von $x + iy$; and further in Note VII, "Zur Interpretation der binären Formen – die Darstellung der komplexen Zahlen auf der heute nach Riemann benannten Zahlenkugel.") This idea, conceived by Klein independently of Riemann, turned out to be of decisive importance, in that it provided, among other things, a link with the theory of forms.

Klein's initial views on the relation between "substitution" theory and geometry are contained in the paper "Über eine geometrische Repräsentation der Resolventen algebraischer Gleichungen" [355] of 1871. He thinks that "the general theory of algebraic equations ... [is] beautifully illustrated by a number of special geometric examples" ("die allgemeine Theorie der algebraischen Gleichungen ... in schönster Weise durch eine Anzahl besonderer geometrischer Beispiele illustriert [wird]") [347, p. 262], and that

* the great value of these examples lies in the fact that they furnish an intuitive sense of the intrinsically abstract conceptions of substitution theory. [347, p. 262]

This view was common at the time. It underlies, for example, the investigations of Clebsch on quintic equations [159], as well as the results of Brioschi, Kronecker, and others, reported in Serret's *Cours d'Algèbre supérieure* and Jordan's *Traité* (II.5). In his paper [355] of 1871 Klein viewed the n roots of an algebraic equation of degree n as elements of an $(n-2)$-dimensional space. To study them analytically, he replaced the permutations of the n roots by linear transformations of a continuous space. Klein described the method he applied in 1871 as follows:

** By means of this representation one establishes a remarkable connection between the theory of equations of nth degree and the theory of covariants of n elements of a space of $n-2$ dimensions, so that each theory can be viewed as an image of the other. [347, p. 262]

In order to advance toward the conception of groups of geometric objects, Klein had to surmount the point of view of the mutual modeling of algebraic and geometric interpretations. For this he required a new idea, the idea of a group-theoretic interpretation of the rotations of the Riemann sphere. Effectively, Klein took this step already in November 1873. (The relevant paper [360] and its predecessor [357] advanced the efforts, going back to Poncelet and Chasles, to delineate the role of the complex in the realm of analytic geometry.[168]) But it was not until 1875 that he carried it out and gave a group-theoretic formulation of the resulting problem in the paper "Über binäre Formen mit linearen Transformationen in sich selbst" [362] (dated June 1875).

Klein here thinks of the Riemann sphere as "surrounding" the regular polyhedra. To each motion of the Riemann sphere there corresponds a fractional linear transformation of a complex variable. Klein describes the object of the study as follows:

*** The following investigations were motivated by our quest to apply the interpretation of $x+iy$ on the surface of the sphere to the theory of binary forms. ... The special task I then embarked on was related to the fact that, for the ordinary metric, the following problem had been solved for a long time: *to construct all finite groups of motions.* It seemed possible to solve the same problem for the general projective metric and to determine, in this way, *all finite groups of linear transformations of a complex argument* $x+iy$, a problem that is bound to be important for algebraic investigations. [347, pp. 275–276]

---

* der hohe Nutzen dieser Beispiele liegt darin, daß sie die an und für sich so eigenartig abstrakten Vorstellungen der Substitutionentheorie in anschaulicher Weise dem Auge vorführen.
** Vermöge dieser Repräsentation wird die Theorie der Gleichungen n-ten Grades in einen merkwürdigen Zusammenhang gebracht mit der Theorie der Kovarianten von n Elementen eines Raumes von $n-2$ Dimensionen, so zwar, daß jede der beiden Theorien geradezu als ein Bild der anderen angesehen werden kann.
(*** see p. 200)

(It should be pointed out, that Klein's use of the term "linear transformations" included fractional linear transformations.)

What follows is a concise description of the course followed by Klein.

Suppose that we wish to determine all finite groups of fractional linear transformations. Then, says Klein, we need only consider transformations "reproduce themselves after a finite number of times. It follows that, in geometric terms, they must be *rotations*" ("nach einer endlichen Anzahl von Malen reproduzieren. Sie müssen daher, geometrisch zu reden, *Rotationen* sein") [347, pp. 277–278], and the rotation angle must be a rational part of 2π. Every such "rational rotation" (rationale Rotation) generates a finite group. To obtain composite groups, we must combine rotations not all of which fix the poles of the Riemann sphere; for a product of two rotations to be a rotation, their axes must intersect. Klein specifically includes rotations that, as he puts it, "interchange the points left fixed under rotation" ("die bei der Rotation festbleibenden beiden Punkte vertauschen") [347, p. 278], that is, are reflections in a plane of a great circle. At this point Klein defines the concept of a "dihedron" (Dieder) as "a figure of zero volume formed of the upper and lower sides of a regular polygon" ("von der Ober- und Unterseite eines regulären Polygons gebildete Figur vom Rauminhalt Null") [347, p. 278].

Having thus described the set of admissible rotations, Klein summarizes the result of his investigations in these words:

* But then there belong here the groups of rotations that bring into coincidence with themselves the *regular solids:* the tetrahedron, the octahedron, and the icosahedron, or, what amounts to the same thing, the tetrahedron, the cube, and the pentagonal dodecahedron. I shall show that *these examples yield all the required groups* [that is, groups of fractional linear transformations of a complex variable (Wussing)]. [347, p. 278]

Thus Klein spelled out the isomorphism of the groups of isometries of the octa-

---

(*** see p. 199)

Die nachstehenden Untersuchungen sind aus dem Streben hervorgegangen, die geometrische Interpretation von $x + iy$ auf der Kugelfläche für die Theorie der binären Formen zu verwerten ... Die spezielle Aufgabe, welche ich weiterhin in Angriff nahm, knüpfte an den Umstand an, daß bei gewöhnlicher Maßbestimmung ein lange erledigtes Problem ist: *alle endlichen Gruppen von Bewegungen zu konstruieren.* Es schien möglich, für allgemeine projektivische Maßbestimmung dasselbe Problem zu lösen und damit also, was für algebraische Untersuchungen von Wichtigkeit sein muß, *alle endlichen Gruppen linearer Transformationen eines komplexen Arguments* $x + iy$ *zu gewinnen.* Es hängt diese Bestimmung auf das genaueste mit der Theorie der regulären Körper zusammen ....

* Dann aber gehören hierher die Gruppen derjenigen Rotationen, welche die *regulären Körper:* Tetraeder, Oktaeder, Ikosaeder, oder, was auf dasselbe hinauskommt: Tetraeder, Würfel, Pentagondodekaeder, mit sich selbst zur Deckung bringen. Ich werde nun zeigen, daß *mit diesen Beispielen alle Gruppen der geforderten Beschaffenheit bereits angegeben sind.*

hedron, the cube, and the pentagonal dodecahedron, respectively. He also gave the orders of these groups.

This turn of his research was so fruitful that in commenting in 1921 on his paper [347] of 1873, Klein spoke of "having struck a vein of ore" ("eine erzführende Ader angeschlagen"). At Easter 1875 Klein moved to the Technische Hochschule in Munich, where he continued this line of research. He maintained scientific contact with Gordan and exchanged letters with Brioschi. In the summer of 1875, L. Fuchs (1833–1902) published his investigations on algebraically integrable linear differential equations, work whose invariant-theoretic component overlapped Klein's investigations on regular solids. Fuchs's results inspired Klein's papers [363] and [364] on algebraically integrable linear differential equations. Gordan also turned [249] to this topic. However, Klein soon returned to the line of research begun in 1873. Already in that year he had realized that there is an intimate connection between the icosahedron and the theory of quintic equations. Now this theory became one of his central concerns. The interpretation of the quintic in terms of the icosahedron and, more generally, of equations in terms of regular polyhedra implied an approach that gave prominence to groups of isometries and considered equations in this context. In this sense the theory of the quintic may be said to have inspired the development of the concept of a transformation group.

Klein gave final form to this line of thought in 1877. The paper "Weitere Untersuchungen über das Ikosaeder" [365] shows that Klein became fully aware of groups whose elements are geometric motions. In the introduction to this paper, Klein again points to the insights attained in 1873, bearing on the connection between the icosahedron and the quintic, and adds that

> * the ease with which it was possible to deduce certain resolvents of sixth and fifth degree that turn up in the theory of quintic equations and to recognize their connection must have been striking even then. But it was Gordan, with whom I discussed these matters in detail, who urged me to invert the question and to try *to deduce the theory of the quintic from consideration of the icosahedron.* While maintaining the steady exchange of ideas with Gordan, I succeeded not only in deducing in a natural way, and from a single source, all of the relevant algebraic theorems and results published—in part without proof—by Kronecker and Brioschi,[169] but also in adding to them new, and I believe, essential contributions. [347, pp. 321–322]

---

* bemerkenswert mußte schon damals die Leichtigkeit erscheinen, mit der es gelang, gewisse in der Theorie der Gleichungen fünften Grades auftretende Resolventen sechsten und fünften Grades abzuleiten und in ihrem Zusammenhange zu erkennen. Aber ich bin erst durch Gordan, mit dem ich diese Gegenstände ausführlich besprach, veranlaßt worden, die Frage umzukehren und zu versuchen, *geradezu die Theorie der Gleichungen fünften Grades aus der Betrachtung des Ikosaeders abzuleiten.* In der Tat gelang es mir — im steten Verkehr mit Gordan — nicht nur sämtliche algebraische Sätze und Resultate, welche Kronecker und Brioschi in dieser Hinsicht — zum Teil ohne Beweis — publiziert haben, aus einer Quelle naturgemäß abzuleiten, sondern ihnen auch neue, und, wie ich glaube, wesentliche Beiträge hinzuzufügen.

The "natural" source of Klein's success was the consistent geometric-group-theoretic viewpoint. In his subsequent papers on the theory of equations Klein emphasized this viewpoint, and in 1884 he derived it from first principles in his famous *Vorlesungen über das Ikosaeder und die Auflösung der Gleichungen vom fünften Grade* [379]. Two years later, he described his method again, in these terms: "In the paper 'Zur Theorie der allgemeinen Gleichungen sechsten und siebenten Grades' [383], I hold to the approach adopted in my 'Icosahedron book' of introducing the relevant algebraic processes by means of geometric constructions" ("Auch in der Abhandlung 'Zur Theorie der allgemeinen Gleichungen sechsten und siebenten Grades' halte ich, wie in meinem 'Ikosaederbuch' daran fest, die in Betracht kommenden algebraischen Prozesse durch geometrische Konstruktionen einzuleiten") [347, p. 440].

The *Vorlesungen über das Ikosaeder* consists of two parts, each subdivided into chapters. The first part is called "Theorie der Gleichungen fünften Grades." A study of chapter I, "Die regulären Körper und die Gruppentheorie," shows that the geometric isometries (Deckabbildungen), rather than their analytic reflections in the form of coordinate transformations, are accepted as group elements.

For the sake of subsequent interpretation on the Riemann sphere, Klein investigates the regular polyhedra "symbolically" ("im übertragenen Sinne"). Specifically, the sphere is circumscribed over the regular polyhedra (including the dihedron), and what is studied on the sphere are the projections, from the center of the sphere, of their edges and faces.

* Thus the direct object of our consideration is a certain subdivision of the sphere. We use the terminology and, in part, the constructions of solid geometry for the sake of a more convenient mode of expression. [379, p. 3]

The terms involved are tetrahedron, octahedron, cube, icosahedron, pentagonal dodecahedron, and dihedron.

In order fully to characterize the chapter in question, I must say that it deals with group theory rather than with geometry. It is basically different from earlier as well as contemporary investigations on regular polyhedra, such as, say, the papers [580, 579, 688] by Reye and Stephanos published in 1883 and dealing with the theory of the cube. I quote Klein:

** To begin with, it must be emphasized that what we are concerned with in the sequel is not really the various figures but rather the *rotations* and *reflections* or, briefly, the *elementary geo-*

---

* Das nähere Object unserer Betrachtung ist also eine bestimmte Kugeltheilung, und nur der bequemeren Ausdrucksweise wegen greifen wir auf die Benennungen und zum Theil auch auf die Constructionen der Raumgeometrie zurück.

(** see p. 203)

metric operations that bring them into coincidence with themselves. *The figures are, for us, only the means of orientation that enables us to survey the totality of certain rotations or other transformations ....*

The problem of studying the relevant rotations, etc., that carry the various configurations into themselves, dictates, as a matter of course, a linkup with the important and comprehensive theory created by *Galois's* pioneering paper and known as *group theory*. This theory grew out of the theory of equations and therefore refers to *permutations* of certain elements. Nevertheless, as has been realized for a long time, this theory subsumes all questions involving a closed system of arbitrary *operations*. We say of arbitrary operations that they form a *group* if the composition of any two of them yields one of the given operations. We thus have immediately the theorem: *The totality of rotations which make a regular solid coincide with itself form a group*. [379, pp. 4–5]

Having formulated the problem of study of geometric motions as a group-theoretic problem, Klein naturally relied on the well-developed theory of permutation groups. Accordingly, he interpolated a section (the second) entitled "Gruppentheoretische Vorbegriffe." There he defined the degree (that is, the order) of a finite group, a subgroup, conjugation of one element by another, a distinguished (that is, normal) subgroup, and—since he had in mind the connection between groups of isometries (Deckabbildungen) and the theory of groups of permutations—an isomorphism of groups. As regards isomorphism, Klein went far beyond Jordan, in that he developed abstractly the idea of the identity of isomorphic groups:

* Two groups are said to be isomorphic if we can establish a correspondence of their operations S, S' such that if S_i corresponds to $S_{i'}$ and S_k to $S_{k'}$, then S_iS_k corresponds to $S_{i'}S_{k'}$. If the isomorphism is one-to-one, then we call it *holoedric*. Then the two groups are abstractly identical and they can differ only in the *meaning* of their respective operations. The subgroups of one group correspond to the subgroups of the other, etc., etc. [379, p. 8]

(** see p. 202)

Es sind nun aber, und dies muss von vornherein hervorgehoben werden, im Folgenden nicht eigentlich die hiermit aufgezählten Figuren selbst, die den Gegenstand unserer Betrachtung ausmachen, vielmehr sind es jene *Drehungen*, oder auch *Spiegelungen*, oder kurz gesagt: *diejenigen elementargeometrischen Operationen*, durch welche die genannten Figuren mit sich selbst zur Deckung kommen. *Die Figuren sind für uns nur das Orientirungsmittel, vermöge dessen wir die Gesammtheit gewisser Drehungen oder sonstiger Umänderungen übersehen ...*

Indem wir uns jetzt die Aufgabe stellen, die in Rede stehenden Drehungen etc. zu studiren, durch welche die genannten Configurationen in sich übergehen, gebietet sich von vornherein der Anschluss an jene wichtige und umfassende Theorie, welche zumal durch *Galois*' bahnbrechende Arbeiten geschaffen worden ist und die man als *Gruppentheorie* bezeichnet. Ursprünglich aus der Gleichungstheorie erwachsen und dementsprechend auf die *Vertauschungen* irgendwelcher Elemente bezüglich, umfasst diese Theorie, wie man seit lange erkannt hat, überhaupt jede Frage, bei der es sich um eine geschlossene Mannigfaltigkeit irgend welcher *Operationen* handelt. Man sagt von beliebigen Operationen, dass sie eine *Gruppe* bilden, wenn je zwei der Operationen zusammengesetzt immer wieder eine Operation unter den bereits gegebenen erzeugen. In diesem Sinne haben wir sofort den Satz: *Die Drehungen, welche einen regulären Körper mit sich selbst zur Deckung bringen, bilden in ihrer Gesammtheit eine Gruppe.*

(* see p. 204)

Following Jordan, Klein called a homomorphism of two groups a meriedric isomorphism. He used the term hemiedric isomorphism for a homomorphism in which to every S' there correspond two S.

While a modern reader would regard Klein's explanations of group-theoretic concepts as verbose, they were actually rather skimpy for their time. That is why he referred the reader to more detailed expositions of group theory such as Serret's *Cours d'Algèbre supérieure* (the fourth French edition of 1879, whose group-theoretic component was essentially unaltered from the third edition's, had been translated into German) and to Jordan's *Traité* of 1870, Netto's *Substitutionentheorie und ihre Anwendungen auf die Algebra* (1882), and the *Gruppentheoretische Studien* (1882, 1883) of W. v. Dyck [180, 181]. (For isomorphism, see also [86].) In this way, Klein consciously established a link with the development of the abstract group concept that was taking place in the eighties (III.4).

The subsequent sections of the *Vorlesungen über das Ikosaeder* are a systematic study of groups of isometries of regular polyhedra. The group of order n of "cyclic rotations" through $2\pi/n$ arises when a sphere, circumscribed about one of the polyhedra mentioned above, rotates about a fixed axis through the center of the sphere (§3); the "group of dihedral rotations" of order $2n$ arises when the dihedron is a regular n-gon (§4). The fifth section is concerned with the "Klein 4-group" (we shall not deal with Klein's definition of this group as a group of isometries (Deckabbildungen) since it has little relevance to our theme), §6 deals with the "group of tetrahedral rotations," §7 with the "group of octahedral rotations," and §8 with the "group of icosahedral rotations."

Already in connection with a cyclic rotation group of order n, Klein stressed the importance of finding isomorphism of groups of symmetries and permutation groups. In this, the simplest of cases, he used the isomorphisms "as the essence of 'cyclic' permutations of n elements taken in a definite order" ("zum Inbegriff der 'zyklischen' Vertauschungen von n in bestimmter Reihenfolge genommenen Elementen") in order "to practice the concept of isomorphism" ("den Begriff des Isomorphismus einüben zu können") [379, p. 9]. This idea appears in later section as well; the well-known isomorphisms of the various groups of isometries of regular polyhedra and of groups of permutations are discovered and ex-

---

(* see p. 203)
Zwei Gruppen heissen isomorph, wenn man ihre Operationen S, S' derart einander zuweisen kann, dass immer $S_i S_k$ dem $S'_i S'_k$ entspricht, sofern S_i dem S'_i, S_k dem S'_k entsprechend gesetzt ist. Die isomorphe Beziehung kann eine wechselseitig eindeutige sein; man spricht dann von *holoedrischem* Isomorphismus. Es sind in diesem Falle die beiden Gruppen abstract genommen überhaupt identisch, und es ist nur die *Bedeutung* der beiderseitigen Operationen, in denen eine Verschiedenheit liegen kann. Die Untergruppen der einen Gruppe liefern also ohne Weiteres die Untergruppen der anderen Gruppe, etc., etc.

plicitly formulated. Thus in the case of the group of the icosahedron we have the statement:

* *We see that the group of 60 icosahedral rotations is holoedrically isomorphic to the group of 60 even permutations of five things.* [379, p. 19]

2 *Differentiation and Scope of the Concept of a Discontinuous Transformation Group*

Historically, the ad hoc extension of the conceptual content of permutation groups to that of transformation groups completed, in principle, in the Erlangen Program (III.2), was followed by the refinement of the concept of a transformation group, in consequence of the diverse ways in which the idea was applied.

In the late sixties Klein and Lie had undertaken, jointly,

** to investigate geometric or analytic objects that are transformed into themselves by *groups of changes.* [379, p. IV]

This is Klein's retrospective (1884) description of their successfully completed program. The division of the work reflected in part the crumbling of their personal friendship. Having described the central aim of their research, Klein continues:

*** This idea has had a decisive effect on our later papers, in spite of the fact that they seem to reflect divergent concerns. While I concentrated largely on groups of *discrete* operations and was led to the investigation of regular solids and their relation to the theory of equations, Lie studied, from the very beginning, the more difficult theory of *continuous* transformation groups, and thus of *differential equations.* [379, p. IV]

In the introduction to the first volume of his *Theorie der Transformationsgruppen* [463] of 1888, Lie gave his view of the division of labor between himself and Klein:

† I noticed [in the early seventies (Wussing)] that *most of the ordinary differential equations, whose integration had been accomplished by the older integration methods, remained invariant under certain easily describable sets of transformations, and that the methods in question were coextensive with the exploitation of this property of the differential equations in question.* In other words, I saw

---

* *Auf solche Art erweist sich die Gruppe der 60 Ikosaederdrehungen mit der Gruppe der 60 geraden Vertauschungen von 5 Dingen holoedrisch isomorph.*

** überhaupt solche geometrische oder analytische Gebilde in Betracht zu ziehen, welche durch *Gruppen von Aenderungen* in sich selbst transformirt werden.

*** Dieser Gedanke ist für unsere beiderseitigen späteren Arbeiten, soweit dieselben auch auseinander zu liegen scheinen, bestimmend geblieben. Während ich selbst in erster Linie Gruppen *discreter* Operationen ins Auge fasste und also insbesondere zur Untersuchung der regulären Körper und ihrer Beziehung zur Gleichungstheorie geführt wurde, hat Hr. Lie von vorneherein die schwierigere Theorie der *continuirlichen* Transformationsgruppen und somit der *Differentialgleichungen* in Angriff genommen.

(† see p. 206)

that, basically, the concept of a differential invariant of a finite continuous group appeared, admittedly *implicitly* and in *special form*, in every textbook on ordinary differential equations. After I had looked at a number of the older integration methods from this general point of view, I undertook the development of a general integration theory for all ordinary differential equations that admit known finite or infinitesimal differential equations. To begin with, it was clear that the transformations in question must, in each case, generate a group.

I formulated and solved this problem by myself. While working on this problem I maintained a lively exchange of ideas with my friend, Professor *Felix Klein*. Klein had set himself a task that, while obviously different from mine, shared with it essential analogies. Specifically, Klein investigated geometric and analytic objects that admitted a *discontinuous* group and, in particular, wished to apply the theory of discontinuous groups of *linear* transformations in the theory of equations. While it is certain that both of us have been influenced by our detailed oral and written exchanges concerning our respective related subjects, I am not in a position to determine the extent of these mutual influences with greater precision. This is due to the fact that our contacts involved general points of view rather than specific theorems. [463, p. V]

While they contain no echoes of the discord that had set in between them,[170] the testimonies of Lie and Klein reflect the division of labor involved in the sharpening of the conceptual content of the word "group" and the extension of its realm of application. I begin this survey of their work with a discussion of Klein's contribution that, in group-theoretic terms, bears on finite or infinite discrete groups.

While the concerns of this book make indispensable a detailed analysis of the concept of a group of transformations—in 1887, when the Vorlesungen über

---

(† see p. 205)

Ich bemerkte, *dass die meisten gewöhnlichen Differentialgleichungen, deren Integration durch die älteren Integrationsmethoden geleistet* wird, *bei gewissen leicht angebbaren Schaaren von Transformationen invariant bleiben, und dass jene Integrationsmethoden in der Verwerthung dieser Eigenschaft der betreffenden Differentialgleichungen bestehen.* Mit anderen Worten: ich sah, dass der Begriff der Differentialinvariante einer endlichen continuirlichen Gruppe im Grunde, wenn auch nur *implicite* und *in specieller Form*, in jedem Lehrbuche über die gewöhnlichen Differentialgleichungen vorkommt. Nachdem ich so eine Reihe von älteren Integrationsmethoden unter einen allgemeinen Gesichtspunkt gebracht hatte, stellte ich mir naturgemäss die Aufgabe, für alle gewöhnlichen Differentialgleichungen, die bekannte endliche oder infinitesimale Transformationen zulassen, eine allgemeine Integrationstheorie zu entwickeln. Dabei war es von vornherein klar, dass die betreffenden Transformationen in jedem einzelnen Falle eine Gruppe erzeugen mussten.

Die eben gekennzeichnete Aufgabe habe ich mir selbständig gestellt und habe sie selbständig erledigt. Während ich damit beschäftigt war, stand ich in einem lebhaften Verkehr mit meinem Freunde Herrn Professor *Felix Klein.* Dieser hatte sich seinerseits eine Aufgabe gestellt, welche allerdings von der meinigen verschieden war, aber doch wesentliche Analogien mit derselben darbot. Er betrachtete nämlich überhaupt geometrische und analytische Gebilde, die eine *discontinuirliche* Gruppe gestatten und wollte insbesondere die Theorie der discontinuirlichen Gruppen von *linearen* Transformationen für die Gleichungstheorie verwerthen. So sicher es nun auch ist, dass unsere eingehenden mündlichen und brieflichen Aussprachen über die uns beschäftigenden verwandten Gegenstände auf uns beide von Einfluss gewesen sind, so bin ich doch ausser Stande, den Umfang dieser gegenseitigen Einflüsse genauer festzustellen; denn bei unserem Verkehr handelte es sich weniger um bestimmte Sätze als um allgemeine Gesichtspunkte.

das Ikosaeder was published, Klein included in this concept groups of coordinate transformations as well groups of geometric motions—they do not require a detailed analysis of the development set in motion by Lie and Klein.[171] It suffices to sketch the range of application of discrete and continuous groups of transformations. For this purpose, we have to delineate the intrinsic power of the group-theoretic method. Indeed, the clarification of the full scope of the concept of a "group of transformations" (Transformationsgruppe) provided the general background and the motivation for the simultaneous emergence of the concept of an *abstract* group, an idea that eventually turned out to be a key concept of mathematics. The recognition of the unifying power of group theory was furthered by the recognition of the multiplicity of its possible—and decisive—applications. Thus, going beyond geometry, it was now possible to achieve a synthesis of earlier results in analysis and in the theory of functions of a complex variable.

As regards the range of application of the theory of discrete groups, the two main areas involved are the theory of equations—especially the theory of the quintic equation—and the theory of automorphic functions.

In 1875 Klein published the paper "Über binäre Formen mit linearen Transformationen in sich selbst" [362]. A technical feature of this paper is the systematic use of homogeneous coordinates. Its principal result is the construction of all finite groups of fractional linear transformations. (The idea behind this achievement is described in the *Vorlesungen über das Ikosaeder* and we dealt with it above; see III.3.1.) This paper marked the conscious inauguration of a development that provided an organic link between the theory of (finite discrete) groups and algebraic invariant theory. In his autobiographical notes [347, pp. 256–257], Klein described the influence that his contacts with Gordan had had on him in this area.

In the seventies, Klein restricted himself to the study of finite groups in a single (nonhomogeneous) variable (in contemporary terminology finite groups of linear substitutions in a single variable). On the other hand, in 1878 Jordan began his investigations [331, 332] of finite groups of linear transformations in finitely many variables, but managed to realize his program only for binary and ternary groups. Extensive research in this area during the eighties and nineties, bound up with transformations of elliptic and hyperelliptic functions, the investigation of algebraic curves and surfaces, and the theory of linear differential equations, yielded insights involving special quaternary, quinary, ... groups (for details see [738]). Eventually, all this activity was submerged in the representation theory of groups by linear homogeneous substitutions. To the extent that it aimed to represent abstract groups by means of transformation groups, and that Frobenius and Burnside used the theory of group characters to prove important results about abstract groups, this theory,[172] developed at the end of the nineteenth and the beginning of the twentieth centuries by G. Frobenius (1849–1917), W. Burnside (1852–1927), T. Molien (1861–1941), and I. Schur (1875–1941), already belongs to the sphere of abstract group theory.

Only in the midseventies, that is, long after the scope of the group concept had been extended to that of a transformation group, did Klein turn his attention to the theory of permutation groups. His appreciation of the full scope of the group concept, hence of a consistently group-theoretic methodology, gave him a deeper insight[173] into the solution of the quintic by means of elliptic modular functions. This solution, based on Galois's approach,[174] was given in 1858 by Ch. Hermite [275].[175] In turn it had inspired, at the end of the fifties, extensive research by, among others, L. Kronecker and F. Brioschi into the generation of resolvents of equations of higher degree. In 1861 [431], Kronecker placed this cluster of problems in a deeper, field-theoretic context. Kronecker distinguished between what Klein [379, p. 157] later (in 1884) called "natural" and "accessory" (natürliche und accessorische) irrationalities.[176] In 1877 [365] Klein, following B. Riemann's elaboration (in his 1858–59 lectures) of the function-theoretic significance of the icosahedron, established connections between the solution of the quintic and the Galois group of the "icosahedron equation" on the one hand, and the geometric interpretation of the generation of resolvents on the other. In discovering this interplay between group theory, the theory of forms, and the solution of the quintic, Klein had relied on isomorphisms between groups of covering transformations and permutation groups. In his *Vorlesungen über das Ikosaeder* of 1884, he had stressed the methodological content of his approach. Klein's earlier papers, as well as the papers [367, 373, 383, 384, 395, 397, 398], all dealing with equations of higher degree, culminated in what Gordan [374, p. 261] jokingly referred to as the "Hypergalois" Program:

* To reduce the solution of equations to the solution of "form problems" connected with finite groups of linear substitutions in the fewest possible variables. [347, p. 258]

After going through certain intermediate stages, this line of Klein's research was subsumed in the study of finite groups of linear substitutions without the intermediary use of the theory of forms.

But all these extremely impressive papers had done little to extend and deepen the new conceptual content of a group. This was because Klein and his fellow workers in this area relied on isomorphisms between groups of isometries and permuta-

---

* *Die Auflösung der Gleichungen auf die mit endlichen Gruppen linearer Substitutionen möglichst geringer Variabelnzahl verknüpften ‚Formenprobleme' zurückzuführen.*

tion groups, and thus stayed within the conceptual environment of (finite) permutation groups.

In the theory of automorphic functions, however, matters took a different turn. Here the transition to infinite discrete groups of transformations made it necessary to extend not only the conceptual content but also the conceptual scope. What is striking—from a historical viewpoint—is that the transition from finite to infinite (discrete) groups was virtually "smooth" and created little stir.[177] At the very least, we can say that this transition involved less serious intellectual difficulties than the transition from permutation groups to transformation groups.

The anchor points of the theory of automorphic functions—as well as of Klein's theory of equations—lay in the transformation theory of elliptic functions.

In his *Traité des fonctions elliptiques et des intégrales euleriennes* of 1825/26, A.-M. Legendre suggested that his division of elliptic integrals into three "normal" types had given the theory of "fonctions elliptiques"—his term for elliptic integrals—its final form. Actually, the theory of elliptic functions had been launched a few years earlier, when N. H. Abel, C. G. J. Jacobi, and C. F. Gauss[178] independently arrived at the decisive idea of considering, instead of the normal integral of the first kind,

$$u = \int_0^\varphi \frac{d\varphi}{\sqrt{1 - k^2 \sin^2 \varphi}},$$

the inverse function $\varphi = \operatorname{am} u$. In 1827 Abel and Jacobi discovered, almost simultaneously, the double periodicity of the three (single-valued) functions sin am u, cos am u, and Δ am u, now designated sn u, cn u, and dn u. Jacobi introduced the so-called theta functions, whose quotients could be used to represent sn u, cn u, and dn u, and showed[179] how one could use them to compute the three normal integrals. It was also Jacobi who discovered that for doubly periodic functions there exists a pair of primitive periods p_1, p_2, with p_1/p_2 complex, such that all periods p can be written as $p = mp_1 + np_2$, that is, that the periods form a module. Set $p_1 = 2\omega$ and $p_2 = 2\omega'$. The pairs of periods p_1, p_2 and $q_1 = 2\dot{\omega}$, $q_2 = 2\dot{\omega}'$, are equivalent—that is, they determine the same lattice points in the complex plane—if and only if there exist integers $\alpha, \beta, \gamma, \delta$ with $D = \alpha\delta - \beta\gamma = \pm 1$ such that

$$\begin{aligned} \dot{\omega} &= \alpha\omega + \beta\omega', \\ \dot{\omega}' &= \gamma\omega + \delta\omega'. \end{aligned}$$

By the middle of the nineteenth century, the further development of the theory of elliptic functions by, among others, K. Weierstrass (1815–97), G. Eisenstein

(1823–52), Ch. Hermite, B. Riemann (1826–66), and R. Dedekind (1831–1916) provided clear indications of a connection between these functions and the theory of linear transformations. Like the σ-function and the ζ-function, the Weierstrass $\wp$-function turned out to be invariant with respect to every rational transformation of the periods. Also, the coefficients g_2 and g_3 of the differential equation $\wp'^2 = 4\wp^3 - g_2\wp - g_3$ and, of course, the quantity $I = g_2^3/\Delta = g_2^3/(g_2^3 - 27_3^2)$, which Klein called an "absolute invariant," turned out to be invariant when a period pair was replaced by an equivalent period pair. Thus if we call τ the ratio of the periods, these "elementary modular forms" are invariant when τ is replaced by $\dot{\tau} = (\gamma + \delta\tau)/(\alpha + \beta\tau)$, where $\alpha, \beta, \gamma, \delta$ are integers such that $\alpha\delta - \beta\gamma = 1$.

The general results of the theory of doubly periodic functions, viewed as pertaining to functions of three arguments, namely, u and the two periods (with complex ratio and positive imaginary part), had prompted Abel and Gauss, and then, among others, Hermite, Eisenstein, Weierstrass and Dedekind[180] to investigate[181] the modular functions.

Klein turned his attention to elliptic modular functions in 1878. There is a close connection between the train of thought in his first substantial paper in this area, "Über die Transformation der elliptischen Funktionen und die Auflösung der Gleichungen fünften Grades" [368], and R. Dedekind's *Erläuterungen zu Riemann's Fragmenten über die Grenzfälle der elliptischen Modulfunktionen* of 1876 [163]. But there is also a close link between the results in this paper and some of his own results, in particular, the construction of all finite[182] groups of linear transformations in one variable and the function-theoretic interpretation of covering mappings of regular polyhedra by means of the Riemann sphere. The group-theoretic approach played a leading role in [368]. In his subsequent papers [369–372, 374, 375, 380–382, 391] on modular functions, Klein tended increasingly to provide a group-theoretic basis for the results.

For Klein, these successes had obvious psychological value and encouraged his further efforts. As for group theory, the fast-moving leading edge of mathematical research provided further proof of the unifying power of the group concept. Klein described the situation in these words:

* [The method of solving the quintic by means of elliptic functions (Wussing)] forced me to concern myself with the transformation theory of elliptic functions. ... In particular, I started with a new approach to a problem raised by Hermite: To write down explicitly, and in the simplest possible manner, the resolvents of degree 5, 7, and 11, which according to a famous theorem of Galois, exist for transformations of order 5, 7, and 11, respectively. By combining the group-theoretic and invariant-theoretic approaches with which I am thoroughly familiar and the in-

(* see p. 211)

tuitive geometric methods of Riemann's theory of functions of a complex variable, I managed to do this in an amazingly simple manner. ... At the same time I obtained a complete overview of the different kinds of modular and multiplier equations found in the literature, as well as a clear insight into the nature of algebraic equations solvable by means of elliptic functions.

In fact, I found in this way a new approach to the systematic study of all relevant questions in the theory of elliptic functions. In abstract terms, the approach consists in putting forward the infinite, discontinuous group of homogeneous linear substitutions

$$\begin{aligned} u' &= \pm u + m_1\omega_1 + m_2\omega_2, \\ \omega_1' &= \alpha\omega_1 + \beta\omega_2, \\ \omega_2' &= \gamma\omega_1 + \delta\omega_2 \end{aligned}$$

(where $m_1, m_2, \alpha, \beta, \gamma, \delta$ are integers with $\alpha\delta - \beta\gamma = 1$) and ordering the material according to the subgroups contained in that group. One recognizes the link with the central idea of the Erlangen Program, which is that one must use group theory as the ordering principle of the confusion of phenomena. [348, p. 3]

The group-theoretic approach had again demonstrated its usefulness. It also suggested the general question of functions that, like the doubly periodic functions, are invariant under a finite or infinite transformation group of one (or more) variables. Finally, Klein stated the following objective:

* To consider geometric or analytic objects that are transformed into themselves by *groups of changes*. [379, p. IV]

(* see p. 210)

Ich wurde so von selbst zur allgemeinen Beschäftigung mit der Transformationstheorie der elliptischen Funktionen ... und damit insbesondere zur Neuaufnahme der Hermiteschen Aufgabe gedrängt: die Resolventen 5-ten, 7-ten und 11-ten Grades, die, einem berühmten Satze von Galois zufolge, bei Transformation 5-ter, 7-ter und 11-ter Ordnung bestehen, in einfachster Form wirklich aufzustellen. Durch Kombination der mir geläufigen gruppentheoretischen, bzw. invariantentheoretischen Ansätze mit den geometrisch-anschaulichen Methoden der Riemannschen Funktionentheorie gelang dies in überraschend einfacher Weise ... Zugleich ergab sich ein voller Überblick über die verschiedenen Arten von Modulargleichungen und Multiplikatorgleichungen, die in der Literatur vorlagen, sowie überhaupt eine klare Einsicht in das Wesen derjenigen algebraischen Gleichungen, welche durch elliptische Funktionen gelöst werden können ...

Hiermit war in der Tat ein neuer Ansatz zur systematischen Behandlung aller einschlägigen Fragen aus der Theorie der elliptischen Funktionen gewonnen. Abstrakt läßt sich derselbe dahin fassen, daß man die unendliche diskontinuierliche Gruppe homogener linearer Substitutionen

$$\begin{aligned} u' &= \pm u + m_1\omega_1 + m_2\omega_2 \\ \omega_1' &= \alpha\omega_1 + \beta\omega_2 \\ \omega_2' &= \gamma\omega_1 + \delta\omega_2 \end{aligned}$$

(wo $m_1, m_2, \alpha, \beta, \gamma, \delta$ ganze, die Gleichung $\alpha\delta - \beta\gamma = 1$ befriedigende, Zahlen bedeuten) an die Spitze stellen und den Stoff nach den Untergruppen ordnen soll, die in der Gesamtgruppe enthalten sind. Man erkennt den Anschluß an die Grundauffassung des Erlanger Programms: die Gruppentheorie als ordnendes Prinzip im Wirrsal der Erscheinungen zu benutzen.

* Überhaupt solche geometrische oder analytische Gebilde in Betracht zu ziehen, welche durch *Gruppen von Aenderungen* in sich selbst transformirt werden.

In the early[183] papers [376–378] of 1882–83, which reflected this more general view, Klein spoke of "single-valued functions with linear transformations into themselves." H. Poincaré (1854–1912), who, without detailed knowledge of the German literature, was led to consider the same questions by papers[184] of L. Fuchs (1833–1902) on the integration of linear differential equations, and through suggestions of Ch. Hermite, spoke of such functions as "functions reproduced by linear transformations" [569]. The term "*automorphic* functions" for functions (especially single-valued functions of one variable) invariant under a finite or infinite transformation group was suggested by Klein only in 1890, specifically [347, p. 549] in the paper *Zur Theorie der allgemeinen Laméschen Funktionen* [386].

It was R. Fricke (1861–1930) who provided the conclusive and explicit group-theoretic foundation of the theory of automorphic functions. In 1897 Fricke, with due attention to the papers[185] of Poincaré (who, like Klein, used group theory and geometric function theory as decisive tools), published, in close collaboration with Klein,[186] the first volume, *Die Gruppentheoretischen Grundlagen* [201], of the *Vorlesungen über die Theorie der automorphen Functionen.* Here Fricke described the central role played in that discipline by group theory, that is, the theory of transformation groups, in these words:[187]

* We are redeeming a pledge made at the end of the second volume of the '*Vorlesungen über die Theorie der elliptischen Modulfunctionen.*' In fact, we had in mind the possibility of following up the exposition of the theory of modular functions with an exposition, based on the same principles, of the theory of single-valued automorphic functions. This plan could not be fully realized in the present volume. Rather, it represents a first important step, the development of the *group-theoretic foundations of the theory of single-valued automorphic functions.* [201, p. III]

** The leading auxiliary theory used in most recnet function theory is group theory, which has been gaining ever greater importance in almost all areas of advanced mathematics. It turns out that certain conceptual constructs, which go back to Riemann and whose further development is the principal task of the present work, are readily amenable to group-theoretic methods. Under these circumstances it is not surprising that the first volume of the lectures on so-called authomorphic functions should be devoted exclusively to the group-theoretic foundations of future investigations. This division is all the more advisable because *the study of discontinuous groups*

---

* Es handelt sich hier um die Einlösung eines Versprechens, welches am Schlusse des zweiten Bandes der ‚Vorlesungen über die Theorie der elliptischen Modulfunctionen' gemacht worden ist. In der That wurde ja daselbst in Aussicht genommen, dass sich an die Behandlung der Modulfunctionen eine nach gleichen Principien angelegte Darstellung der Theorie der eindeutigen automorphen Functionen anschliessen sollte. Aber allerdings hat dieser Plan im vorliegenden Bande noch nicht im ganzen Umfange zur Ausführung gelangen können. Vielmehr ist hier nur erst ein erster wichtiger Schritt geschehen: es sind die *gruppentheoretischen Grundlagen der Theorie der eindeutigen automorphen Functionen* zur Entwicklung gebracht.
(** see p. 213)

of linear substitutions of one variable, to be presented here, has, in recent times, acquired to an ever larger extent the position of an independent discipline—a position further enhanced by its inner connections with geometry and number theory. [201, p. 1][187]

The *Vorlesungen über die Theorie der automorphen Functionen* of Fricke and Klein became a model, in particular with regard to the role of group theory, which strongly influenced subsequent monographs such as G. H. Halphen's *Traité des fonctions elliptiques*, volume 3, published in Paris in 1900 [259] and G. Fubini's *Introduzione alla teoria dei gruppi discontinui e delle funzioni automorphe*, published in Pisa in 1908 [216].

3 Lie and the Differentiation of the Concept of a Continuous Transformation Group: Classification of Transformation Groups

The general concept of a transformation group was explicitly introduced by Lie in "Über eine Klasse geometrischer Transformationen" [471], the German continuation of his Norwegian dissertation "Over en Classe geometriske Transformationer" [470].[188] Earlier, Lie had written a few papers on point transformations, on the role of the imaginary in geometry and, in collaboration with Klein, the paper "Sur une certaine famille de courbes et de surfaces" [350], which demonstrated his perfect command of projective and other coordinate transformations. Now, in [471], Lie developed the concept of contact transformations (Berührungstransformationen), and thus embarked on the study, inaugurated earlier by Plücker, of transformations other than point transformations. I quote:

* It is well known that in addition to point transformation there exists an extensive class of transformations of space that carry touching surfaces into touching surfaces. Such a transformation, to be called a contact transformation, can be given by means of five equations connecting x, y, z, p, q and X, Y, Z, P, Q. [304, p. 420]

(** see p. 212)
Unter den Hilfstheorien, welche die neueste Functionenlehre für ihre Zwecke verwertet, steht die in fast allen Teilen der höheren Mathematik sich mehr und mehr Geltung verschaffende Gruppentheorie im Vordergrunde. Es erweisen sich namentlich gewisse functionentheoretische Ideenbildungen, die in ihrem ersten Ursprung auf Riemann zurückgehen und deren Weiterentwicklung eine Hauptaufgabe des vorliegenden Werkes ist, den gruppentheoretischen Methoden leicht zugänglich. Bei dieser Sachlage wird es nicht überraschen, wenn der erste Band der Vorlesungen über die sogen. automorphen Functionen einzig den gruppentheoretischen Fundamenten der künftigen Untersuchungen gewidmet ist. Es erschien diese Einteilung um so mehr rätlich, als sich die *Lehre von den discontinuierlichen Gruppen linearer Substitutionen einer Veränderlichen*, welche wir hier zu entwickeln haben, in neuester Zeit mehr und mehr zu einer besonderen Disciplin herangebildet hat, die namentlich auch durch ihre innigen Beziehungen zur Geometrie und Zahlentheorie ein selbständiges Interesse für sich in Anspruch nimmt.

* Ausser den Punkt-Transformationen giebt es bekanntlich eine ausgedehnte Klasse räumlicher
(* continued on p. 214)

This is how Lie himself announced the new concept in 1874. His deeper study of contact transformations was based on the new insight [304, p. 421] that the system of two equations $F_1(x, y, z, p, q) = 0$, $F_2(X, Y, Z, P, Q) = 0$ determines a contact transformation linking line geometry to a geometry of 3-dimensional space whose elements are spheres. Thus Lie's dissertation and its sequel, while certainly not group theoretic in nature, show the pursuit of his program[189] (elaborated jointly with Klein) of searching for invariants under transformations. Naturally, this leads to the investigation of the transformations themselves:

* We can pose the problem of determining all groups of linear transformations. I note[190] that a set (of) continuous or discontinuous transformations forms a group if a combination of a number of these transformations is always equivalent to some transformation of the set. [459, pp. 208–209]

This statement marks the moment[191] when Lie, citing Klein and Jordan's investigations of groups of motions, chose his own direction of work.

Lie's papers [470–471] of 1871 illustrate a characteristic feature of all of his work, namely the close connection between geometry and the theory of differential equations; in this Lie followed Monge and Plücker. Lie's general concern, set out with considerable clarity in these papers, is the same as Klein's: the search for geometric or analytic invariants in "closed system of transformations" ("geschlossenen Systemen von Transformationen"). This is the joint formulation by Klein and Lie found in [351].

Between 1871 and 1874, and especially during the winter of 1873–74, Lie worked hard to develop a theory of transformation groups, spurred[192] by the multiplicity of applications that he envisioned. He realized that the theory of transformation groups was an unbelievably powerful tool in geometry and in the theories of ordinary and partial differential equations. Throughout his work (which I shall not analyze in detail) Lie emphasized this function of the theory of transformation groups. For example, in the introduction to his paper *Klassifikation der Flächen nach der Transformationsgruppe ihrer geodätischen Kurven* [481], which he thought of as a program for the application of transformation groups to differential geometry [459, p. 358], he states that, prior to the development of the theories that

---

(* continued from p. 213)
Umformungen, bei denen Flächen, die sich berühren, in ebensolche übergehen. Jede solche Transformation, die eine Berührungs-Transformation heissen soll, drückt sich durch fünf Gleichungen zwischen x, y, z, p, q und X, Y, Z, P, Q aus.

* Es läßt sich die Aufgabe stellen, alle Gruppen linearer Transformationen anzugeben. Ich sage dabei, daß eine kontinuierliche oder diskontinuierliche Schar [von] Transformationen eine Gruppe bildet, wenn die Kombination einiger dieser Transformationen jedesmal mit einer Transformation der gegebenen Schar äquivalent ist.

he saw with his mind's eye, he must "first ... develop an extensive auxiliary theory, the theory of transformation groups" ("zuerst eine umfangreiche Hilfstheorie, die Theorie der Transformationsgruppen ... entwickeln") [459, p. 358].

This tells us about Lie's main motives for his systematic study of transformation groups und their resulting classification. Documented details are missing, however, because Lie's letters written to Klein in 1873 and in 1874 were not preserved [461, p. 584]. Some light is thrown on the matter by a few letters that Lie wrote to A. Mayer (1839–1908), who, at the time, worked effectively in the area of partial differential equations. Thus, in 1873, Lie wrote to Mayer,

* I have obtained extremely interesting results and except a great deal more. What is involved is an idea whose origin goes back to earlier papers of Klein and myself, namely, the introduction of substitution theory into the theory of differential equations. [461, p. 584]

The letter ends as follows:

** If I guess right, and I cannot be all wrong, then I may be able to construct a rational classification of partial differential equations in n variables of arbitrary order. Each class is characterized by a canonical form obtainable by means of a contact transformation. I shall provide a general method for the reduction of any equation to its canonical form. These are mere conjectures. ...—At the moment I am very excited; I hope that this will not all end in disappointment. [461, p. 585]

At the end of 1874, Lie published the programmatic paper "Über Gruppen von Transformationen" [472], which is directly related to transformation groups. Here he defined the r-parameter (continuous) transformation group as follows:

*** The concept of a group of transformations, first developed in number theory and substitution theory, has recently been applied in geometric and general analytic investigations, respectively. We say that a set of transformations

$$x_i' = f_i(x_1, \ldots, x_n, \alpha_1, \ldots, \alpha_r)$$

(the x are the original variables, the x' are the new variables, and the α are parameters that are always supposed to vary continuously) is an r-parameter group, when the composition of any two transformations of the set yields a transformation belonging to the set, that is, when

* Ich habe höchst interessante Resultate erhalten und ich erwarte sehr viel mehr. Es handelt sich um eine Idee, deren Ursprung sich in früheren Arbeiten von Klein und mir findet, nämlich die Begriffe der Substitutionstheorie in die Theorie der Differentialgleichungen einzuführen.
** Sind meine Vermuthungen begründet, und ganz leer sind sie nicht, so werde ich eine rationelle Klassifikation partieller Differentialgleichungen beliebiger Ordnung mit n Variabeln aufstellen können. Jede Klasse ist charakterisiert durch eine kanonische Form, die sie durch eine Berührungstransformation erhalten kann. Ich werde eine allgemeine Methode geben, um jede Gleichung auf ihre kanonische Form zu bringen. Doch dies letzte sind nur Vermuthungen ... – Im Augenblicke bin ich sehr exaltirt; möchte ich nicht zu sehr enttäuscht in meinen Hoffnungen werden.
(*** see p. 216)

the equations

$$x_i' = f_i(x_1, \ldots, x_n, \alpha_1, \ldots, \alpha_r),$$
$$x_i'' = f_i(x_1', \ldots, x_n', \beta_1, \ldots, \beta_r)$$

imply that

$$x_i'' = f_i(x_1, \ldots, x_n, \gamma_1, \ldots, \gamma_r),$$

where γ denotes quantities that depend on α, β alone. [461, p. 1]

The relatively simple results obtained by Lie in this first paper involve a single variable x. In the case of 1-parameter groups there is just one "type"—as Lie puts it—of transformation group, namely, the translation type $x' = x + \alpha$. In the case of 2-parameter groups there is again just one type, namely, the type $x' = \gamma x + \delta$ of a linear transformation obtained by combining a translation with a similarity transformation. The 3-parameter transformation group is represented by the totality of linear transformations

$$x' = \frac{\alpha x + \beta}{\gamma x + \delta}.$$

For one variable, there are no transformation groups with more than three (independent) parameters. Lie's results required only elementary calculations. He summarized them in the following theorem:

If a group of transformations in one variable contains only a finite number of parameters, hen this number does not exceed three. Such a group is similar to the group of all linear transormations or to a subgroup of that group. [461, p. 5]

(*** see p. 215)

Der Begriff einer Gruppe von Transformationen, welcher zunächst in der Zahlentheorie und in der Substitutionstheorie seine Ausbildung fand, ist in neuerer Zeit verschiedentlich auch für geometrische, resp. allgemeine analytische Untersuchungen verwendet worden. Man sagt von einer Schar von Transformationen:

$$x_i' = f_i(x_1, \ldots, x_n, \alpha_1, \ldots, \alpha_r)$$

(wo die x die ursprünglichen, die x' die neuen Variabeln und die α Parameter bedeuten. die im folgenden stets kontinuierlich veränderlich gedacht werden), daß sie einer r-gliedrige Gruppe bilden, wenn irgend zwei Transformationen der Schar zusammengesetzt wieder eine der Schar angehörige Transformation ergeben, wenn also aus den Gleichungen:

$$x_i' = f_i(x_1, \ldots, x_n, \alpha_1, \ldots, \alpha_r)$$
und: $$x_i'' = f_i(x_1', \ldots, x_n', \beta_1, \ldots, \beta_r)$$
hervorgeht: $$x_i'' = f_i(x_1, \ldots, x_n, \gamma_1, \ldots, \gamma_r)$$

unter den γ Größen verstanden, die nur von den α, β abhängen.

* Soll eine Gruppe von Transformationen einer Veränderlichen nur eine endliche Zahl von Parametern enthalten, so kann diese Zahl nicht größer sein als drei, und es ist die Gruppe als dann entweder mit der Gruppe aller linearen Transformationen oder mit einer in dieser enthaltenen Untergruppe ähnlich.

Lie defined two transformation groups as "similar" (ähnlich) [461, p. 1–2] if the analytic form of the transformations in one of the groups can be changed to the analytic form of the transformations in the other by the introduction of a suitable coordinate system.

At the end of the paper, following the relatively straightforward results on transformation groups in one variable, Lie turns to the "significantly more complicated" ("bedeutend komplizierteren") transformation groups in two variables. He considers immediately the problem of transformations that go beyond point transformations, that is, the problem of contact transformations[193] (which he defines as transformations "in which x_1', x_2', and $p' = dx_1'/dx_2'$ are expressed in terms of x_1, x_2, and $p = dx_1/dx_2$ ("die x_1', x_2' und $p' = dx_1'/dx_2'$ durch x_1, x_2 und $p = dx_1/dx_2$ ausdrücken") [461, p. 5]. He then refers to his earlier results on contact transformations and alludes to the connection between the results on groups of contact transformations and the theory of differential equations[194] on the one hand, and the Erlangen Program on the other.[195] The paper concludes with the hint that the introduction of transformation groups in more than two variables makes possible "similar considerations for partial differential equations," and that it must be possible directly to convert the resulting "fruitful direction of research" ("fruchtbare Untersuchungsrichtung") [461, p. 8] into an integration theory of differential equations.

This review of the paper Lie wrote toward the end of 1874 shows that it was essentially programmatic. Early in 1876, Lie began to write a numbered series of papers on the theory of transformation groups. His enthusiasm for the unifying power[196] of the group concept shines through his sober style. To give the reader an idea of this enthusiasm, I quote from the introduction to Lie's definitive paper [474] on the theory of transformation groups:[197]

> * This is the first of a proposed series of papers in which I intend to develop a new theory, which I shall call the theory of transformation groups. The reader will note that the relevant investigations touch on many mathematical disciplines, such as substitution theory, geometry and modern manifold theory [that is, *n*-dimensional geometry (Wussing)] and, finally, the theory of differential equations. To some extent, these investigations will unify these so far unconnected disciplines. The importance and scope of the new theory will be presented in future papers. [461, p. 9]

---

* In einer Reihe von Abhandlungen, unter denen die nachstehende die erste ist, beabsichtige ich, eine neue Theorie, die ich die Theorie der Transformationsgruppen nennen werde, zu entwickeln. Die betreffenden Untersuchungen haben, wie der Leser bemerken wird, viele Berührungspunkte mit mehreren mathematischen Disziplinen, insbesondere mit der Substitutionstheorie, mit der Geometrie und der modernen Mannigfaltigkeitslehre und endlich auch mit der Theorie der Differentialgleichungen; sie werden gewissermaßen einen Zusammenhang zwischen diesen früher getrennten Disziplinen zustandebringen. Im übrigen muß ich es mir für spätere Arbeiten vorbehalten, die Wichtigkeit und Tragweite dieser neuen Theorie darzulegen.

The assumptions about the number of independent parameters are the same in this paper as in [472]. What is different is the consideration of the *structure* of transformation groups in one variable; what is new is the presence of existence proofs for identities and for inverses of elements belonging to 1-parameter [461, p. 14], 2-parameter [461, pp. 19–20], and 3-parameter [461, pp. 33–34] groups of transformations, respectively. In this way Lie determines all transformation groups in one variable or, to use his new definition, all transformation groups of a 1-dimensional manifold. He writes,

* *Every transformation group of a simply extended manifold is similar to a linear group, and thus contains at most three parameters.* [461, p. 41]

When it comes to the definition of a transformation group, there is no difference between this paper and [472]. The group property is stated as follows: "*The succession of two transformations of the set* (must be) *equivalent to a single transformation in that set*" ("*die Sukzession zweier Transformationen in der Schar mit einer einzigen Transformation derselben Schar equivalent*" sein muss) [461, p. 10].

In the second paper on the theory of transformation groups [475], Lie deals with general theorems on transformation groups "of an arbitrarily extended manifold," that is, groups depending on n variables $x_1, \ldots, x_n$. Here too the sufficient group property continues to be closure under composition of transformations, and the existence of an identity transformation and of inverses is proved. For computational purposes, Lie gives the following analytical definition of a group (it is an analytical extension of the formulation given in [474]):

** *The n equations*

$$x_i' = f_i(x_1, \ldots, x_n;\ a_1, \ldots, a_r) = f_i(a)$$

determine a transformation group if for every i we have a relation of the form

$$f_i(f_1(a), \ldots, f_n(a),\ b_1, \ldots, b_r) = f_i(x_1, \ldots, x_n,\ c_1, \ldots, c_r),$$

where $c_1, \ldots, c_r$ are certain functions of the a and b that do not depend on i. [461, p. 43]

The third paper [478] on the theory of transformation groups appeared in 1878,

* *Eine jede Transformationsgruppe einer einfach ausgedehnten Mannigfaltigkeit ist ähnlich mit einer linearen Gruppe und enthält somit höchstens drei Parameter.*

** Die *n Gleichungen:*

$$x_i' = f_i(x_1, \ldots, x_n;\ a_1, \ldots, a_r) = f_i(a)$$

bestimmen eine Transformationsgruppe, wenn für jedes i eine Relation der Form:

$$f_i(f_1(a), \ldots, f_n(a),\ b_1, \ldots, b_r) = f_i(x_1, \ldots, x_n,\ c_1, \ldots, c_r)$$

stattfindet; dabei vorausgesetzt, daß $c_1, \ldots, c_r$ gewisse Funktionen der a und b sind, die von der Zahl i unabhängig sind.

after a long delay due to the need for extensive computations. Here Lie determined all groups of a doubly extended manifold. The fourth paper [479] gives all the contact transformations in the plane.

A progressively more dominant feature of these papers, which culminated in the distinction between finite and infinite[198] continuous groups elaborated by Lie in 1883, was the investigation of transformation groups with the aid of infinitesimal transformations. Already in [475] Lie showed that every r-parameter group contained r independent infinitesimal transformations. Using notation somewhat different from the modern, Lie gave the following definition of an infinitesimal transformation:[199]

* A transformation is said to be infinitesimal if it can be written in the form

$$x_i' = x_i + \delta t \, X_i(x_1, \ldots, x_n),$$

where δt is an infinitesimal magnitude. We usually write the latter equations as

$$\delta x_i = \delta t \cdot X_i(x_1, \ldots, x_n). \qquad [461, \text{p. } 45]$$

In terms of Lie's symbolism, the connection between infinitesimal transformations and contact transformations is realized as follows. The infinitesimal transformation

$$\delta x = \xi(x, y, p)\, \delta t, \qquad \delta y = \eta \, \delta t, \qquad \delta p = \pi \, \delta t$$

is a contact transformation of the plane with coordinates x and y if

$$\frac{\delta}{\delta t}(dy - p\, dx) = \varphi(x, y, p)\,(dy - p\, dx).$$

After the fifth paper on transformation groups [480], dating from 1879, the treatment of transformation groups by means of infinitesimal transformations became—in Lie's words—a "direct method" (eine direkte Methode). This method gave rise to the concept of a "continuous group" (Kontinuierliche Gruppe) and to the distinction between finite and infinite continuous groups. Lie used the term "continuous group" for the first time in a review, written for the *Jahrbuch der Fortschritte der Mathematik* [484, p. 641], of one of his own papers of 1882.

The distinction between finite and infinite continuous transformation groups—

* Eine Transformation heißt infinitesimal, wenn sie die Form:

$$x_i' = x_i + \delta t \, X_i(x_1, \ldots, x_n)$$

erhalten kann, und dabei δt eine infinitesimale Größe bezeichnet. Wir schreiben im allgemeinen die letzten Gleichungen folgendermaßen:

$$\delta x_i = \delta t \cdot X_i(x_1, \ldots, x_n).$$

where, of course, the important groups are the groups of infinite order—was stated by Lie in detail in the paper "Über unendliche kontinuierliche Gruppen" [485] of 1883. This paper begins as follows:

* A class of operations forms a group if the succession of two of its operations is equivalent to a single operation of that class. A group is said to be continuous if all of its operations are generated by infinite repetition of infinitesimal operations, and discontinuous if all of its operations are finitely [rather than infinitesimally (Translator)] different.

A discontinuous group is finite or infinite according as the number of its operations is bounded or unbounded. There is an analogous division of continuous groups into two main categories according as their operations depend on variable parameters or on variable functions. I call a continuous group a finite continuous group if its operations depend only on variable parameters. I shall call a continuous group infinite if its operations depend on arbitrary functions. [461, p. 314]

To justify this definition, which "may at first [appear] unnatural" ("möglicherweise im ersten Augenblicke nicht naturgemäß [erscheint]") [461, p. 314], Lie quoted the result, obtained with the aid of infinitesimal transformations, that the two types of groups differ by the number of (linearly) independent infinitesimal transformations. In this way he arrived at the following definition, which was basic for his subsequent research:

** *A continuous group is called finite or infinite according as the number of its independent infinitesimal transformations is bounded or unbounded.* [461, p. 315]

Obviously, in addition to continuous and discontinuous transformation groups there is a third category to which there belongs, for example, the group of motions and reflections of the plane. Of course Lie realized this fact and noted it, in connection with his division of transformation groups, in a footnote on page 315 of [416], from which I took the quotation above.

---

* Eine Schar von Operationen bildet eine Gruppe, wenn die Sukzession zweier Operationen der Schar mit einer einzigen Operation derselben äquivalent ist. Kontinuierlich heißt eine Gruppe, deren sämtliche Operationen durch unendlichmalige Wiederholung von infinitesimalen Transformationen erzeugt sind; diskontinuierlich heißt dagegen eine Gruppe, deren Operationen sämtlich endlich verschieden sind.

Eine diskontinuierliche Gruppe heißt endlich oder unendlich, je nachdem die Zahl ihrer Operationen begrenzt oder unbegrenzt ist. Dementsprechend können auch die kontinuierlichen Gruppen in zwei Hauptkategorien geteilt werden, je nachdem ihre Operationen von variablen Parametern oder von variabeln Funktionen abhängen. Eine kontinuierliche Gruppe, deren Operationen nur von variablen Parametern abhängen, nenne ich eine endliche und kontinuierliche Gruppe; ein einfaches Beispiel bilden alle Bewegungen einer Ebene. Dagegen werde ich eine kontinuierliche Gruppe unendlich nennen, wenn ihre Operationen von arbiträren Funktionen abhängen.

** *Eine kontinuierliche Gruppe heißt endlich oder unendlich, je nachdem die Zahl ihrer unabhängigen infinitesimalen Transformationen begrenzt oder unbegrenzt ist.*

In addition to the classification of continuous transformation groups in terms of the number of their independent infinitesimals, Lie introduced a classification, inducing the same division, based on the connection of these groups with differential equations. The latter classification was worked out to an extent that reflected Lie's development of the theory of integration of differential equations on the basis of the theory of transformation groups. If the transformations $x_i' = f_i(x_1, \ldots, x_n; a_1, \ldots, a_r)$, $i = 1, \ldots, n$, form a finite continuous group, then differentiation with respect to the x_i and elimination of the parameters yield a system of differential equations that defines the functions f_i, and thus has as solutions the functions f_i with the integration constants a_ϱ. The group property implies that, together with the solution systems

$$x_i' = f_i(x_1, \ldots, x_n; b_1, \ldots, b_r) \quad \text{and} \quad x_i' = f_i(x, a),$$

the system

$$x_i' = f_i(f_1(x, a), \ldots, f_n(x, a); b_1, \ldots, b_r), \qquad i = 1, \ldots, n,$$

is also a solution system.

Now if a class of transformations is defined by means of a system of differential equations

$$W_k\left(x_1', \ldots, x_n'; \frac{\partial x_1'}{\partial x_1}, \ldots, \frac{\partial^2 x_1'}{\partial x_1^2}, \ldots\right) = 0, \qquad k = 1, 2, \ldots, n,$$

which, as above, has the property that, together with the solutions $x_i' = f_i(x_1, \ldots, x_n)$ and $x_i' = g_i(x_1, \ldots, x_n)$, the system $x_i' = g_i(f_1(x), \ldots, f_n(x))$ is always a solution; and if, in distinction to the previous case, the general solution depends not just on a finite number of arbitrary constants but—in Lie's words—

* on arbitrary elements of a higher kind such as, for example, arbitrary functions, [then] the totality of transformations satisfying the differential equations in question obviously forms a group that, in general, is continuous but, rather than being finite, is of a type that we designate as an *infinite continuous* group. [463, p. 6]

We see that 1883 may be regarded as the year in which the classification of transformation groups was completed[200]—discontinuous groups had been classified earlier by Klein. At this point it is useful to make a number of observations about the subsequent development of the theory of transformation groups.

---

* „von willkürlichen Elementen höherer Art, wie zum Beispiel von willkürlichen Functionen" abhängt, dann „bildet der Inbegriff von allen Transformationen, welche den betreffenden Differentialgleichungen genügen, offenbar wiederum eine Gruppe und zwar im Allgemeinen eine continuirliche, aber nicht mehr eine endliche, sondern eine, die wir als *unendliche continuirliche* bezeichnen.

In collaboration[201] with F. Engel (1861–1941), later the editor of his collected works, Lie published a comprehensive account of the theory of transformation groups that he had developed as a tool for the treatment of differential equations. This work, called *Theorie der Transformationsgruppen*, consisted of three volumes [463–465], published in 1888, 1890, and 1893,[202] respectively. One of Lie's students in Leipzig was G. Scheffers (1866–1945). In collaboration with Scheffers, Lie published in 1896 the *Geometrie der Berührungstransformationen* [467]. It was not granted him to complete a planned comprehensive account of the theory of integration of differential equations by means of transformation groups, announced in the third volume of his *Theorie der Transformationsgruppen* [466, pp. xvii–xviii].

In connection with differential equations, Lie had earlier formulated an exact analogue of Galois's theory of algebraic equations. In a letter written to A. Mayer early in 1874, Lie stated that

> * Before Galois, the question asked in the theory of algebraic equations was, Is the equation solvable in radicals, and how is it solved? One of the questions asked after Galois is, What is the simplest way of solving the equation in radicals? ... I believe that the time has arrived for a similar advance in the theory of differential equations. [461, p. 586]

In the late seventies and early eighties, Lie pursued energetically the difficult search within the theory of continuous groups for concepts analogous to the ideas of the theory of permutation groups that had proved relevant in Galois theory. But the actual formulation of a "Galois theory of differential equations" is due not to Lie but to E. Picard (1856–1941) and E. Vessiot (1865–1952). The relevant work of Picard dates back to 1883 and 1887, and the conclusive work of Vessiot to 1892.

In spite of the fact that Lie had emphasized the decisive role of the group concept—a theme he never tired of—since the early seventies, his advance was judged only in terms of his geometric and analytic results. In this sense, the support he obtained can only be described as indirect. In geometry, such support came from G. Darboux, with whom Lie had closer contact in Paris (III.2), and in differential equations it came largely from L. Cremona (1830–1903), A. Meyer, and C. Jordan. Recognition of the group-theoretic objectives of Lie's investigations came only in the early eighties, from Picard and Vessiot. And yet these fundamental advances in differential equations were due to the inner shift in the evaluation

---

* In der Theorie der algebraischen Gleichungen stellte man vor Galois nur die Fragen: Ist die Gleichung auflösbar durch Wurzelzeichen und wie löst man sie auf? Seit Galois stellt man u. a. auch die Frage: Wie löst man die Gleichung am einfachsten durch Wurzelzeichen auf? ... Die Zeit ist gekommen, glaube ich, um in Differentialgleichungen einen ähnlichen Fortschritt zu machen.

of the advances associated with the transition from permutation groups to transformation groups.

The subsequent development of the theory of continuous groups is associated with the name of M. Poincaré (1854–1912) and, above all, with that of E. Cartan (1869–1951), who was undoubtedly the most important of Lie's followers. A description of Cartan's development of the theory of Lie groups lies outside the aims of the present work,[203] but certain historically significant effects of his work are important for our theme. Thus between 1893 and 1894 Cartan published four papers [92–95] on the "structure" of finite continuous transformation groups, describing this structure as the true object of investigation.[204] These four papers were followed by the programmatic paper "Sur la structure des groupes de transformations finis et continus" [96], the famous "Paris thesis" of 1894, which so exhaustively analyzed the open questions associated with the contemporary theory of finite continuous transformation groups that it produced a shift in interest from finite to infinite continuous groups. Interest in the finite case revived, however, in the 1920s, through the rise of relativity theory and the associated issue of differential invariants.

The evolution of the concept of a (finite) continuous group to that of a topological group is associated with an extremely fruitful development of its content. In 1900, at the International Congress of Mathematicians, D. Hilbert presented a famous programmatic address, whose "Fifth Thesis" proposed the investigation of the necessity of the differentiability conditions associated with Lie's formulation of the concept of a finite continuous transformation group. This, and the set-theoretic formulation of the abstract group concept, paved the way for O. Schreier's (1901–29) establishment of the theory of topological groups in 1925 [645]—but this too lies outside the period covered by my account.

4 Lie and the Evolution of the Group Axioms for Infinite Groups

Sophus Lie played an important role not only in the classification of transformation groups but also in the axiomatic formulation of the group concept.

From an abstract point of view, the definition of a permutation group by Klein and of a transformation group by Lie, that is, the definitions of a group used until the eighties, shared with the modern definition of a group the single requirement of closure under the rule for combining the elements. The associative law was used for permutation groups with or without explicit reference to the calculus of permutations; the same was true of the existence of an identity and of the inverse permutation. Lie's merit lies in his having recognized that the remarkable difference between finite (in terms of order) and infinite groups is that in the latter case the existence of an inverse is one of the essential group properties. This insight came to him not in a sudden flash but gradually, through his study of transformation groups. Whereas 1883 was the year in which the classification of transformation groups was completed, the formulation of group axioms for transformation groups (including permutation groups) continued into the nineties.

As mentioned above, the early papers on the theory of (finite) continuous groups dating back to 1876–78 established, by calculation, the existence of (at least) one identity transformation and of (at least) one transformation inverse to a given transformation. On the other hand, the validity or application of the associative law was not mentioned.

Lie's change of attitude in the matter of making explicit the defining assumptions for groups of transformations, a change of special interest in connection with the working out of the abstract group concept (III.4), occurred in 1888. Having established a vast number of results in geometry and analysis, Lie began at that time the conceptual elaboration of the theory of transformation groups, the splendid tool he had created. Here he was clearly under the influence of the early evolution of the axiomatic method and of the development of mathematical logic. In the short paper "Zur Theorie der Transformationsgruppen" [486], Lie stated that "the so-called associative law" ("das sogenannte assoziative Gesetz") [461, p. 553] holds, in all generality, for transformations because (!) we may "regard transformations as operations" ("Transformationen als Operationen auffassen") [461, p. 553]. This is a remarkable argument for the justification of the associative law. It is based on the abstract interpretation, attained at the time by mathematicians such as G. Peano and R. Dedekind, of such fundamental concepts of mathematics as "element," "composition," "operation" (Element, Verknüpfung, Operation) and others. From the associative law, valid for transformation groups in general and for r-parameter continuous groups in particular, Lie deduced various other assertions about such groups—for example, a more convenient formulation of the statement of isomorphism of two finite continuous transformation groups.

A year later, in 1889, Lie published another short paper, "Ein Fundamentalsatz in der Theorie der unendlichen Gruppen" [487], which lists as a group requirement "the possibility of grouping the transformations of our class in mutually inverse pairs" ("daß die Transformationen unserer Schar sich paarweise als inverse zusammenordnen lassen") [461, p. 558]. This explicit assumption was necessitated by Lie's definition of the group property of a class of transformations in terms of the requirement—explained above—that the iteration of two solutions of a system of differential equations also be a solution of the system.

In the first paper, "Die Grundlagen für die Theorie der unendlichen Kontinuierlichen Transformationsgruppen. I. Abhandlung" [488] of 1891, Lie took a decisive step in the elaboration of the group axioms. It was a step away from the abstract definition of a group in the sense that he defined an infinite continuous group in terms of the solutions of a system of (partial) differential equations.[205] Lie's definition was to the effect that a class of transformations

$$\mathfrak{x}_i = F_i(x_1, \ldots, x_n), \qquad i = 1, \ldots, n,$$

is called an infinite continuous group if the F_i are the most general solutions of a system of partial differential equations

$$W_k\left(x_1, \ldots, x_n, \mathfrak{x}_1, \ldots, \mathfrak{x}_n, \frac{\partial \mathfrak{x}_1}{\partial x_1}, \frac{\partial^2 \mathfrak{x}_1}{\partial x_1^2}, \ldots\right) = 0, \qquad k = 1, 2, \ldots,$$

with the following properties:

1. The most general solutions of the system of differential equations must not depend on finitely many arbitrary constants alone.
2. If $\mathfrak{x}_i = F_i(x_1, \ldots, x_n)$ and $\mathfrak{x}_i = \Phi_i(x_1, \ldots, x_n)$ are any two solutions of the system of differential equations, then so is $\mathfrak{x}_i = \Phi_i(F_1(x), \ldots, F_n(x))$. In Lie's words: "*Successive application of two arbitrary transformations of the class defined* [in this manner] must *always yield a transformation of that class*" ("*es sollen zwei beliebige Transformationen der* [solcherart] *definierten Schar nacheinander ausgeführt stets wieder eine Transformation der Schar ergeben*"). [462, p. 300].

In §2, entitled "Generalities" ("Allgemeine Bemerkungen"), the section that follows this definition, Lie noted that while there exist infinite continuous groups not definable by means of differential equations, the restriction implied by his definition was justified by the difficulties associated with an alternative approach and by the applications in the theory of differential equations.

Lie added another restriction in the form of the following requirement:

* To simplify the theory we introduce an additional assumption. We shall consider only the infinite continuous groups whose transformations are pairwise mutually inverse. This implies that such groups contain the identity transformation. In fact, successive application of two mutually inverse transformations of a group yields a transformation of the group, namely, the identity transformation. [462, p. 302]

---

* Zur Vereinfachung der Theorie führen wir noch eine zweite Voraussetzung ein, wir wollen nämlich nur solche unendliche kontinuierliche Gruppen betrachten, deren Transformationen paarweise zueinander invers sind. In dieser Voraussetzung liegt dann zugleich, daß die betreffenden Gruppen die identische Transformation enthalten; denn führen wir zwei zueinander inverse Transformationen einer Gruppe aus, so bekommen wir wieder eine Transformation der Gruppe, eben die identische Transformation.

The progress in the evolution of the group axioms here consisted in the fact that the existence of inverses was made part of the group definition and was thereby made explicit. Lie went on to show that the infinite continuous transformation groups that he considered always satisfied this assumption. In fact, the group elements were solutions of a system of differential equations and yielded under iteration other elements of the same class. Thus for the concrete groups in question Lie could claim that

* the above assumption looks like a restriction but is not. In fact, it is possible to prove that every infinite continuous group that can be defined by differential equations [of the above form (Wussing)] ... contains the identity transformation and consists of pairwise mutually inverse transformations. [462, pp. 302–303]

By distinguishing between a postulate and a condition that can always be satisfied, Lie had opened the way to the distillation of the abstract content of the group concept. To put it differently, the working out of the group axioms for infinite (in terms of the number of elements) groups was for Lie a matter of providing a setting that would accommodate the concept of a transformation group; the abstract formulation of the group concept came later.

Lie undertook the elaboration of the abstract group-theoretic ingredients only three years after the publication of the paper [488] just discussed. He was motivated in part by symbolic considerations. Galois had been a student in the Paris Ecole Normale Supérieure, founded in 1795. In commemoration of the 100th anniversary of the school's founding, a *Festschrift* [152] was published in Paris in 1895, and Lie contributed a paper, dated 17 November 1894, entitled "Influence de Galois sur le développement des mathématiques" [490].[206] While not actually an historical account, this paper is important for the axiomatic formulation of the group concept. At the same time, it is an enthusiastic expression of Lie's reverence for Galois's genius as well as an especially beautiful and profound assessment[207] of the group concept held by the mathematicians of the end of the nineteenth century.

In fact, the finest tribute to Galois was the description of the many areas of mathematics in which the group concept turned out to play a fundamental role.

** Everyone knows today that the notion of a group, discovered by Galois, relates not only to substitutions but, on the contrary, admits of generalizations in many directions. [462, p. 597]

---

* die eben eingeführte Voraussetzung sieht aus wie eine Beschränkung, ist es aber nicht. Es läßt sich nämlich beweisen, daß jede unendliche kontinuierliche Gruppe, die sich durch Differentialgleichungen ... definieren läßt, die identische Transformation enthält und aus paarweise inversen Transformationen besteht.

** Toute le monde sait aujourd'hui que la notion de groupe, due à Galois, ne se rapporte pas seulement aux substitutions, mais peut au contraire être généralisée dans des directions multiples.

The anniversary afforded Lie the chance to analyze the abstract content that resulted from the enlargement of the idea of a group. He starts in this paper with Jordan's permutation-theoretic concept and proceeds to replace it with a sharper one, which is nevertheless far inferior to the abstract definition of a group that had been found in the meantime (III.4):

* The following is the most general definition that has been made until now: A set of operations is said to form a group if the product of two of these operations is equivalent to a single operation belonging to the set. [462, p. 597]

In relating this definition to that of a transformation group, Lie states that changing "operation" (opération) to the more special "transformation" (transformation) and of "product" (produit) to "succession" (succession) may lead to the "possibly definitive" ("peut-être définitive") [462, p. 597] definition of a transformation group:

** In practice, we make this definition specific in each particular case. For example, we divide groups into continuous, discontinuous, and mixed ones. Each of these classes consists of finite and infinite groups. In the case of continuous groups it seemed convenient to set aside the groups not defined by differential equations. Finally, it has not yet been shown, in a manner entirely satisfactory from a theoretical point of view, whether it is necessary to add to the definition of continuous groups that among their transformations there is also the identity transformation.

When we extend the definition of a group to arbitrary operations we do not have the right to conclude that the definition given above [the one quoted by us above (Wussing)] implies the division of the operations of the group into pairs of mutually inverse operations. Also, the principle of associativity is not valid for all groups of operations. [462, pp. 597–598]

Thus Lie requires—admittedly in implicit form— all the modern group axioms. Nevertheless, this is not the set-theoretic and axiomatic formulation of the group concept. Neither Lie[208] nor Klein took this step.[209] In fact, the opposite is true. Klein, who lived to see the evolution of the axiomatic method and the transition to the so-called "Modern Algebra" in the early twenties, objected to the loss of

* La définition la plus générale que l'on ait posée jusqu'ici est la suivante: Un ensemble d'opérations forme un groupe, si le produit de deux de ces opérations est équivalent à une seule opération appartenant à l'ensemble considéré.

** Pratiquement, on précise dans chaque cas particulier cette définition. Par exemple, on partage les groupes en groupes continus, discontinus, et mixtes, et chacune de ces classes comporte deux divisions, les groupes finis et les groupes infinis. Parmi les groupes continus, il a paru convenable de laisser de côté ceux qui ne sont pas définis par des équations différentielles. Enfin, on n'a pas encore montré, d'une facon entièrement satisfaisante au point de vue purement théorique, s'il était nécessaire d'ajouter à la définition des groupes continus que parmi leurs transformations se trouve la transformation identique. En étendant la notion de groupe à des opérations quelconques, on n'a pas le droit de conclure que de la définition donnée plus haut résulte le partage des opérations du groupe en couples d'opérations inverses. Il arrive aussi que le principe d'associativité n'est pas valable pour tous les groupes d'opérations.

concrete representations of the elements of a group. On the other hand, in his "Vorlesungen über die Entwicklung der Mathematik" (these were lectures given between 1915 and 1919; part 1 [403] appeared in print in 1926), he emphasized the importance of group theory:

* *Group theory* appears as a distinct discipline throughout the whole of modern mathematics. It penetrates the most varied areas as an ordering and clarifying principle. [403, p. 334]

Klein evaluated correctly the historical role of his own and Lie's contributions to the working out of the group concept. He then went on to say:

** Then [after the publication of Jordan's *Traité* (Wussing)] Lie and I decided to elaborate the significance of group theory for different areas of mathematics. We stated that a group is a class of unique operations $A, B, C, \ldots$ such that the combination of any two operations A, B again yields an operation C of the class

$$A \cdot B = C.$$

In the course of further investigations on infinite groups Lie found it necessary to require that for each A the group should contain its inverse A^{-1}. Modern mathematicians employ a far paler but more precise definition. One speaks of a system of things or elements $A, B, C, \ldots$ rather than operations. It is then postulated that

1. The "product" or combination $A \cdot C = B$ belongs to the system (the requirement of closure).
2. The associative law holds, that is,

$$(AB) \cdot C = A \cdot (BC).$$

3. There exists an identity E such that

$$AE = A \text{ and } EA = A.$$

4. There exists an inverse, that is, the equation

$$Ax = E$$

is solvable.

Thus there is a complete loss of the appeal to the imagination. On the other hand, the logical skeleton is carefully prepared. ... This abstract formulation is excellent for the working out of proofs but it does not help one find new ideas and methods. Rather, it represents the conclusion of an advance. It therefore seems to ease the learning process, for it helps one prove known theorems in a flawless and simple manner. On the other hand, it makes matters far more difficult for the mind of the student, for he confronts something closed, does not know how one arrives at these definitions, and can imagine absolutely nothing. In general, the disadvantage of the method is that it fails to encourage thought. All one must beware of is that one does not violate the four commandments. [403, pp. 335–336]

---

* Die *Gruppentheorie* zieht sich als besondere Disziplin durch die ganze neuere Mathematik. Sie greift als ordnendes und klärendes Prinzip in die verschiedensten Gebiete ein.

** Als dann Lie und ich es unternahmen, die Bedeutung der Gruppentheorie für die verschiedensten Gebiete der Mathematik herauszuarbeiten, da sagten wir: ‚Gruppe' ist der Inbegriff (** continued on p. 229)

The extension of the group concept to that of a transformation group was inaugurated by Klein and Lie. In the seventies and eighties, this development contributed to the evolution of the group concept and produced a tremendous expansion of the range of applications of group theory. But the abstract formulation of the group concept would require input from other areas of mathematical progress.

(** continued from p. 228)
von eindeutigen Operationen A, B, C, ... derart, daß irgend zwei der Operationen A, B kombiniert wieder eine Operation C des Inbegriffes ergeben:

$$A \cdot B = C.$$

Bei seinen weiteren Untersuchungen über unendliche Gruppen sah sich Lie genötigt, ausdrücklich zu verlangen, daß neben A auch die Inverse A^{-1} in der Gruppe vorhanden sein solle. Bei den neueren Mathematikern tritt eine abgeblaßtere Definition auf, die aber präziser ist. Man spricht nicht mehr von einem System von Operationen, sondern von einem System von Dingen oder Elementen A, B, C, ... Dann wird postuliert, daß

1. das ‚Produkt' oder die Verknüpfung $A \cdot C = B$ selbst dem System angehört (Abgeschlossenheit des Systems),
2. das assoziative Gesetz gilt, also
$$(AB) \cdot C = A \cdot (BC),$$
3. eine Einheit E existiert, so daß
$$AE = A$$
und $$EA = A$$
ist,
4. die Inverse existiert, d. h. daß die Gleichung
$$Ax = E$$
lösbar ist.

Der Appell an die Phantasie tritt also hier völlig zurück. Dafür wird das logische Skelett sorgfältig herauspräpariert ... Diese abstrakte Formulierung ist für die Ausarbeitung der Beweise vortrefflich, sie eignet sich aber durchaus nicht zum Auffinden neuer Ideen und Methoden, sondern sie stellt vielmehr den Abschluß einer voraufgegangenen Entwicklung dar. Daher erleichtert sie den Unterricht äußerlich insofern, als man mit ihrer Hilfe bekannte Sätze lückenlos und einfach beweisen kann; andererseits wird die Sache für den Lernenden dadurch innerlich sehr erschwert, daß er vor etwas Abgeschlossenes gestellt wird und nicht weiß, wieso man überhaupt zu diesen Definitionen kommt, und daß er sich dabei absolut nichts vorstellen kann. Überhaupt hat die Methode den Nachteil, daß sie nicht zum Denken anregt; man hat nur aufzupassen, daß man nicht gegen die aufgestellten vier Gebote verstößt.

Already in Galois's work we find a statement that may be viewed as an early stage of the abstract view of a group, a faint allusion to isomorphism between permutation groups.

The paper "Mémoire sur les conditions de résolubilité des équations par radicaux" [229], in which we find the definition of the "group of the equation" (II.3) contains the lemma "Scolie II. —Les substitutions sont indépendantes même du nombres des racines" [221, p. 40]. Tannery's new inspection[210] of Galois's manuscripts [222], begun in 1908, showed that this theorem was followed by another—which, since it was crossed out, Picard [221] had ignored. It was to the following effect:

* This is what characterizes a group: One can start with any one of its permutations. [228, p. 8]

Actually, it was C. Jordan who, in his *Traité* of 1870, first made clear the essential significance of the isomorphism and homomorphism of permutation groups (II.5).

1 The Group as a System of Defining Relations on Abstract Elements

Another development that contributed to the shaping of the abstract group concept was launched by Cayley. His short paper, "Note on the Theory of Permutations" [138] of 1849, shows that it was then, or perhaps earlier, that he consciously linked his own ideas on permutations with Cauchy's (II.2). Five years later Cayley, referring specifically to Galois,[211] began to use the term "group". Cayley published two papers [140, 141], "On the Theory of Groups as depending on the Symbolic Equation $\theta^n = 1$," in which he presented a remarkable conception of groups. At a time when the Galois theory of equations had just begun to be known (II.4), and when the only explicit groups under study were permutation groups, Cayley recognized the generalizability of the group concept as well as the implicitly group-theoretic nature of many contemporary ideas. Clearly, Cayley's orientation toward an abstract view of groups was due to his close contact with the abstract position of G. Boole (III.1), whose *An Investigation of the Laws of Thought, on which are Founded the Mathematical Theories of Logic and Probabilities* [50] appeared in 1854.

Cayley introduces a general operation symbol Θ applied to a system of quantities

---

* Ce qui caractérise un groupe: On peut partir d'une des permutations quelconques du groupe.

$x, y, \ldots$. He writes

$$\Theta(x, y, \ldots) = (x', y', \ldots).$$

Here $x', y', \ldots$ are arbitrary functions of $x, y, \ldots$ whose number is not necessarily the same as that of the $x, y, \ldots$. Cayley continues as follows:

In particular x', y', &c. may represent a permutation of x, y, &c., Θ is in this case what is termed a substitution; and if, instead of a set $x, y, \ldots$, the operand is a single quantity x, so that $\Theta x = x' = fx$, Θ is an ordinary functional symbol. It is not necessary (even if this could be done) to attach any meaning to a symbol such as $\Theta \pm \Phi$, or to the symbol Θ, nor consequently to an equation such as $\Theta = 0$, or $\Theta \pm \Phi = 0$; but the symbol 1 will naturally denote the operation which (either generally or in regard to the particular operand) leaves the operand unaltered, and the equation $\Theta = \Phi$ will denote that the operation Θ is (either in generally or in regard to the particular operand) equivalent to Φ, and of course $\Theta = 1$ will in like manner denote the equivalence of the operation Θ to the operation 1. [130, p. 123]

The symbol $\Theta\Phi$ denotes the "compound operations." It is "obvious" that, in general, $\Phi\Theta$ and $\Theta\Phi$ are different. The validity of the associative law is expressed in the following theorem:

But the symbols Θ, Φ ... are in general such that $\Theta \cdot \Phi\chi = \Theta\Phi \cdot \chi, \ldots$, so, that $\Theta\Phi\chi$, $\Theta\Phi\chi\omega$, &c. have a definite signification independent of the particular mode of compounding the symbols. [130, p. 123]

This is followed shortly [130, p. 124] by the statement that "these symbols are not in general convertible but are associative."

We note that Cayley's view of the rule for combining abstract symbols anticipates that of Kronecker in 1870 (I.3). The agreement of the two views extends even to the emphasis on the formal standpoint.

Cayley defines a group and introduces the group table:

A set of symbols

$$1, \alpha, \beta, \ldots$$

all of them different, and such that the product of any two of them (no matter in what order), or the product of any one of them into itself, belongs to the set, is said to be a *group*.[212] It follows that if the entire group is multiplied by any one of the symbols, either as further or nearer factor, the effect is simply to reproduce the group; or what is the same thing, that if the symbols of the group are multiplied together so as to form a table, thus:

Further factors

Nearer factors	1	α	β	...
1	1	α	β	...
α	α	α^2	$\beta\alpha$	
β	β	$\alpha\beta$	β^2	
	$\vdots$			

that as well each line as each column of the square will contain all the symbols 1, α, β, ... It also follows that the product of any number of the symbols, with or without repetitions, and in any order whatever, is a symbol of the group. [130, p. 124]

Cayley's 1854 papers contain the group tables of special permutation groups. But given my present concerns, what must be emphasized is Cayley's realization, not shared by his contemporaries, that his group concept extended to particular calculi, such as the multiplication of matrices and the theory of quaternions, which were of great interest in England at the time due to the publication (in 1853) of W. R. Hamilton's *Lectures of Quaternions* [260]. It is obvious that Cayley's clear recognition of noncommutativity[213] was due to these calculi.

Cayley's 1854 advance in the direction of the abstract group concept was, in historical terms, premature. The conditions favoring appreciation of the abstract, that is formal, approach had not yet fully developed. Also, as long as permutation groups were the only groups under investigation, there was no interest in generalizing the group concept, nor reason to do so. Thus Cayley's 1854 papers had no immediate impact on the move toward abstraction.

Matters were different with regard to Cayley's group-theoretic papers of 1878. These became the main inspiration of a process of abstraction consciously launched by W. v. Dyck (1856–1934) and completed in 1882 in his "Gruppentheoretische Studien" [180, 181]. One of v. Dyck's achievements was to reduce the concept of a transformation group to an instance of the abstract case. Cayley's 1878 papers also inspired a number of other fundamental group-theoretic works. For example, in his 1893 paper "Die Gruppen der Ordnungen p^3, pq^2, pqr, p^4" [291], based entirely on the abstract group concept, O. Hölder (1859–1937) referred explicitly to the problem posed by Cayley, in [133, p. 403],[214] of constructing all groups of a given order.

Cayley's return to group theory in 1878 was due to the rapidly growing recognition of its importance. In that year he published four papers [148–151].[215] One of these, called "The Theory of Groups," discusses fundamental problems of the subject, including the graphical representation of groups. The paper begins as follows:

Substitutions, and (in connection therewith) groups, have been a good deal studied; but only a little has been done towards the solution of the general problem of groups. [133, p. 401]

Cayley uses "function symbols" α, β, ..., "each operating upon one and the same number of letters and producing as its result the same number of functions of these letters" [133, p. 401]. In other words he works, quite generally, with mappings of a finite set onto itself. These operations, he says, can be iterated and composed, are in general noncommutative, are invariably associative, and include an identity, that is, a function symbol that keeps the letters fixed. As for the connection with permutation groups, Cayley states that "The functional symbols *may* be substitutions ..." [133, p. 402]. Having made this acknowledgment, he pursues the more abstract approach and considers groups whose elements are the "function symbols":

> A set of symbols α, β, γ, ..., such that the product $\alpha\beta$ of each two of them (in each order, $\alpha\beta$ or $\beta\alpha$), is a symbol of the set, is a group. ... A group is defined by means of the laws of combination of its symbols. [133, p. 402]

This generalized view of permutation groups suggested to Cayley, who was obviously concerned with finite groups alone, the problem of constructing all groups of a given order n. He realized that this problem is identical to the problem of constructing all permutation groups of order n:

> But although the theory as above stated is a general one, including as a particular case the theory of substitutions, yet the general problem of finding all the groups of a given order n, is really identical with the apparently less general problem of finding all the groups of the same order n, which can be formed with the substitutions upon n letters. [133, p. 403]

This is the content of Cayley's theorem that asserts that every finite group can be represented as a group of permutations. In proving this theorem, Cayley used the method of group tables, introduced by him in 1854.

2 *The Evolution of the Abstract Group Concept*

We saw that acceptance of Cayley's premature abstraction of 1854 failed, in the first place, through the lack of a body of concrete representations accepted in mathematical practice. To ascertain the significance of other factors we must look at developments in the seventies and eighties.

In view of the fact that Cayley's abstract group conception of 1854 was, not only in the chronological but also in the historical sense of the term, a mere anticipation (though in very many respects) of a development that was to come to fruition a quarter of a century later, it would not be reasonable to designate 1854 as the year of transition to the abstract conception of a group, let alone the year in which abstract group theory was born—though if assessment weakens the historical role of Cayley's contribution of 1854, it points up more strikingly his personal achievement. It also poses two questions relevant to our concerns.

The first of these bears on the other deep changes in mathematics that coincided in time with the process of abstraction in group theory and gave this process *causal* support. I shall show later that these factors acted in two ways, supplying to the developments in group theory both a methodological model and a natural analogy.

The second question has to do with the historical fact that the abstract group concept emerged not from a single act or the creation of a single scholar but was rather the outcome of a process of abstraction with certain discernible steps. These were not essentially logical in nature; rather, they represented the variable extent to which the process of abstraction was carried through. These steps depended largely on the degree of insight into the basic role of the group concept and are consequently a function of the historical interplay between the progressive abstraction itself and the recognition of that fundamental role. This observation will greatly influence my manner of presentation, simplifying it to the extent that typical steps can be characterized by a few examples—the only way to survey the wealth of relevant contributions.

The first noteworthy step within this progressive abstraction was the amalgamation, with the help of the abstract theory of abelian groups, of the theory of permutation groups on the one hand and, on the other hand, the implicit group-theoretic patterns of thought in number theory (I.3)—in particular, the theory of quadratic forms.

The most important paper of this kind was "Ueber Gruppen von vertauschbaren Elementen" by G. Frobenius (1849–1917) and L. Stickelberger (1850–1936), published in 1879. This represents a conscious effort to achieve the amalgamation alluded to above and serves as a kind of program for the subsequent development.

In the introduction the authors refer to the theory of power residues developed by Euler and Gauss, the solution of algebraic equations by Lagrange and Abel, and the composition of quadratic forms by Gauss, Schering, and Kronecker (I.3). This material suggests to the authors what we now call the basis theorem for (finite) abelian groups. Its formulation by the authors rests on the concept of irreducibility of a group:

* Here, just as in the case of the study of complex integers, the main difficulty was the transformation of the concepts of elementary number theory. For example, in the latter a prime is an integer whose only divisors are 1 and itself. On the other hand, in the case of groups we must call a group irreducible if it cannot be decomposed into two factors without one of them being the whole group. [207, p. 218]

The authors formulated the basis theorem[216] as follows:

* Die Hauptschwierigkeit bei dieser Untersuchung bestand, ähnlich wie bei der Lehre von den complexen ganzen Zahlen, in der Umformung der Begriffe, welche die elementare Zahlentheorie darbietet. Während man z. B. dort eine Zahl eine Primzahl nennt, wenn sie nur durch 1 und sich selbst theilbar ist, mussten wir hier eine Gruppe irreductibel nennen, wenn sie nicht in zwei Factoren zerfällt werden kann, ohne dass einer derselben gleich der ganzen Gruppe ist.

* A group that is not irreducible can be decomposed into purely irreducible factors. As a rule, such a decomposition can be accomplished in many ways. However, regardless of the way in which it is carried out, the number of irreducible factors is always the same and the factors in the two decompositions can be so paired off that the corresponding factors have the same order. The proof of this statement ... and the precise characterization of the irreducible factors of a decomposable group ... form the main subject of our investigation. [207, p. 221]

As regards methodology, the authors relied on Dedekind's theory of modules. Since they worked in a commutative setting, they could reduce multiplication to addition. From the very beginning they adopted an abstract standpoint:

** In order to present the abstract development in the most convenient and understandable manner we rely on the investigation of classes of numbers which are incongruent relative to a given modulus and relatively prime to it, and make no use of the special properties of these elements. We treat the theory of these number classes briefly [in the last section of the paper (Wussing)] as an application of the general investigation. [207, p. 218]

In order to define a group, the authors start with a complete residue system modulo M. These are "the elements of our investigation" ("die Elemente unserer Untersuchung"):

*** A number of these elements forms a (finite) group if the product of any two of them is contained among them. [207, p. 218]

Infinite groups are referred to in footnotes.[217] The cited examples of such groups include the groups of units of number fields (provided they are not all roots of 1) and the group of all roots of unity.

Clearly, the point of view taken by Frobenius and Stickelberger is very close to that of Kronecker in 1870 (I.3). The direct and extremely fruitful influence of Kronecker is also particularly clear in the textbook *Substitutionentheorie und ihre Anwendungen auf die Algebra* [54] published by his student E. Netto (1846–1919) in 1882. Netto's textbook was, of course, very strongly under the influence of the

---

* Eine Gruppe, die nicht irreductibel ist, kann in lauter irreductible Factoren zerlegt werden. Eine solche Zerlegung ist in der Regel auf viele verschiedene Weisen möglich. Wie man sie aber auch ausführen mag, man erhält doch stets die gleiche Anzahl von irreductiblen Factoren, und dieselben können einander in zwei verschiedenen Zerlegungen so zugeordnet werden, dass die entsprechenden Factoren von gleicher Ordnung sind. Der Beweis dieser Behauptung ..., sowie die genaue Charakterisierung der irreductiblen Factoren einer zerlegbaren Gruppe ... bildet den Hauptgegenstand der folgenden Untersuchung.

** Um die abstracte Entwicklung möglichst bequem und fasslich darstellen zu können, knüpfen wir sie an die Untersuchung der Klassen von Zahlen an, die in Bezug auf einen gegebenen Modul incongruent und relativ prim zu demselben sind, ohne dabei von den speciellen Eigenschaften dieser Elemente Gebrauch zu machen. Die Theorie dieser Zahlenklassen haben wir dann als Anwendung der allgemeinen Untersuchung kurz abgehandelt.

*** Eine Anzahl dieser Elemente bildet eine (endliche) Gruppe, wenn das Product von je zweien derselben wieder unter ihnen enthalten ist.

excellent, and by then classical, presentation of the theory of permutation groups by J.-A. Serret and the superb *Traité des substitutions* of C. Jordan (II.5). Comparison of texts shows that in matters of content and notation Netto followed Jordan closely[218] and thus took over the line of development represented by Vandermonde, Lagrange, Ruffini, Abel, and Galois. On the other hand, whenever Netto broke from this line and developed an abstract conception of a group, he was under the direct and explicitly recognized influence of Kronecker.[219] A few passages from the introduction show that Netto was aware of the novelty of his presentation, and found it necessary to justify his approach:

* The presentation of substitution theory given in this book differs in many nontrivial points from the standard one. In this connection I must comment on certain decisive viewpoints.

There is no doubt that the circle of applications of an algorithm increases if one manages to remove from its foundations and development all but the indispensable assumptions, and if one enables it to penetrate many areas by virtue of the generality of the objects it manipulates. The fact that the theory of group formation admits such a presentation is an argument in favor of its far reaching significance and in favor of its future. [541, p. III]

Netto's definition of a group and his reliance on the study of the valuedness of entire rational functions of n variables $x_1, x_2, \ldots, x_n$ for pedagogical and methodological motivation were admittedly conventional, that is, they reflected the viewpoint associated with permutation groups:

** The substitutions

$$s_1 = 1, s_2, s_3, s_\alpha, s_r$$

that leave unchanged a given function $\varphi(x_1, x_2, \ldots, x_n)$ form ... a closed group in the sense that their totality is unchanged as a result of the multiplication of the constituent substitutions by each other. The term "group" will always denote a complex of substitutions possessing the ... characteristic property of reproduction of the complex through multiplication of its members. [541, pp. 25–26]

---

* Die im vorliegenden Buche durchgeführte Darstellung der Substitutionentheorie weicht in mehreren nicht unwesentlichen Punkten von den bisher üblichen ab. Hierbei waren Gesichtspunkte massgebend, welche hervorgehoben werden müssen.

Es ist unzweifelhaft, dass der Kreis der Anwendungen eines Algorithmus sich ausdehnen wird, wenn es gelingt, die Grundlagen und den Aufbau desselben von allen nicht unbedingt geforderten Voraussetzungen zu befreien, und ihm durch die Allgemeinheit der Objekte, mit denen er arbeitet, auch die Möglichkeit des Eingreifens in die verschiedensten Gebiete zu geben. Dass die Theorie der Gruppenbildung eine solche Darstellung zulässt, spricht für ihre weitgreifende Bedeutung und für ihre Zukunft.

** Diejenigen Substitutionen

$$s_1 = 1, s_2, s_3, \ldots, s_\alpha, \ldots, s_r,$$

welche eine gegebene Funktion $\varphi(x_1, x_2, \ldots, x_n)$ ungeändert lassen, bilden ... in der Hinsicht eine in sich geschlossene Gruppe, dass sich ihre Gesamtheit durch Multiplikation der einzelnen zuge-

(** continued on p. 237)

On the other hand, Netto went beyond the tradition of permutation groups in his treatment of abelian groups, and especially in the derivation of the basis theorem. In §131, Netto quotes almost verbatim Kronecker's 1870 paper [432] on the class number of ideal complex numbers, which we analyzed in detail in I.3:

* We shall now concern ourselves with groups of commuting substitutions. But we can use a more general approach [that is, an approach which is more abstract and dispenses with permutations (Wussing)], which makes the results obtained far more applicable. [541, pp. 143 to 144][220]

This is followed by a passage from Kronecker's paper cited above (in I.3), virtually unchanged: "Let θ', θ'', θ''', ... be finitely many elements ...."

Netto derives (in §133) the basis theorem for abelian groups from Kronecker's result:

** In our case, the elements θ must be replaced by commuting substitutions. The number n of elements θ becomes the order r of the group. We thus have Theorem IV: If the substitutions of a group commute with each other, then there exists a fundamental system of substitutions $s_1, s_2, s_3, \ldots$ such that the expression

$$s_1^{h_1} s_2^{h_2} s_3^{h_3} \cdots \qquad (h_i = 1, 2, \ldots, r_i)$$

represents each of the substitutions of the group exactly once. The numbers $r_1, r_2, r_3, \ldots$ are the orders of $s_1, s_2, s_3, \ldots$ and have the property that each is equal to or divisible by the next. The product $r_1 r_2 r_3 \cdots$ of the orders is equal to the order r of the group. [541, p. 146]

It is clear that when it comes to the use of the abstract group concept, Netto's position is not consistent: "Group" is still synonymous with "permutation

(** continued from p. 236)
hörigen Substitutionen untereinander nicht ändert. Mit diesem Namen einer Gruppe soll stets ein Komplex Substitutionen bezeichnet werden, welcher die ... charakteristische Eigenschaft der Reproduktion des Komplexes durch Multiplikation seiner Individuen besitzt.

* Wir wollen uns jetzt noch mit solchen Gruppen beschäftigen, deren Substitutionen untereinander vertauschbar sind. Doch können wir hierbei eine allgemeinere Behandlung eintreten lassen, welche die erlangten Resultate vielfach verwendbar macht.

** In unserem Falle müssen die Elemente θ durch Substitutionen ersetzt werden, die mit einander vertauschbar sind. n, die Anzahl der Elemente θ, geht in die Ordnung r der Gruppe über. Wir haben demnach:

Lehrsatz IV. Sind die Substitutionen einer Gruppe unter einander vertauschbar, so giebt es ein Fundamentalsystem von Substitutionen $s_1, s_2, s_3, \ldots$, welches die Eigenschaft besitzt, dass der Ausdruck

$$s_1^{h_1} s_2^{h_2} s_3^{h_3} \ldots \quad (h_i = 1, 2, \ldots, r_i)$$

sämtliche Substitutionen der Gruppe und zwar jede nur einmal darstellt. Dabei sind die Zahlen $r_1, r_2, r_3, \ldots$ die Ordnungen von $s_1, s_2, s_3 \ldots$ und so beschaffen, dass jede derselben durch die folgende teilbar oder ihr gleich ist; das Produkt dieser Ordnungen $r_1 r_2 r_3 \ldots$ ist gleich der Ordnung r der Gruppe.

group." Netto's formulation of the basis theorem is a relapse, the surrender of a newly attained level of abstraction. A similar state of affairs prevails in the case of his extensive investigations of isomorphic and homomorphic mappings[221] of groups. The investigations and conclusions do not really depend on the properties of permutations, and one would have liked to see the word "substitution" replaced by "element."

In his *Substitutionentheorie* of 1882, Netto used the abstract group concept to combine, admittedly in an imperfect and inconsistent way, two of the three historical roots of abstract group theory, namely, the already explicitly developed and largely independent group theory of permutations and the group ideas present implicitly in number theory. *At that time* he had not subsumed the concept of a transformation group under his version of abstract groups.

What we have here is a typical intermediate stage, soon to be left behind, which strikes us as a mere episode. But it should be pointed out that Netto was not the only representative of this position. Another was H. Weber who, also in 1882, interpreted Kronecker's implicit system of axioms for abelian groups (quoted by Netto) in group-theoretic terms in his paper "Beweis des Satzes, dass jede eigentlich primitive quadratische Form unendlich viele Primzahlen darzustellen fähig ist" [713], whose content closely follows Gauss, Dirichlet, and Kronecker. Like Netto, Weber used the group concept in his paper without the inclusion of transformation groups. Weber recognized the essential sameness of the rules of combination of Kronecker's "elements" on the one hand and of permutations on the other and therefore used the respectable and accepted term "group," formulated Kronecker's postulates in the language of group theory, attained in this way an axiomatically based abstract group concept[222] and—unlike Netto—managed to adhere to the abstract approach throughout his paper. But only later (see below) did Weber use the abstract group concept in connection with the theory of transformation groups.

So far we have mentioned the year 1882 as marking a transitional stage in the evolution of the abstract group concept. But one may also justifiably consider 1882 as decisive for that concept's full elaboration. For in that year there appeared a paper that consciously combined all three of the concept's historical roots—the theory of algebraic equations, number theory, and geometry. Moreover, infinite groups, in particular, infinite transformation groups, were consciously subsumed under the axiomatically based group concept here set forth. The one case not included was that of the continuous transformation groups —but in fact Lie in 1882 (III.3) was forced to make strenuous efforts to gain recognition anywhere for his theory of these groups.

The decisive paper alluded to above was due to W. v. Dyck (1856–1934). In

1882 there appeared the first part [180], and in the following year the second part [181] of his "Gruppentheoretische Studien." Both works deserve detailed analysis, for here v. Dyck not only completes the elaboration of the abstract group concept but also refers to the intellectual sources of his approach in an effort to justify it. Moreover, his statement of motivation [180, p. 2] is remarkable for its historical objectivity.

Actually v. Dyck had been within range of all the influences necessary for the working out of a comprehensive abstract group concept. This applies to the influences due to the development of group theory itself as well as to the tendencies, which were becoming more explicit at the end of the nineteenth century, to provide for mathematics an abstract foundation formulated by means of axioms and justified through mathematical logic.

The inclusion of (discrete) transformation groups in the abstract group concept goes back to the close contact between v. Dyck and Klein. In 1879 v. Dyck obtained his doctoral degree [179] under Klein, and he was Klein's assistant in Leipzig between the summer of 1871 and winter of 1882–83, and thus also during the decisive years 1882 and 1883.

There can be little doubt that Cayley's 1878 advance toward an abstract group concept, contained in a number of papers, and his stress on the role of generators of groups, triggered v. Dyck's paper. In fact, v. Dyck was so strongly influenced by Cayley that he adopted his group definition ("A group is defined by means of the laws of combination of its symbols") as the motto of his own paper.

Cayley's abstract group concept was met in 1854 with silence, but in 1878 its reception was vastly different. At that time there was general recognition of the need for a deeper view of mathematics and of the necessary methods of mathematical logic. Boole's papers on formal algebra (III.4) were no longer read by just his friends, including Cayley, but were accepted on the whole European continent. Following the contribution of his son C. S. Peirce (1839–1914) to mathematical logic [550–553], B. Peirce (1809–80) of the United States developed far-reaching approaches [549] to general associative linear algebra.

In Germany there was recognition of the deeper significance of Grassmann's *Ausdehnungslehre* [253] of 1862, and the work was pointed to as a methodological model. In fact, the retrospective recognition of Grassmann provided the direct motive[223] for G. Peano's (1858–1932) development of the postulates of arithmetic in 1889. Through G. Frege's (1848–1925) *Begriffsschrift* [196] of 1879 and E. Schröder's (1841–1902) numerous publications [659–661] of the seventies and eighties on the calculus of logic and on, as Schröder called it, "absolute" algebra, mathematical logic began to develop in Germany as an independent area of research.

An advantageous development was the acceptance in Germany of the viewpont, modeled on Lejeune-Dirichlet and Kronecker, that aimed at working out the laws of combination of "magnitudes." Here the pacesetting works were H. Hankel's (1839–73) *Theorie der complexen Zahlensysteme* [262] of 1867, R. Dedekind's "Sur la théorie des nombres algébriques" [162] of 1876–77, and his "Supplemente" to the third edition (1879) of Lejeune-Dirichlet's *Vorlesungen über Zahlentheorie* [455].

These were the conditions under which v. Dyck wrote his "Gruppentheoretische Studien." Thanks to Klein, v. Dyck was familiar with all three areas —the theory of algebraic equations, number theory, and geometry—of application of group theory and was able to identify, among contemporary ideas, the elements most significant for group theory. Obviously he was best acquainted with the relevant German literature, but he relied substantially on Cayley's paper [150] of 1878 and on W. R. Hamilton's *Elements of Quaternions* [261] of 1866, as well as on Netto's *Substitutionentheorie* [541] of 1882. For his concept of an abstract group, he drew especially on the work of Grassmann, Hankel,[224] and Schröder. In particular, Schröder had more than once [659–661] discussed the position of group theory in what he referred to as "absolute" algebra.[225]

Accordingly, v. Dyck arrived at the following arrangement of group theory in the general scheme of algebra—a rather modern formulation in the sense of contemporary "universal algebra":

* If we ... view the group-theoretic operations quite formally, then their position in a formal development of all analytic operations is quite clear. They are multiplication operations that obey the associative but *not* the commutative principle. Also, *certain multiplication rules* give these operations the character of a *special* circle of operations, which includes an infinite or finite group of operations. [180, p. 2]

He formulated the following general problem, aimed at the abstract definition of a group:

** *To define a group of discrete operations that are applied to a certain object, while ignoring the particular form of representation of the individual operations, regarding them as given only in terms of the properties essential for group building.* [180, p. 1]

---

* Indem wir ... die gruppentheoretischen Operationen rein formal auffassen, zeigt sich deutlich ihre Stellung in einer formalen Entwicklung analytischer Operationen überhaupt. Es sind Multiplicationsoperationen, welche das associative, *nicht* aber das commutative Princip befolgen. Dabei wird diesen Operationen *durch gewisse Multiplicationsregeln* ... der Charakter eines *speciellen* Operationskreises ertheilt, der eine unendliche oder auch eine endliche Gruppe von Operationen umfasst.

** *Eine Gruppe von discreten Operationen, welche auf ein gewisses Object angewandt werden, zu definiren, wenn man dabei von einer speciellen Darstellungsform der einzelnen Operationen absieht, diese vielmehr nur nach den zur Gruppenbildung wesentlichen Eigenschaften als gegeben voraussetzt.*

Following Cayley's example, v. Dyck used the method of generators to solve the problem. He starts,

* From certain *generating operations* $A_1, A_2, A_3, \ldots$ about whose *particular* nature no assumptions are made.

Then it is possible to specify every group that can be built by means of iteration and combination of these operations through the knowledge of certain relations which occur in connection with the composition of the original operations. [180, p. 1]

In modern terms, v. Dyck is constructing the free group on the elements $A_1, A_2, \ldots$. The passage quoted above is identical with the theorem that asserts that every group with a finite number of generators is a factor group of a free group of finite rank. Just as we do today, v. Dyck constructs expressions

$$A_1^{\mu_1} A_2^{\mu_2} \cdots A_1^{\nu_1} A_2^{\nu_2} \cdots,$$

now called "words." He writes the given initial relations in the form

$$F_h(A_1) = 1.$$

The associative law is assumed rather than proved. On the other hand, v. Dyck stresses that

** in this way all *holoedrically isomorphic* groups are included in *a single* group" ... [and] ... the *essence* of a group is no longer expressed by a particular form of its operations but rather by their mutual relations. [180, p. 2]

We have determined v. Dyck's intellectual starting point. In the first part [180] of his "Gruppentheoretische Studien" he investigates the free group on m generators $A_1, A_2, \ldots, A_m$, which he denotes by G. By introducing defining relations $F_h = 1$, $h = 1, 2, \ldots, n$, he arrives at a group $\bar{G}$ that —to use modern terms—he shows to be [180, p. 12] a factor group of G.[226]

The first part of the "Gruppentheoretische Studien" is also a link to Klein's function-theoretic interpretation of infinite discrete groups (III.3). But, as v. Dyck states explicitly, he regards this "geometric representation ... *only as a transition to the development of abstract group-theoretic views*" ("geometrische Darstellung ...

---

* Von gewissen *erzeugenden Operationen* $A_1, A_2, A_3, \ldots$ aus, über deren *speciellen* Charakter keinerlei Annahmen gemacht werden.

Dann kann man jede Gruppe, welche durch Iteration und Combination dieser Operationen sich bilden lässt, individualisiren, durch die Kenntnis gewisser Relationen, die bei der Zusammensetzung dieser ursprünglichen Operationen auftreten.

** damit werden alle *holoedrisch isomorphen* Gruppen in *eine einzige* Gruppe begriffen ... [und] ... das *Wesen* der Gruppe drückt sich nicht mehr an einer speciellen Darstellungsform ihrer Operationen aus, sondern lediglich in der gegenseitigen Beziehung derselben zu einander.

nur als ein Durchgangstadium zur Entwickelung der abstracten gruppentheoretischen Anschauungen") [180, p. 3].

v. Dyck links his theory to the study of known "concrete" groups by means of the clearly formulated problem (which he solves for finite groups) of determining these groups by means of "suitable modifications of the relations $F_h = 1$" ("zweckmässige Umformungen der Relationen $F_h = 1$") [180, p. 25]. In this way he obtains, in the second part of his "Gruppentheoretische Studien" [181], the theories of permutation groups and of polyhedral groups. He discusses extensively his position toward "concrete" group-theoretic investigations, that is, the theory of permutation groups, number-theoretic groups, and transformation groups. These arguments, some lengthy, some short, and some in the form of footnotes, represent v. Dyck's effort to justify his approach. We quote a lengthy passage that illustrates very clearly his conscious attempt to unify the three historical roots of group theory by means of the abstract group concept:

* *The following investigations aim to continue the study of the properties of a group in its abstract formulation. In particular, this will pose the question of the extent to which these properties have an invariant character present in all the different realizations of the group, and the question of what leads to the exact determination of their essential group-theoretic content.*

I wish to emphasize here, at the very outset, that this approach does not aim to surrender the *individual* advantages that may derive from a particular formulation of each particular problem. For each specific problem we have at our disposal a treasure of specific information—it seems hardly necessary to mention the algebraic, function-theoretic and number-theoretic questions connected with group-theoretic problems—that can be used to advantage in any given case. *But it is precisely these special connections that call for a discussion of the extent to which they are based on purely group-theoretic as against other properties of the posed problem.*

These considerations gave rise to the following investigations. Thus, while I make no claims of originality for the *subject matter* of the following lines, I believe that the form of the question allows *the known properties* of a group to be seen from a new point of view and defines them with greater clarity. [181, pp. 70–71]

---

* *Es ist die Absicht der hier folgenden Untersuchungen, die bisher studirten Eigenschaften einer Gruppe in ihrer abstracten Formulirung zu verfolgen. Dabei wird insbesondere die Frage zur Geltung kommen, in wie weit diese Eigenschaften durch alle verschiedenen Erscheinungsformen der Gruppe einen invarianten Charakter tragen, was zur exacten Fixirung ihres eigentlichen gruppentheoretischen Inhaltes führt.*

Es sei dabei gleich hier hervorgehoben, dass durch diese Darstellung keineswegs ein Aufgeben derjenigen *individuellen* Vortheile angestrebt wird, die für jedes specielle Problem gerade in seiner speciellen Darstellungsform liegen können. Für jedes einzelne Problem steht uns ein Schatz individueller Kenntnisse zur Verfügung – ich brauche kaum an algebraische, an functionentheoretische und zahlentheoretische Fragen zu erinnern, die mit gruppentheoretischen Problemen in Verbindung stehen – die wir in jedem gegebenen Falle mit Vortheil verwenden. *Aber gerade solche specielle Beziehungen fordern zur Discussion der Frage, in wie weit sie in rein gruppentheoretischen und in wie weit in anderen Eigenschaften des gestellten Problems ihre Begründung finden.*
(* continued on p. 243)

A second point to be mentioned in connection with v. Dyck's "Gruppentheoretische Studien" is that in the second part [181, pp. 73–75], the author provides a link to the development of axiomatics. What is especially noteworthy is the explicit requirement of the existence of an inverse:

* *We require for our considerations that a group that contains the operation T_k must also contain its inverse T_k^{-1}.* [181, p. 74]

We see that v. Dyck's concept of a group fulfills all the requirements demanded of a fully developed abstract approach, namely, an axiomatic definition and the subsuming of all "concrete" group definitions.

In the years immediately following its publication, v. Dyck's book was frequently quoted, less because of its results than because of its formulation of the abstract group concept. He had managed to set forth in masterly fashion a development whose presentation had been attempted more or less consciously but more casually by many other authors.

The abstract group concept spread rather rapidly during the eighties and nineties, accompanied by a proliferation—especially in France—of papers in the areas of permutation and transformation groups, although, for methodological reasons, many textbook authors kept it out of their books or introduced it inductively. One very good and popular textbook of this type is W. Burnside's *Theory of Groups of Finite Order*, published in 1897.

My summary of the "spread of the abstract group concept" during the eighties and nineties does not do justice to the genuine complexities of the historical situation and requires additional distinctions. As before, I rely on the method of choosing typical representatives of a particular level of abstraction.

In connection with the application of the abstract group concept we encounter the typical desire to see to what extent theorems from "concrete" group theory are independent of the specific nature of the group elements, in order possibly to include familiar theorems of, say, the theory of permutation groups in a final presentation of an abstract *theory* of groups. As an example of such tendencies, I cite Frobenius's paper [208] of 1887,[227] in which he proved Sylow's theorem to the effect that every (finite) group whose order is divisible by the νth power of

---

(* continued from p. 242)

Aus diesen Ueberlegungen sind die nachfolgenden Untersuchungen entstanden. Macht also der *stoffliche* Inhalt der folgenden Zeilen keineswegs den Anspruch auf Neuheit, so glaube ich doch, dass gerade die Form der Fragestellung *die bekannten Eigenschaften* einer Gruppe unter einem neuen Gesichtspunkt erscheinen lässt und deren Definition in grösserer Klarheit und Schärfe ergiebt.

* *Wir setzen nun für unsere Betrachtungen noch fest, dass eine Gruppe, welche die Operation T_k enthält, stets auch ihre inverse Operation T_k^{-1} mit einbegreifen soll.*

a prime p must contain a subgroup of order p^ν. Frobenius notes that Cauchy had already proved a special case of this theorem for permutation groups[228] and goes on to state the theorem proved by Sylow [632] for permutation groups in 1872. He admits that the fact that every finite group can be represented by a group of permutations proves that Sylow's theorem must hold for all finite groups but nevertheless wishes to establish the theorem abstractly.

* However, in view of the nature of the theorem to be proved [which is abstract, that is, not referring to a permutation group (Wussing)] we shall not use this interpretation in what follows. [208, p. 179]

To this end, following Kronecker [432] and Weber [713], Frobenius sets down a system of postulates such that a set of arbitrary "elements" satisfying these postulates must have the group property. The postulates in question are, (1) the result of combining two elements is a unique third element, (2) $AC = BC$ and $CA = CB$ each imply that $A = B$, (3) the associative law holds, and (4) the number of elements is finite. Frobenius then proves the theorem by induction (on ν).

In addition to efforts aimed at extracting the essential group-theoretic content of "concrete" group-theoretic investigations we encounter studies and conceptual constructions based on an abstract foundation. Most of the latter tendencies are due to North American and German mathematicians. Thus, starting from an abstract basis, Dedekind [167] and G. A. Miller (1863–1951) [520] investigated in the late nineties the so-called hamiltonian groups, that is, nonabelian groups whose subgroups are invariably normal. Again, Dedekind [167] and G. A. Miller [518] introduced abstractly the concepts of a commutator and of a commutator subgroup. Finally, O. Hölder [291] and E. M. Moore (1862–1932) [534, 535] introduced abstractly the concept of an automorphism of a group, an extremely important methodological research tool.

The concept of a factor group, one of the core ideas of classical group theory, as well as the second half of the Jordan-Hölder theorem on the isomorphism of the prime factors of two different composition series of a group, were deductions from the abstract group concept. O. Hölder, citing v. Dyck's "Gruppentheoretische Studien" of 1882, placed at the beginning of his 1889 paper "Zurückführung einer beliebigen algebraischen Gleichung auf eine Kette von Gleichungen" [287] an abstract definition of a group. Specifically, §1, called "Die definierenden Eigenschaften der Gruppen," states that

* Indessen will ich gerade in Anbetracht des zu beweisenden Satzes diese Auffassung im Folgenden nicht zugrunde legen.

* our ... theorems hold for all *finite* groups of operations. The nature of the operations is irrelevant. All we assume is the group property. The latter can be described under the following headings:

1) The result of combining (multiplying) two operations in a given order must be a uniquely determined operation that also belongs to the class of operations.
2) The composition of operations must obey the associative law but need not obey the commutative law.
3) Each of the symbolic equations

$$AB = AC, \qquad BA = CA$$

must imply that

$$B = C.$$

These assumptions, together with the finiteness of the number of operations, imply that there exists a unique so-called *identity* operation J that is the only operation leaving all others unchanged under multiplication, and that for each operation A there is a unique inverse operation A^{-1} such that

$$AA^{-1} = A^{-1}A = J. \qquad \text{[287, pp. 28–29]}$$

Hölder's other fundamental contributions to classical group theory rely on the same group concept in a different formulation.[229] The papers in question are [290] and [291] of 1892–1893, devoted to simple groups, and [293] of 1895, devoted to composite groups. The latter may be regarded as an early stage of Schreier's theory of group extensions [653, 655–658], dating from the 1920s.

3 *The Group as a Fundamental Concept of Algebra*

In the introduction to his *Substitutionentheorie* [541, p. IV], Netto regrets the fact[230] that he had not been able to make full use of Kronecker's "Grundzüge der arith-

---

* die ... entwickelten Sätze gelten für alle Gruppen, die aus einer *endlichen* Anzahl von Operationen bestehen. Die Art der Operationen ist dabei gleichgültig. Es wird nur die Gruppeneigenschaft vorausgesetzt, welche in den folgenden Bestimmungen zusammengefasst werden kann:

1) Je zwei Operationen sollen in bestimmter Aufeinanderfolge zusammengesetzt (multiplicirt) eine eindeutig bestimmte Operation ergeben, welche gleichfalls derselben Gesammtheit angehört.
2) Für die Zusammensetzung der Operationen soll das associative Gesetz gelten, während das commutative nicht erfüllt zu sein braucht.
3) Aus jeder der beiden die Operationen A, B, C enthaltenden symbolischen Gleichungen

$$AB = AC, \qquad BA = CA$$

soll geschlossen werden können, dass

$$B = C$$

ist.

Eine Folge dieser Bestimmungen im Zusammenhang mit der Endlichkeit der Operationenzahl ist es, dass eine sogenannte *identische* Operation J vorhanden ist, und zwar eine einzige, welche alle anderen bei der Multiplication unverändert lässt, und dass zu jeder Operation A eine eindeutig bestimmte umgekehrte Operation A^{-1} sich findet, so dass

$$AA^{-1} = A^{-1}A = J$$

ist.

metischen Theorie der algebraischen Grössen," also published in 1882, in the theory of substitutions and its applications. More than all the discussions[231] of the "arithmetization of mathematics" ("Arithmetisierung der Mathematik"), this monograph of Kronecker exerted a strong methodological influence on the further development of algebra and on the further—and eventually conclusive—amalgamation of the permutation-theoretic group concept and the group-theoretic approaches of field theory and number theory. Another strong influence was that of R. Dedekind—in particular, his "Supplementen" to Lejeune-Dirichlet's *Vorlesungen über Zahlentheorie* [455]. The abstract group concept owes its elevation to the position of a central concept of algebra to the encounter of the theory of permutation groups with the two directions that may be identified with the names of Kronecker and Dedekind, respectively.

The publications of H. Weber (1842–1913), a student of Dedekind, typify the convergence of these three developments. Moreover, his often reprinted *Lehrbuch der Algebra* became the standard German textbook in this field and played a major role in shaping the views of the next generation of mathematicians.

The voluminous series of four papers on the "Theorie der Abel'schen Zahlkörper" [715–718] published by Weber in 1886–87 was the first systematic treatment of commutative number fields based explicitly on permutation groups and consciously related to the works of Schering [637] and Kronecker [432] mentioned above (I.3). As regards the group-theoretic foundations, Weber referred to the work of Frobenius and Stickelberger [207] and to his own paper [713], in which he developed the main theorems on abelian permutation groups.

The series on abelian number fields signified the conclusive amalgamation of "concrete" field theory with the theory of permutation groups. In this work the term "group" is synonymous with "permutation group." At just one point, in connection with the proof of the theorem that two abelian (permutation) groups with the same invariants (the same number of basis elements of the same order) are isomorphic, Weber refers—in a footnote—to the abstract group concept. I quote:

* If we separate the group concept entirely from the particular meaning that its elements have in each specific case and rely on a formal definition, then we can identify isomorphic groups and say, in this sense, that the group is completely determined by the invariants. [718, p. 110]

Just a few years later, in 1893, we witness a radical change in Weber's position.

---

* Löst man den Begriff der Gruppe gänzlich ab von der besonderen Bedeutung, welche die Elemente derselben in jedem einzelnen Falle haben, und fasst die Definition nur formal, so kann man isomorphe Gruppen auch schlechthin als identisch bezeichnen und in diesem Sinne sagen, dass durch die Invarianten die Gruppe vollständig bestimmt sei.

He had earlier dismissed abstract groups with a casual reference; now he refers to the group as a fundamental structure of algebra. In the introduction to the paper "Die allgemeinen Grundlagen der Galois'schen Gleichungstheorie" [720], Weber spoke of the scope and limits of the "formal" method of algebra in terms reminiscent of the 1920s:[232]

* What follows is an attempt to present the Galois theory of algebraic equations so that it possibly includes all cases in which the theory has been applied. In this presentation the theory is a direct consequence of the group concept extended to the field concept, independent of the numerical significance of the elements involved. ... The effect of this viewpoint is that the theory appears as a pure formalism that acquires content and life only after the individual elements are assigned numerical values. On the other hand, this form can be applied in all imaginable cases that satisfy the required assumptions. These extend on the one hand to the theory of functions and on the other hand to number theory. [720, p. 521]

After this introduction, Weber describes the group concept in §1 of his paper. He considers this necessary in spite of the fact that "the concepts that mathematicians associate today with this word [group (Wussing)] are rather uniform and clear" ("die Vorstellungen, die Mathematiker heutigen Tags mit diesem Worte verbinden, ziemlich übereinstimmend und klar sind") [720, p. 521]. What is new in Weber is the consistently abstract axiomatic definition of a group, a definition that includes infinite groups:

** A system $\mathfrak{S}$ of finitely many or infinitely many arbitrary things (elements) forms a *group* if it satisfies the following conditions.

1) *A rule is given that assigns to a first and second elements of the system a definite third element of that system.* [720, p. 522]

Weber introduces the symbolic notation $AB = C$ and $C = AB$ and refers to C as "composed" of A and B:

*** We do not assume that this composition always satisfies the commutative law ..., but

---

* Im Folgenden ist der Versuch gemacht, die Galois'sche Theorie der algebraischen Gleichungen in einer Weise zu begründen, die soweit möglich alle Fälle umfasst, in denen diese Theorie angewandt worden ist. Sie ergiebt sich hier als eine unmittelbare Consequenz des zum Körperbegriff erweiterten Gruppenbegriffs, als ein formales Gesetz ganz ohne Rücksicht auf die Zahlenbedeutung der verwendeten Elemente ... Die Theorie erscheint bei dieser Auffassung freilich als ein reiner Formalismus, der durch Belegung der einzelnen Elemente mit Zahlwerthen erst Inhalt und Leben gewinnt. Dagegen ist diese Form auf alle denkbaren Fälle, in denen die gemachten Voraussetzungen zutreffen, anwendbar, die einerseits in die Functionentheorie andererseits in die Zahlentheorie hinübergreifen.

** Ein System $\mathfrak{S}$ von Dingen (Elementen) irgend welcher Art in endlicher oder unendlicher Anzahl wird zur *Gruppe*, wenn folgende Voraussetzungen erfüllt sind.

(1) *Es ist eine Vorschrift gegeben, nach der aus einem ersten und einem zweiten Element des Systems ein ganz bestimmtes drittes Element desselben Systems abgeleitet wird.*

(*** see p. 248)

2) *We assume the associative law ....*
3) *We assume that if $AB = AB'$ or if $AB = A'B$, then, necessarily, $B = B'$ or $A = A'$.*
   If $\mathfrak{S}$ contains a finite number of elements, then the group is called a *finite* group and the number of its elements is called its *degree*. In the case of finite groups (1), (2), and (3) imply that
4) *If two of three elements A, B, C are taken arbitrarily from $\mathfrak{S}$ then the third can always be uniquely determined* so that

$$AB = C. \quad [720, \text{p. } 522]$$

After proving (4), Weber continues:

* This conclusion need not hold for infinite groups. *Thus in the case of infinite groups we shall include property* (4) *as one of the required axioms.* [720, p. 524]

Weber goes on to deduce consequences of this definition. In particular, he proves the existence of an identity. He also introduces additional definitions, such as those of commutativity and of isomorphism.

An additional reason for our interest in Weber's paper is his detailed comments on the nature of the abstract group concept:

** We can ... combine all isomorphic groups into a single class of groups that is itself a group whose elements are the generic concepts obtained by making one general concept out of the corresponding elements of the individual isomorphic groups. The individual isomorphic groups are then to be regarded as different representatives of the generic concept, and it makes no difference which representative is used to study the properties of the group. [720, p. 524]

Weber goes on to give examples of groups, including the additive group of complex numbers, the group of polyhedral rotations (here he refers specifically

---

(*** see p. 247)
Bei dieser Composition wird im Allgemeinen nicht das commutative Gesetz vorausgesetzt ..., dagegen wird
2) *das associative Gesetz vorausgesetzt ...*
3) *Es wird vorausgesetzt, dass, wenn $AB = AB'$ oder $AB = A'B$ ist, nothwendig $B = B'$ oder $A = A'$ sein muss.*
Wenn $\mathfrak{S}$ eine endliche Anzahl von Elementen umfasst, so heisst die Gruppe eine *endliche* und die Anzahl ihrer Elemente ihr *Grad.* Bei endlichen Gruppen ergibt sich aus 1) 2) 3) die *Folgerung*
4) *Wenn von den drei Elementen A, B, C zwei beliebig aus $\mathfrak{S}$ genommen werden, so kann man das dritte immer und nur auf eine Weise so bestimmen, dass*

$$AB = C$$

ist.
* Für unendliche Gruppen ist dieser Schluss nicht mehr zwingend. *Für unendliche Gruppen wollen wir also die Eigenschaft 4) noch als Forderung in die Begriffsbestimmung mit aufnehmen.*
** Man kann ... alle unter einander isomorphen Gruppen zu einer Classe von Gruppen, die selbst wieder eine Gruppe ist, zusammenfassen, deren Elemente die Gattungsbegriffe sind, die man erhält, wenn man die entsprechenden Elemente der einzelnen isomorphen Gruppen zu einem Allgemeinbegriffe zusammenfasst. Die einzelnen unter einander isomorphen Gruppen sind dann als verschiedene Repräsentanten eines Gattungsbegriffs aufzufassen, und es ist gleichgültig, an welchem Repräsentanten man die Eigenschaften der Gruppe studirt.

to Klein), permutation groups, the additive group of residue classes with respect to an arbitrary modulus, and the group of classes of binary forms combined by means of Gauss's rule of composition of forms. Weber's examples of finite and infinite groups are striking in the sense that they are associated with the decisive stages in the evolution of the abstract group concept. Only Lie groups are missing—probably because the theory of Lie groups had not at this time gained general acceptance.

Finally Weber makes clear the relation between groups and fields. He emphasizes the primacy of the group concept as against that of the (commutative[233]) field concept.

* A group becomes a field if it can have *two kinds of composition*, of which the first is called *addition* and the second *multiplication*. [720, p. 526]

There follow the field postulates. What is new in Weber's approach is the fact that the field concept is applied in particular to finite fields—the existence of which is emphasized.

Weber commented frequently on the role of the group as a fundamental structure of algebra. He did this for the most part in his popular textbooks. Thus in the so-called "Kleine Ausgabe" ("pocket edition") of his *Lehrbuch der Algebra* [729] of 1912 we read,

** For the most part, modern algebra is dominated by two general concepts. The existence and significance of these concepts could not be recognized until such time as algebra had reached a certain level of completeness and became the property of mathematicians. Only then could one see in them the connecting and leading principle.

The two concepts which we are about to explain are those of group and field. The more general concept is that of a group, and we begin with it. [729, p. 180]

Weber's paper of 1893, discussed above, exerted a substantial influence on algebra through its abstract presentation of the fundamental algebraic concepts of group and field. Along with many other authors, E. Steinitz (1871–1928) referred to it—specifically in the introduction to his "Algebraische Theorie der Körper" [686] of 1910. This fact is historically significant because Steinitz's paper, together with Hilbert's "Zahlbericht" [279] of 1897, marked the conclusion of the axiomatization of algebra and served as the linchpin[234] of the "Modern

---

* Eine Gruppe wird zum Körper, wenn in ihr *zwei Arten der Composition* möglich sind, von denen die erste *Addition*, die zweite *Multiplication* genannt wird.

** Es sind hauptsächlich zwei große allgemeine Begriffe, von denen die moderne Algebra beherrscht wird. Die Existenz und Bedeutung dieser Begriffe konnte allerdings erst erkannt werden, nachdem die Algebra bis zu einem gewissen Grad fertig und zum Eigentum der Mathematiker geworden war. Erst dann konnte in ihnen das verbindende und führende Prinzip erkannt werden.

Es sind die Begriffe der G r u p p e und des K ö r p e r s, zu deren Erklärung wir jetzt fortschreiten. Der allgemeinere Begriff ist der der Gruppe, mit dem wir also beginnen.

Algebra" of the twenties associated with the names of, among others, E. Noether (1882–1935), H. Hasse (1898–1979), E. Artin (1898–1962), O. Schreier (1901–29), and B. L. van der Waerden (b. 1903).

Since it is intended for beginners, the first volume of the *Lehrbuch der Algebra* does not go beyond the group concept in the sense of permutation group. However, it does emphasize this concept's importance:

> * It was my intention to write a textbook that would introduce readers with a modest background to modern algebra, including the higher and more difficult parts in which the subject comes truly to life. In order to make the presentation largely independent of other textbooks I intend to develop the required elementary and advanced tools in the book itself.
>
> Two things have acquired very special significance for the newest development of algebra. One is the ever dominant group theory whose ordering and clarifying influence is felt everywhere and the other is the intervention of number theory. [721, p. v]

In the first volume, Weber uses elements of the theory of permutations to introduce, among other things, Galois theory. In this connection he mentions Dedekind [721, p. vii], in particular Dedekind's Göttingen lecture on Galois theory given in the winter of 1857–58, and he also acknowledges the stimulus provided by his "colleague F. Klein."

During the planning of the second volume of his textbook Weber was in Göttingen. Klein was also in Göttingen at the time and his influence is reflected in the content[235] of that volume. Indeed, in addition to the introduction of the abstract group concept, Weber here sets forth the theory of groups of isometries, the theory of groups of linear substitutions, and—following Klein—the correspondence between group theory and the theory of forms as well as the theory of equations of degrees 5 and 7 (III.3).

In Weber's words, the second volume of his *Lehrbuch* is to

> ** present the general theory of finite groups, the theory of linear substitution groups, and applications to a variety of specific problems, and is to close with the theory of algebraic numbers, where an effort is made to unify the various viewpoints from which the theory has been considered until now. [721. p. vi]

---

* Es war meine Absicht, ein Lehrbuch zu geben, das, ohne viel Vorkenntnisse vorauszusetzen, den Leser in die moderne Algebra einführen und auch zu den höheren und schwierigeren Partien hinführen sollte, in denen das Interesse an dem Gegenstande erst recht lebendig wird. Dabei sollten die erforderlichen Hülfsmittel, die elementaren sowohl als die höheren, aus dem Gange der Entwickelung selbst abgeleitet werden, um die Darstellung von anderen Lehrbüchern möglichst unabhängig zu machen.

Zwei Dinge sind es, die für die neueste Entwickelung der Algebra ganz besonders von Bedeutung geworden sind; das ist auf der einen Seite die immer mehr zur Herrschaft gelangende Gruppentheorie, deren ordnender und klärender Einfluss überall zu spüren ist, und sodann das Eingreifen der Zahlentheorie.

(** see p. 251)

Actually, the second volume of the *Lehrbuch* opens, with the definition of the "general group":

* In connection with permutations, we were introduced in the first volume to the concept of a group and we used it for significant algebraic applications. Our next task will be to formulate this concept, so important in all of newer mathematics, in a more general manner and to learn the relevant laws. [722, p. 1]

The definition of an abstract group is very similar to the definition employed by Weber in his paper [720] of 1893. It starts with a "system of arbitrary [elements]" ("System [von Elementen] irgend welcher Art") [722, p. 3] and involves the four familiar postulates stated in a manner close to that used by Weber two years before.

Weber's *Lehrbuch der Algebra* was distinguished by its excellent presentation of the material and was reprinted a number of times. One should mention in particular Weber's considerable influence on group-theoretic *terminology*, the result of his effort to develop a unified system of concepts and to carry it through various areas of algebra. For example, the German "Normalteiler" (normal subgroup) goes back to Weber [721, p. 511] and is possibly linked to Dedekind's term "Teiler" (divisor), which was employed in ideal theory.

4 *The First Monographs on Abstract Group Theory*

In the eighties and nineties, the unmistakable incentive for the transition to the abstract group concept based on axioms was the variety of applications of group theory and the resulting disclosure of the vast reach of the group concept as a central idea of all mathematics. The transition to the abstract group concept was also the partial cause, as well as the partial effect, of the growing general acceptance of the "axiomatic method" in Hilbert's sense of the term.

It is safe to say that the evolution of the abstract group *concept* came to an end in the eighties. But this was not accompanied by a general acceptance of the associated method of presentation in papers, textbooks, monographs and lectures (III.3). Group-theoretic monographs based on the abstract group concept did

---

(** see p. 250)
die allgemeine Theorie der endlichen Gruppen, die Theorie der linearen Substitutionsgruppen und Anwendungen auf verschiedene einzelne Probleme bringen, und soll abschliessen mit der Theorie der algebraischen Zahlen, wo der Versuch gemacht ist, die verschiedenen Gesichtspunkte, unter denen diese Theorie bisher betrachtet worden ist, zu vereinigen.

* Wir haben im ersten Bande bei den Permutationen den Begriff einer Gruppe kennen gelernt und wichtige algebraische Anwendungen von ihm gemacht. Es muss nun unsere nächste Aufgabe sein, diesen in der ganzen neueren Mathematik so überaus wichtigen Begriff allgemeiner zu fassen und die dabei herrschenden Gesetze kennen zu lernen.

not appear until the beginning of the twentieth century. Their appearance marked the birth of abstract group *theory*.

The earliest monograph on group theory was due to the French scientist J.-A. de Séguier (1862–1937). De Séguier managed to isolate himself from the "concrete" tradition, which, at the time, was nourished in France by steady production in the areas of permutation groups and transformation groups. His *Eléments de la théorie des groupes abstrait* [605] of 1904 represents a conscious attempt to make the detailed abstract group concept of Dedekind, Frobenius, Stickelberger, v. Dyck, Hölder, Moore, Miller, and Weber the basis of a comprehensive account of group theory. De Séguier also used the then very popular fundamental textbooks on the theory of permutation groups, such as Jordan's *Traité* and Burnside's *Theory of Groups*,[236] but he presented the results on permutation groups as concrete applications of abstract group theory obtained by suitable specialization. (Obvious exceptions are concepts defined for permutation groups alone).

The introduction to his book characterizes de Séguier's view of group theory:

* The idea of the abstract group, that is to say, the group considered by itself, independently of its elements, had necessarily to result from the different specific groups encountered in algebra, analysis, and geometry. Many past investigations from different areas have been combined into a more general theory that has not ceased to develop. [605, p. I]

There is originality in de Séguier's conception of an abstract group and in his construction of an algebraic monograph. This manifests itself at the very beginning of his book, in the famous set-theoretic foundation based on the work of G. Cantor. The independence of the usual four group postulates is demonstrated by means of counterexamples. The *Eléments* sets forth the abstract definition of a group and its consequences (chapter I), theorems on subgroups, normal divisors, isomorphisms, homomorphisms and automorphisms, as well as theorems on factor groups and the Jordan-Hölder theorem (chapter II), the theory of abelian groups including the basis theorem and the theory of Hamiltonian groups (chapter III), and, finally, the theory of p-groups (chapter IV). "Concrete" groups, some related to Jordan's groups of motions, are investigated in special notes in the appendix.

De Séguier's book was devoted largely to finite groups. The first abstractly based group-theoretic monograph that dealt with infinite groups first, treating finite groups as special cases relegated to special chapters, was the work of O. Ju.

* Des divers groupes particuliers rencontrés en Algèbre, en Analyse et en Géométrie devait nécessairement se dégager l'idée du groupe abstrait, c'est-à-dire du groupe considéré en lui-même, indépendamment de la nature de ses éléments. Beaucoup de recherches déjà faites dans divers domaines vinrent dès lors se fondre en une théorie plus générale qui depuis n'a cessé de se développer.

Schmidt (1861–1956), founder of the Russian-Soviet school of group theory. His *Abstract Group Theory* (in Russian) [644] was published[237] in 1916 in Kiev. Through his algebraic papers published at the beginning of the twentieth century, Schmidt's teacher D. A. Grave (1863–1939) maintained close contact with the French school of mathematics. Grave published (in Russian) textbooks called, respectively, *Theory of Finite Groups* (1908) [254] and *Elements of Higher Algebra* (1914) [255]. These were the first comprehensive Russian accounts of the main areas of classical algebra, including the Galois theory of equations. Part II, chapter 1, of Grave's *Theory* contains the elements of the theory of "general," that is, abstract, groups including the axiomatic definition, and chapters on polyhedral groups and groups of linear substitutions as well as other material.

Schmidt's *Abstract Group Theory* of 1916 begins with the abstract group concept and devotes the first four chapters to group properties common to finite and infinite groups. Discussion of finite groups is postponed until chapter V. The book contains the first Russian account of the theory of group characters. Many of the proofs are due to Schmidt himself, including the partial proofs of Remak's theorem [578] on the decomposition of finite groups into a direct product of irreducible factors, proofs that he obtained [642] in 1912. In 1928 Schmidt extended these results to infinite groups with the finite chain condition [645].

5 *Group Theory around 1920*

The following list of theorems and concepts suggest the "inventory" of group theory around 1920: the basis theorem for finite abelian groups, Sylow's theorem, the Jordan-Hölder theorem; direct product, subgroup and normal divisor, isomorphism, homomorphism, automorphism. Group theory at this time also included extensive investigations of permutation groups and of groups of linear homogeneous substitutions, in particular of representation theory.

Already at the beginning of the twentieth century we encounter applications of group theory to physics (mechanics, crystallography, and relativity). Nevertheless these were merely hints of the decisive application of group-theoretic methods in modern physics.

The investigations of group axiomatics,[238] in particular the studies of "weakest" systems of group axioms, date back to the first years of the twentieth century. Viewed in historical terms, these beginnings represent only the germ of the subsequent development of group axiomatics and of the generalizations[239] of the group concept.

In spite of the many approaches that hindsight shows to have contained the seeds of later developments, the year 1910 marks a temporary pause in the development of group theory, and thus represents a natural boundary for the investigation undertaken in this book.

This pause in the development of group theory was due to the historically conditioned distinction—artificial from the point of view of the theory itself—between finite and infinite groups. At the beginning of the twentieth century, in spite of the full development of an abstract group concept that included both cases, there was as yet no organic connection between them. (A single exception was O. Ju. Schmidt's *Abstract Group Theory*.) Although the question of finiteness conditions was brought up as early as 1900,[240] the systematic study[241] of such conditions was undertaken only in the late thirties.

The amalgamation of the theories of finite and infinite groups into so-called "general" group theory in the late thirties had a twofold effect. On the one hand, it focused attention on classes of groups with special properties—including finiteness—and, on the other hand, it opened the way to stubborn problems in the theory of finite groups and in the process suggested a great many decisive new questions and methods.

This development of group theory was also part—both cause and effect—of a process of restratification within algebra. The universal acceptance of the axiomatic method and of the set-theoretic foundation of mathematics brought about, in the thirties, the transition from "modern algebra" in the sense of van der Waerden to a theory of mathematical structure in which a group is just an instance of an algebraic structure with a binary operation.

Epilogue

The history of the evolution of the abstract group concept is the history of the earliest case of the elaboration and conscious study of an axiomatic algebraic structure.

Obviously such a statement could only be made in the middle of the twentieth century, when the study of mathematical structures is recognized and stressed as a fruitful direction of research. But the emphasis on structure must not be regarded as the characteristic difference between nineteenth- and twentieth-century mathematics. That the genesis of the abstract group concept was completed in the nineteenth century points to a fact that is sometimes overlooked today, namely, that the modern "mathematics of structures" belongs, not only in origin but even effectively to the nineteenth century. What is more, the research directions and methods of "classical mathematics"—the mathematics of numbers, functions and geometric quantities—obviously continue to be developed in the twentieth century. The assertion that the objects of investigation of "classical mathematics" are essentially just models of the abstract structures also fails to provide a basis for the distinction between the mathematics of the nineteenth and twentieth centuries. It is conceivable that the essential difference should perhaps be sought in the direction of the new social function of the mathematical sciences—a suggestion I shall not go into here.

It will be necessary to discuss another question bearing on the history of nineteenth-century mathematics, one directly connected with the evolution of the group concept, namely, the historical role and the inner mathematical function of thinking in terms of structures. Although the question is difficult to state and its precise formulation must await the publication of similar problem-historical works, and this very fact cautions against rash generalizations, the historical facts point emphatically to the conclusion that there is, at all times, a historically determinable correspondence between the development of structural thinking and the level of maturity of the various mathematical disciplines pursued with "classical" methods toward "classical" objectives.

Group theory went through implicit stages of varied duration, in number theory, in geometry, in the theory of algebraic equations, and even in analysis. When solutions of problems were sought in the form of numbers of transformations of variables, it was quite natural to study situations and to employ methods equivalent to the treatment of concrete group-theoretic models. The attitudes of Euler, Lagrange, Ruffini, Abel, and Möbius (I.2, 3) show clearly that their primary interest was the study of the framework of a number manifold or the connections between transformations, rather than the study of structure in the true sense of the word. In this respect it is perhaps Kronecker's position that is most revealing. As a result of his contact with Galois theory, Kronecker was familiar with the permutation-theoretic group concept since the early fifties (II.4). In 1856 Kronecker, consciously following Galois, used the term "group" to denote groups of relatively prime residue classes. Nevertheless, even in 1870 he did not establish any connection between the concept of a (permutation) group, which by that time he used consistently in the theory of equations, and his implicit axiomatization of the group concept (for finite abelian groups) (II.4). This connection was established by his student Netto (III.3). It is clear that consciously following Gauss's theory of composition of forms and the work of Schering, Kronecker had essentially succeeded in describing the inner framework of the concrete material. Consequently, at that time and in that connection, he was not particularly interested in naming the axiomatically determined structure; it was essentially a matter of using axiomatization as a tool for a more convenient and understandable presentation of structured material. This is also the point of departure of the later, idealistically falsified, so-called formalist view of mathematics. On the other hand, twelve years later (1882) Netto urges that the investigation of group structure be recognized as an independent direction of research.

The new orientation in terms of objectives can already be seen clearly in the work of Gauss, Galois, and Eisenstein (I.3), all of whom wished to go beyond the computational manipulation of concrete group-theoretic collections and to penetrate their inner framework, their structure. We are indebted to Gauss for detailed anticipations, largely in number theory, of essential ideas and methods of the subsequent development of algebra; we may say with certainty that Gauss possessed significant insight into the content of future developments. In the case of Galois, the reflections on the new kind of mathematics went hand in hand with the conceptual determination of the group structure—which, until his time, had been studied only implicitly.

The evolution and specification of the conceptual content of the term "group" in the sense of a group of permutations gave rise, in the fifties and sixties of the nineteenth century, to a correspondence between the group concept and those areas of mathematics that had produced it and in which it was explicitly applied. To this incompletely evolved structural concept of "group" corresponded, logically and historically, the contemporary state of the theory of algebraic equations and the fact that this was the only area in which group theory was explicitly applied. That this assertion is not just a pleonasm is attested by Cayley's "premature" transition, in 1854 (III.4) to a group concept that was abstract in the sense of being an extension of the conceptional scope and a refinement of the conceptional content of a permutation group. In contrast to Cayley's second advance in this direction (1878) his first attempt was historically without effect. This was because in 1854 he could not possibly specify either the mathematical disciplines or the circumstances that might offer the abstract group concept genuine needs and opportunities for effective application—that is, needs and opportunities tied to the abstract formulation. In particular, such circumstances could not be inferred from the state of geometry in the middle of the nineteenth century (I.2). Cayley's own activities in geometry and invariant theory were instrumental in making geometry receptive to structural thinking and helped create the preconditions for the use of group-theoretic tools in geometry. Thus Cayley prepared the way for Klein's Erlangen Program of the early seventies by means of his contributions to geometry as well as by his extension of the group concept.

In the seventies and eighties of the nineteenth century group theory rapidly acquired essential significance in geometry, in number theory, and in various areas of analysis. The fruitfulness of group-theoretic concepts and methods then became apparent—not only through the mathematicians' growing subjective insight into their role but also because the new inner conditions of these disciplines offered objectively more favorable conditions for their application.

The extension of the explicit range of application of the group concept beyond the theory of algebraic equations resulted in modifications reflecting the requirements of these new areas of application (III.2 and III.3). A recognition of the decisive role of groups and the penetration into mathematics of the modes of thought of incipient mathematical logic (III.4) brought about the transition to the abstract group concept. That is why the process of evolution of the first algebraic structure was completed before the end of the nineteenth century.

Klein's attitude toward the abstract group concept and toward the axiomatization of mathematics in general (III.4) makes it clear that accepting the study of group structure as a research program was a genuinely forward step. This phase of the building and consolidation of abstract group theory involved considerable methodological difficulties. That the last step could be taken so early—by de Séguier in 1904—was undoubtedly due to the fact that, with continued strong productivity in the areas of permutation and transformation groups, the adoption of the abstract group concept enhanced the application of group theory to a great variety of mathematical disciplines. Thus, already at the turn of the century, the first conceptualized structure in the history of algebra played an active role in the development of mathematics.

The transition to "modern algebra" in the 1920s was accomplished against the background of the penetration of set-theoretic thinking into all of mathematics.

Abstract group theory was embedded in that process and provided a methodological model of provable historical influence. Finally, abstract group theory appears as the earliest component of the "universal algebra" of our time.

To some extent, this study of the genesis of the earliest algebraic structure implies an overall evaluation of the role of nineteenth-century mathematical thinking in terms of structures. Taking into consideration my sketch (not to say portrayal) of the genesis of other algebraic structures, and the special transparency of the first manifestation of a historical phenomenon, my overall assessment of structural thinking is that its role in the nineteenth century—measured in terms of the actual development of contemporary mathematics, that is, the dynamic of mathematical development associated with the mastery of problems—was of secondary importance in terms of its mode of origin, though anything but secondary in terms of its significance. Arising from the solution of problems in the classical mathematics of numbers, functions, and geometric quantities, structural thinking actively furthered classical mathematics in the last third of the nineteenth century even before the study of abstract algebraic structures became fashionable.

The unity of the historical and logical elements in the evaluation of the historical function of structural thinking in the nineteenth century is implied by the fact, obvious to every mathematician, that basically it is not the structure itself that is the object of investigation, but rather the structure made more specific by the impress of additional properties or the consideration of special requirements. This way of achieving inner mathematical readiness for applications selects in the course of the historical process the fruitful structures from among all those conceivable in mathematics. In this sense the group is an algebraic structure that combines singular logical transparency with genuine poverty of content. Since it lends itself to specificity and has been given specific forms, it has been, and is today, one of the most difficult and comprehensive structures.

Within the limits of my study, and to the extent to which these limits bear on the history of the structural concept of "group," the connections between structural thinking and classical mathematics are relatively clear. The very advanced systematization of algebraic structures within contemporary mathematics, that is, the existence of "universal algebra", suggests analogous studies of the genesis of other algebraic structures. But in view of the absence of methodological models and the state of modern mathematics, one is likely to encounter far greater difficulties in the study of ordered and topological structures.

Obviously the circle of problems associated with the historical relation between structural thinking and classical mathematics, problems suggested by the example of the relation between group theory and classical mathematics, is just one of the many aspects that would have to be investigated by studies of the overall nature of the development of mathematics. It seems, however, that this circle of problems may be particularly important for an evaluation of mathematical development over the last 200 years. The task of mathematical historiography is the discovery of regularities valid for all time or at least for certain periods of time and the construction of corresponding historical criteria. Such objective knowledge could then provide the basis for a new attempt to understand mathematics in its historical context.

It appears that we are still relatively far from a satisfactory solution to these problems. When it comes to the nineteenth and twentieth centuries, we have not yet even sifted or concentrated the historical material. We have no vigorous methodology of mathematical historiography determined by its aims and adequate to the comprehensive task of creating a history of mathematics as a special and extremely important form of social awareness. Nor do we as yet have a terminology that would measure up to the multifaceted nature of the problems. On the other hand, urgent demands are now being made on mathematical historiography. It is therefore important to dispense with the idea of the history of mathematics as a subject for benevolent amateurs and to direct the writing of the history of the mathematical sciences toward the study of fundamental problems. The history of the sciences, and not least the history of mathematics, is an integrating component of the varied, objectively necessary, and significant efforts to establish a science of science.

Notes

1. So explosive and thoroughgoing that nineteenth-century mathematicians felt that geometry had come to occupy a truly dominant position. Regarding the judgment of those contemporary with that development, the explanations in the introduction to the *Encyklopädie der mathematischen Wissenschaften* are particularly illuminating. While the preponderance of the other mathematical disciplines, already apparent around 1900, is noted, it is clearly regarded as upsetting the entire tradition, and an appeal is made to revive geometry and to restore it to the position of the leading mathematical discipline.

2. In order to provide at least one relatively recent example in support of this assertion, we refer the reader to the introduction of T. Reye's three-volume reference work *Geometrie der Lage*, Leipzig 1866, [581]. This work went through several editions and remained in print well into this century. It contains a systematic and comprehensive development of projective geometry, including the results of Steiner and von Staudt.

3. C.-L. Brianchon: *Mémoire sur les lignes de second ordre*, Paris, 1817; Pappos, *Collectiones*, 7, theorem 129.

4. G. S. Klügel, *Mathematisches Wörterbuch*, Vol. 1, Leipzig 1803. There we find the following item: "COORDINATES of a curve with simple curvature consist of *two continuous sequences of straight lines*. One of these is a sequence of straight lines lying on a given straight line that lies at or near the curve and starts from a fixed point. The other sequence is connected with the first one, proceeding from the same point under a constant (right or oblique) angle" ("COORDINATEN einer krummen Linie von einfacher Krümmung sind zwey stetige Folgen gerader Linien, von welchen die eine Folge auf einer gegebenen geraden Linie in oder neben der krummen Linie von einem gegebenen Puncte an genommen, und die andere Folge mit jener unter einem unveränderlichen Winkel, rechten oder schiefen, verbunden wird.") [406, p. 556].

5. In this connection see [562, 177].

6. According to Kant—in *Kritik der reinen Vernunft*, 2nd edition, 1787—space is neither an empirical concept nor part of what objectively underlies nature, but (like time) a form of an a priori perception. "Space is nothing other than the ... subjective condition of physical existence, and it is the only condition that makes external perception possible" ("Der Raum ist nichts anderes als nur die ... subjektive Bedingung der Sinnlichkeit, unter der allein uns äußere Anschauung möglich ist.") [342, p. 60]. Since (again in Kant's view) all mathematical judgments, and thus also geometric theorems, are synthetic a priori, and since all a priori knowledge can contain only that which the thinking subject can produce from within himself, the conclusion that must be drawn from Kant's system is that geometry in its contemporary form (that is, three-dimensional, euclidean geometry) must, in view of its very existence, be the only possible geometry. What supports this conclusion is the fact that the theorem on the angle sum in a triangle, which is mathematically equivalent to the fifth axiom of Euclid, was specifically cited by Kant as having a priori certainty. Naturally, Kant did not himself argue explicitly against noneuclidean geometries. For one thing, they did not yet exist. For another, he could not even have conceived noneuclidean geometries without contradicting himself. Klein infers specifically the three-dimensionality of space: "This perception [of space (Wussing)] must occur to us a priori, that is, before the observation of any object, and must therefore be a pure, non-empirical perception. For the geometric theorems are altogether apodictic, that is, they are linked to the awareness of their necessity, for example, space has just three dimensions; such theorems cannot be empirical or experiential judgments, nor can they be derived from them" ("Aber diese Anschauung muß a priori, d. i. vor aller Wahrnehmung eines Gegenstandes in uns angetroffen werden, mit-

hin reine, nicht empirische Anschauung sein. Denn die geometrischen Sätze sind insgesamt apodiktisch, d. i. mit dem Bewußtsein ihrer Notwendigkeit verbunden, z. B. der Raum hat nur drei Abmessungen; dergleichen Sätze aber können nicht empirische oder Erfahrungsurteile sein, noch aus ihnen geschlossen werden.") [342, p. 60]. Equally "apodictically," that is, here, in his awareness of the necessity of his perception of the concept of space and of geometry, Kant arrives at the following conclusion: "Thus ours [that is, Kant's (Wussing)] is the only explanation that makes the possibility of geometry as synthetic a priori knowledge conceivable" ("Also macht allein unsere [also Kant's (Wussing)] Erklärung die Möglichkeit der Geometrie als einer synthetischen Erkenntnis a priori begreiflich.") [342, p. 60]. It is interesting that in his precritical period Kant took into consideration entirely different geometries, and, in this connection, expressed penetrating and far reaching ideas. In the youthful paper "Gedanken von der wahren Schätzung der lebendigen Kräfte ..." (1746), he precedes the critical analysis of the debate between followers of Leibniz and Descartes on the measurement of what we now call kinetic energy with a few fundamental remarks on the interaction between soul and body motion, and goes on to dispute Leibniz's pseudoproof of the three-dimensionality of space. On the other hand, he thinks it likely that the three-dimensionality is the result of the fact that in the world around us forces are inversely proportional to the squares of distances. Kant thinks that in other worlds—to which we admittedly have no access, but whose possible existence must be conceded in view of God's omnipotence—there may operate another law of interaction of forces, and thus another geometry. He goes on to say that "a science of all such possible space forms would undoubtedly be the most sublime geometry that finite reason could pursue" ("Eine Wissenschaft von allen diesen möglichen Raumesarten wäre ohnfehlbar die höchste Geometrie, die ein endlicher Verstand unternehmen könnte.") [341, p. 23]. This shows that the intensification of the idealistic component, the transition to critical philosophy, deprived Kant of an important intellectual approach to the question of the relation between space and geometry. By the way, similar approaches can be found in the thought of D. Diderot.

7. Recent investigations [340, pp. 9–10] indicate that Lobachevski was born in 1792.

8. In 1827, in his *Der barycentrische Calcul* (2. Abschnitt), Möbius, propelled, as it were, by mathematics, arrived spontaneously at the idea of a fourth dimension, but lacked the intellectual courage to pursue it. Two oppositely congruent coplanar figures can be made to coincide only by going out of the plane. Similarly, the only motion that can make two oppositely congruent solids coincide is a half turn in four-space. Möbius goes on to say that "since such a space is inconceivable, the coincidence is, in this case, impossible" ("Da aber ein solcher Raum nicht gedacht werden kann, so ist auch die Coinzidenz in diesem Falle unmöglich.") [523, p. 172]. Thus in Möbius's case also the concepts of physical space and geometric space coincide.

9. In his *Geometrie der Lage* of 1803, Carnot states that "the geometry of position is mostly concerned with the connection between the positions of the different parts of a given figure and their relative values" [90, p. 43]. Thus for Carnot it is not yet the transformation of figures that is the main object of study but rather the modification of figures due to the transformation.

10. This name of the method applied by Carnot is due to Chasles and is found in his "Aperçu historique" of 1837.

11. See the account of P. Baltzer in Vol. 1 of the collected works of Möbius [523, pp. V–XX] and the account of C. Reinhardt, in Vol. 4 of the collected works, devoted to Möbius's *Nachlass*: "Ueber die Entstehungszeit und den Zusammenhang der wichtigsten Schriften und Abhandlungen von Möbius" [526, pp. 699–728].

12. See *Mathematisches Wörterbuch* [741, p. 715].

13. The term "analytic geometry," in its modern sense rather than in the sense of the analysis of a geometric figure in the manner of Euclid, appears for the first time in a work of S. F. Lacroix.

The term is *Géométrie analytique*, the work is *Traité du calcul différentiel et du calcul intégral*, Paris, 1797–1800, and it means the application of algebra to geometry. This development was due to the lasting influence of J.-L. Lagrange (especially to his *Mécanique analytique*), who aimed at the elimination of constructions from geometry and mechanics and their replacement with computations.

14. In their *Lehrbuch der analytischen Geometrie* of 1905, L. Heffter and C. Köhler call homogeneous rectangular coordinates "Hessian" coordinates [267, p. 201].

15. Möbius refers ([523], p. 8) in a footnote to A. G. Kästner, *Anfangsgründe der Analysis endlicher Größen*, 3rd ed., p. 248.

16. C. Reinhardt's account of Möbius's *Nachlass* implies that Möbius wrote *Der barycentrische Calcul* in 1821 during a period when he could not work at practical astronomy, a period that lasted until the completion (in 1821) of the Leipzig observatory. At that time, what was considered was just the permutation-theoretic approach to the theory of algebraic equations—a subject whose relations with geometry Möbius could not possibly have recognized at the time.

17. "In these developments I have used the barycentric calculus. This involved the use of new terms. Since both of these factors might deter some from familiarizing themselves more deeply with my studies, and since the results I found may be not without value in geometry, I thought it expedient to treat the above mentioned topics anew in the sequel using the ordinary, and thus generally understood, method of coordinates. An additional reason for my doing so is that, apparently, line-geometric studies habe been lately pursued, especially by French mathematicians, with a kind of predilection. ... To prevent this essay from taking up too much space, I have restricted myself to plane figures" ("Da ich mich aber bei diesen Entwickelungen der barycentrischen Rechnung bedient habe, und hierdurch sowie durch die neuen dabei gebrauchten Ausdrücke, mancher von einem näheren Eingehen in meine Untersuchungen zurückgehalten werden mag, gleichwohl aber die gefundenen Resultate für die Geometrie nicht ohne einigen Werth sein dürften, so hat es mir zweckmässig geschienen, die gedachten Gegenstände im Vorliegenden mittelst der gewöhnlichen Coordinatenmethode, und somit allgemein verständlich, von neuem zu behandeln, und dieses um so mehr, da die linealgeometrischen Untersuchungen in neuester Zeit, besonders von französischen Mathematikern, mit einer Art Vorliebe betrieben zu werden scheinen ... Um übrigens diesem Aufsatze einen nicht zu grossen Umfang zu geben, habe ich mich bloss auf ebene Figuren beschränkt.") [523, p. 450].

18. After sifting the Möbius *Nachlass*, C. Reinhardt was able to show that these investigations grew out of Möbius's studies of polyhedra of a more complex type. Incidentally, these studies led Möbius to the discovery of the Möbius strip (or sheet, as it was originally called), which was to become famous later. In this connection, see [524, pp. 515–559].

19. In this connection see Gauss's letter to Schumacher, dated 15 May 1843, in which Gauss, in connection with his review of Möbius's *Der barycentrische Calcul*, expresses his opinion of "new calculi in general" ("neue Calculs ueberhaupt"). Quoted in [523, p. XII].

20. In connection with the history of the origin of line geometry, see A. Clebsch's commemorative address on J. Plücker, "Zum Gedächtnis an Julius Plücker," *Göttinger Abhandlungen* **15** (1872). Reprinted in [559, pp. IX–XXXV].

21. This is not to say that Plücker was the first or the only person who, independently, arrived at the idea of line coordinates or plane coordinates. In 1827, Möbius (*Der barycentrische Calcul*, §6) showed that all planes whose distances from four base points, measured parallel to some direction, satisfy a linear equation pass through the same point. However, the concept of plane coordinates is missing. (In this connection, see also Möbius's letter to W. Fiedler, printed in Salmon-Fiedler, *Analytische Geometrie der Kegelschnitte*, Leipzig, 1860, p. 574ff.)—M. Chasles used

plane coordinates almost at the same time as Plücker and independently of him; see his letter to A. Quetelet, dated 10 December 1829 [*Corr. math. phys.* **6** (1830), 81–84, and for a relevant remark pp. 85–87]. Further, see "Aperçu historique," 2nd ed., 1875, pp. 628–639. Also, see G. Darboux: *Bull. Sc. math.* **6** (1874), 113–115, and E. Kötter [417, p. 355]. Also, the nature of the "tangential coordinates" of J. Booth (*On an Application of a New Analytic Method of the Theory of Curves and Surfaces*, Liverpool, 1843) is the same as that of plane coordinates.

22. See [744].

23. [19], for example, is also typical of the retrospective valuation of the advance in number theory due to Gauss in explicitly group-theoretic terms.

24. See [298].

25. A. Speiser, among others, has pointed out the group-theoretic core of Euler's paper [193] in the anthology *Klassische Stücke der Mathematik*, Zürich, 1927 [628], which contains a partial translation of [193].

26. The paragraph numbers are based on Euler's original paper [193]. The quoted text is taken from the partial translation in [628].

27. Euler puts it as follows: "Sic residuum $p + r$ aequivalet residuo r. ... Itaque omnia haec residua $r \pm np$ pro eodem residuo r reputantur" [194, pp. 494–495]. What is here being quite clearly prepared is the later concept of equivalence, of partition into classes, so decisively elaborated by Gauss in the *Disquisitiones arithmeticae* of 1801.

28. Euler puts it as follows: "Unde haec demonstratio magis naturalis videtur" [194, p. 510].

29. See [433].

30. The third edition (1830) of Legendre's book on number theory contains many of the results in Gauss's *Disquisitiones arithmeticae* but their style is strictly that of Legendre. For this edition Legendre changed the title to *Théorie des nombres*. Regarding the method, Lagrange wrote about his relation to Gauss in the introduction to the third edition: "One would have liked to enrich this Essay with more of the excellent material which makes up the work of *Gauss*. However, the methods of this author are so unique, that one could not profit from his discoveries without making large detours and becoming a mere translator" ("On aurait désiré enricher cet Essai d'un plus grand nombre des excellents matériaux qui composent l'ouvrage de Mr. *Gauss*: mais les méthodes de cet auteur lui sont tellement particulières qu'on n'aurait pu, sans des circuits très-étendus, et sans s'assujetir au simple rôle de traducteur, profiter de ses autres découvertes.") (quoted in [454, p. 5]).

31. Section 3, "De residuis potestatum," articles 45–93. The quotations in the sequel are taken from H. Maser's German edition [237] of the *Disquisitiones arithmeticae*.

32. We refer, in particular, to G. J. Rieger, "Die Zahlentheorie bei C. F. Gauss" [238, pp. 37–77], and R. Kochendörffer, "Gauss' algebraische Arbeiten" [238, pp. 79–91]. Kochendörffer's paper contains a special section entitled "Ansätze zu modernen algebraischen Begriffsbildungen." Note also the small item [176] by J. Dieudonné dealing with Gauss's scientific contributions and the collection "Materialien zu einer wissenschaftlichen Biographie von C. F. Gauss." In the latter, note, in particular, [19] and [498].

33. Article 343 is entitled "The totality of roots Ω is subdivided into certain classes (periods)" ("Sämtliche Wurzeln Ω werden in gewisse Klassen (Perioden) eingeteilt").

34. "To simplify printing" ("zur Erleichterung des Druckes") Gauss writes [1], [2], ... for r, r^2, ..., that is, quite generally, $[\sigma]$ for r^σ. Using this notation, Gauss states that "This readily implies that the roots $[1], [g], [g^2], \ldots, [g^{n-2}]$ coincide with Ω. A similar argument shows, more generally, that if λ is a whole number not divisible by n, then $[\lambda], [\lambda g], [\lambda g^2], \ldots, [\lambda g^{n-2}], \ldots$ coincide

with Ω. Also, since $g^{n-1} \equiv 1 \pmod{n}$, it is easy to see that two roots $[\lambda g^{\mu}]$ and $[\lambda g^{\nu}]$ are the same or different according as μ, ν are congruent or incongruent modulo $n - 1$" ("Hieraus folgt unmittelbar, dass die Wurzeln [1], $[g]$, $[g^2]$, ..., $[g^{n-2}]$ mit Ω zusammenfallen, und ganz ebenso werden allgemeiner $[\lambda]$, $[\lambda g]$, $[\lambda g^2]$, ..., $[\lambda g^{n-2}]$ mit Ω zusammenfallen, wenn λ irgend eine ganze durch n nicht teilbare Zahl bezeichnet. Ferner sieht man leicht, da $g^{n-1} \equiv 1$ (mod. n) ist, dass die beiden Wurzeln $[\lambda g^{\mu}]$, $[\lambda g^{\nu}]$ identisch oder verschieden sind, je nachdem μ, ν nach dem Modul $n - 1$ congruent oder incongruent sind.") [237, p. 405].

35. Using his notation, Gauss writes

$$[\lambda] + [\lambda h] + [\lambda h^2] + \cdots + [\lambda h^{f-1}].$$

36. That is, [1], in Gauss's notation.

37. See F. Klein's obituary [394] of E. Schering.

38. The lectures of Kummer and Kronecker were given on 1 December 1870. In the *Monatsberichte*, Kronecker's paper begins with this sentence: "Herr Kronecker appended to Herr Kummer's lecture the following explanation of certain properties of the class number of ideal complex numbers" ("Herr Kronecker knüpfte an den Vortrag des Hrn. Kummer die folgende Auseinandersetzung einiger Eigenschaften der Klassenanzahl idealer complexer Zahlen."). When it was reprinted in Kronecker's *Werke*, Vol. 1, Leipzig, 1895, this sentence was relegated to a footnote.

39. We do not propose to go into the philosophical debate that flared up in this connection beyond noting that Kronecker defended the "arithmetical and formalist" position, and that it foreshadowed the debates about set theory and intuitionism during the so-called foundations crisis in mathematics at the turn of the century. In this connection, see, for example, P. du Bois-Reymond, *Allgemeine Functionentheorie*, Tübingen, 1882 [46]. In this work, Du Bois-Reymond rejects the "arithmetization" of mathematics—a term coined by Kronecker, which later became the watchword of his followers—as formalist: It reduced mathematics to mere playing with symbols. See also Klein's 1895 lecture "Über die Arithmetisierung der Mathematik" [392].

40. The reference is to E. Schering's paper [637].

41. At this point Kronecker remarks, "Instead of multiplication we can also use addition, which *Gauss* preferred, for obvious reasons, when he introduced a symbolism for the composition of quadratic forms" ("Anstatt der Multiplikation kann auch die Addition gebraucht werden, welcher *Gauss* bei Einführung einer Symbolik für die Composition der quadratischen Formen aus leicht erkennbaren Gründen den Vorzug gegeben hat.") [432, p. 275].

42. Kronecker took over the phrase "belonging to the exponent e" ("zum Exponenten e gehörend") from Gauss, *Disquisitiones arithmeticae*, article 52 [237, p. 35].

43. Since the meaning of "permutation" fluctuates even today (see *Mathematisches Wörterbuch* [741]), we note that, unless otherwise stated, the term, as used here, refers to finite permutations and denotes the operation as well as its outcome. Also, apart from quotations, "permutation" is used even where an author may have used "substitution."

44. Lagrange almost always prefaced his papers with historical notes. These are gems of problem-historical mathematical literature.

45. "Algebraic solution of equations" ("algebraische Auflösung von Gleichungen") is the term used at that time for the solution of equations by radicals.

46. Gauss was one of the first to spell out the unsolvability by radicals of equations of degree higher than four. He did so when he realized that the problem was reducible to pure equations. He did not manage to produce a proof. In the section on cyclotomy, article 359 of the *Disquisitiones arithmeticae* (1801), Gauss states, "It is well known that, so far, the efforts of the most eminent geometers [that is, mathematicians (Translator)] to solve equations of degree higher than the

fourth, or (to define more precisely what one wants to do) to reduce mixed equations to pure equations, have all been in vain. There is hardly any doubt that this problem not only exceeds the powers of modern analysis, but also represents an attempt to achieve the impossible" ("Bekanntlich sind alle Bemühungen der grössten Geometer, die allgemeine Auflösung der Gleichungen welche den vierten Grad übersteigen, oder (um genauer zu definieren, was man will) die **Reduction der gemischten Gleichungen auf reine Gleichungen** zu finden, bisher stets vergeblich gewesen, und es bleibt kaum zweifelhaft, dass dieses Problem nicht sowohl die Kräfte der heutigen Analysis übersteigt, als vielmehr etwas Unmögliches erreichen will.") [237, p. 433]. Gauss had made a similar pronouncement earlier (1799) in article 9 of his dissertation "Demonstratio nova theorematis omnem functionem algebraicam rationalem integram. ..."

47. In 1808, Lagrange published another large paper on the solution of algebraic equations: *Traité de la résolution des équations numériques de tous les degrés*, Paris, 1808. Reprinted in *Oeuvres de Lagrange*, publiées par les soins de M. J.-A. Serret, Tome huitième, Paris 1879, pp. 11–131. See note 51.

48. The interpretation of the term "resolvent" advocated here by Lagrange is noteworthy in that it signified his disregard of the then prevailing view that the resolvent must be "simpler" than the given equation. In other words, Lagrange admitted the possibility of the degree of the resolvent being consistently higher than the degree of the initial equation. In Lagrange's case, what matters is that "it [the resolvent (Wussing)] can be solved like a quadratic equation" ("qu'elle peut se résoudre comme celles du second degré") [446, p. 208].

49. Lagrange writes, "The problem of the solution of equations of degree higher than the fourth is one of those which have not yet been solved; on the other hand, nothing about it indicates that its solution is impossible" ("Le Problème de la résolution des équations des degrés supérieurs au quatrième est un de ceux dont on n'a pas encore pu venir à bout, quoique d'ailleurs rien n'en démontre l'impossibilité.") [446, p. 305].

50. In a lecture given in the eighties, Kronecker stated that "Vandermonde's paper, presented to the Paris Academy in 1770 ... started a new flourishing of algebra. The conceptual depth so clearly expressed in this paper is downright astonishing" ("Mit Vandermonde's im Jahre 1770 der Pariser Academie vorgelegten Abhandlung ... beginnt der neue Aufschwung der Algebra; die Tiefe der Auffassung, welche sich in dieser Arbeit in so klaren Worten ausspricht, erregt geradezu unser Erstaunen."). Kronecker's statement was handed down by C. Itzigsohn, who published a German translation of Vandermonde's papers on the solvability theory in 1888 [702, introduction by C. Itzigsohn].

51. Lagrange did not return to the investigation of algebraic equations with "the aid of the combinatorial calculus." The second edition of his *Traité de la résolution des équation numériques de tous les degrés* [449], dated 1808, includes 14 notes. Note XIII contains a summary of the "Réflexions" whose only new feature is the modernized treatment of the roots of unity, inspired by Gauss's *Disquisitiones arithmeticae*, which had appeared in the meantime. At the end of note XIII, Lagrange, now familiar with Vandermonde's "beautiful paper" ("le beau travail"), explains briefly its objective and arrives at the following conclusion: „Since the method of Vandermonde derives from a principle based on the nature of equations, and is thus more direct than that presented in this Note, one can regard the common results of both methods as necessary consequences of the general theory of equations" ("Comme la méthode de *Vandermonde* découle d'un principe fondé sur la nature des équations, et qu'à cet égard elle est plus directe que celle que nous avons exposée dans cette Note, on peut regarder les résultats communs de ces méthodes sur la résolution générale des équations qui passent la quatrième degré, comme des consequences nécessaires de la théorie générale des équations.") [449, p. 274].

52. It is very surprising that, so far, this fact has been completely overlooked even by H. Burkhardt [69].

53. That is why the content of Lagrange's "Réflexions," and especially his procedure for resolvent construction—modified by the use of the subsequently developed calculus of permutations—appears in many algebra textbooks until the middle of the nineteenth century. For example, the first edition of J.-A. Serret's *Cours d'Algèbre supérieure* (1849), the leading textbook of its time, is one of the books containing unaltered the complete content of the third section of the "Réflexions."

54. Ruffini's other papers deal, among other things, with the alleged method, due to the Polish mathematician H. Wronski [742], of solution by radicals of equations of arbitrary degree [601], with the numerical solution of equations [595, 600], and with the roots of the cyclotomic equation [602]. In this connection see also [85].

55. In contrast to J. Hecker's doctoral dissertation [265] of 1886, which considers only Ruffini's last paper of 1813 and thus ignores his deeper results.

56. In this connection see also the introduction and notes of E. Bortolotti, the editor of Ruffini's mathematical works [590, 591], which contain a great many bibliographic references to the secondary literature on Ruffini and items pertaining to the reactions inspired by Ruffini.

57. The concept "permutazione" is used only implicitly in the *Teoria* of 1799. The explicit definition of a group by means of the term "permutazione" was first given by Abbati [1, p. 471] and after that was also used by Ruffini.

58. According to Burkhardt [69, p. 134], all that is missing is the subgroup S_5, generated by the cycles (12345) and (132).

59. See note 65.

60. Of special interest is the introduction in which Cauchy describes the reasons and scope of his investigations and devotes special attention to the theory of quadratic forms (I.3): "I shall now examine with special care a certain variety of alternating symmetric functions that turn up by themselves in very many analytic investigations. It is by means of these functions that one expresses the general values of the unknows involved in several equations of the first degree. They also turn up whenever one forms conditional equations, for example, in the general theory of elimination. *Laplace* and *Vandermonde* have considered them from this point of view in the Memoirs of the Academy of Sciences (1772), and *Bezout* has since reexamined them from the same point of view in his Theory of equations. *Gauss* has used them to advantage in his analytic investigations to discover the general properties of quadratic forms, that is, quadratic polynomials in two or more variables. He called these functions *discriminants* (*déterminans*). I shall retain this name, which gives me a simple means of stating the results. I shall only note that one sometimes calls the functions of which we are speaking the *resultants* (*résultantes*) of two or more letters" ("Je vais maintenant examiner particulièrement une certaine espèce de fonctions symétriques alternées qui s'offrent d'elles-mêmes dans un grand nombre de recherches analytiques. C'est au moyen de ces fonctions qu'on exprime les valeurs générales des inconnues que renferment plusieurs équations du premier degré. Elles se représentent toutes les fois qu'on a des équations de condition à former, ainsi que dans la théorie générale de l'élimination. MM. *Laplace* et *Vandermonde* les ont considerées sous ce rapport dans les Mémoires de l'Académie des sciences (année 1772), et M. *Bézout* les a encore examinées depuis sous le même point de vue dans sa Théorie des équations. M. *Gauss* s'en est servi avec avantage dans ses Recherches analytiques, pour découvrir les propriétés générales des formes du second degré, c'est-à-dire, des polynômes du second degré à deux ou à plusieurs variables; et il a désigné ces mêmes fonctions sous le nom de *déterminans*. Je conserverai cette dénomination qui fournit un moyen facile d'énoncer les résultats; j'observerai seulement qu'on donne aussi quelquefois aux fonctions dont il s'agit le nom de *résultantes* à deux ou à plusieurs lettres.") [100, p. 51].

61. For the second time in this paper, Cauchy speaks of the results of Ruffini: "These two propositions have been proved by *Paolo Ruffini* in the Memoirs of the Italian Society, Vol. XII, and in his Theory of equations. My research in number theory has led me to consider the theory of combinations and I have arrived at a proof of a more general theorem, which includes the two preceding theorems ..." ("Ces deux propositions ont été demontrées par M. *Paolo Ruffini*, dans les Mémoires de la Société italienne tome XII, et dans sa Théorie des équations. Ayant été conduit par des recherches sur les nombres, à m'occuper de la théorie des combinaisons, je suis arrivé à la démonstration d'un théorème plus général qui renferme les deux précédens ...") [99, pp. 8–9].

62. In 1845 J. Bertrand obtained a better result. Specifically, in [28, p. 123], it is shown that if $n > 4$, then there is no group whose index is >2 and $<n$. The same theorem turns up later, in the work of Cauchy in 1845 [111, p. 1101] and in the work of Serret in 1848 [611, p. 147]; and it is once more improved by Serret in 1849 [612, p. 135]. All the corresponding proofs use exclusively permutation theory. Proofs based on the simplicity of the alternating group A_n for $n > 4$ were given by J. König in 1879 [415, p. 215] and by L. Kronecker in 1879 [434, p. 211].

63. In any case, the author cannot agree with R. Taton when the latter describes Cauchy's 1844 exposition of permutation theory as a transition to *abstract* group theory [286, p. 13]. What is involved at all times are systems of conjugate permutations, and thus permutation groups.

64. As well, Galois considered "arithmetic substitutions" before Cauchy, and went further than Cauchy. What is involved is the following. One starts with a permutation of order n on n letters. Its powers form a cyclic group of order n. Its elements can be represented by means of the congruences

$$x' \equiv x + c \pmod{n}.$$

Similarly—and this had been noted by Galois—the p^μ elements, represented by the congruences

$$x_1' \equiv x_1 + c_1, \quad x_2' \equiv x_2 + c_2, \quad \ldots, \quad x_\mu' \equiv x_\mu + c_\mu$$

with respect to a prime p form a group of order p^μ.

65. The papers involved are [108] and [110–128].

66. See note 62.

67. Recently, L. Nový considered Cauchy's role in the formation of an independent theory of groups [543]. Like the present author, Nový arrives at the conclusion that in Cauchy's case the link with the theory of equations is partly broken and that, as a result, permutation theory as developed by Cauchy hampered the formation of abstract group theory.

68. Cauchy's last paper [128] in the series published in 1845–46 comes closest to a group-theoretic formulation. In this paper Cauchy proves a result stated in group-theoretic terms, which contains many individual results. The theorem is stated as follows: "Let us form with n variables

$$x, y, z, \ldots,$$

two systems of conjugate substitutions, and let

$$(1) \qquad 1, P_1, P_2, \ldots, P_{a-1},$$

$$(2) \qquad 1, Q_1, Q_2, \ldots, Q_{b-1},$$

be these two systems, the first of order a and the second of order b. Further, let h and k be any two whole numbers, let I be the number of substitutions R satisfying the symbolic equations

$$(3) \qquad RP_h = Q_kR,$$

and let us put for brevity

$$N = 1 \cdot 2 \cdot 3 \ldots n.$$

Then the numbers N and I yield the same remainders upon division by the product ab" ("Formons avec n variables

$$x, y, z \ldots$$

deux systèmes de substitutions conjuguées; et soient

(1) $1, P_1, P_2, \ldots, P_{a-1},$

(2) $1, Q_1, Q_2, \ldots, Q_{b-1},$

ces deux systèmes, le premier de l'ordre a, le second de l'ordre b. Soient d'ailleurs h, k deux nombres entiers quelconques, nommons I le nombre des substitutions R pour lesquelles se vérifient des équations symboliques de la forme

(3) $RP_h = Q_k R,$

et posons, pour abréger

$$N = 1 \cdot 2 \cdot 3 \cdots n.$$

Les nombres N et I fourniront le même reste lorsqu'on les divisera par le produit *ab*.") [128, p. 631].

69. Just as, say, Abel, so too Cauchy uses the term "le groupe" to denote a system, a complex of permutations, or other elements. This occurs, for example, in [99, p. 21].

70. In particular, see the articles by G. J. Rieger, "Die Zahlentheorie bei C. F. Gauss" [238, pp. 37–77] and R. Kochendörffer, "Gauss' algebraische Arbeiten" [238, pp. 79–91].

71. See note 84.

72. See note 46.

73. For example, there is a direct causal connection between the letter to Hansteen, dated 29 March 1826 (which Abel wrote in Dresden on his way through that city) [3, pp. 263–265], in which Abel develops his views on the progress of analysis, and especially on the study of convergence of series, and the paper [8] on the binomial series, which is of fundamental significance for increasing the precision of the concept of convergence of an infinite series. This paper served as a test piece for the clarification of his public views. Abel was aware that his insistence on precision, on the rejection of uncontrollable intuition and on existence proofs, placed him in the same category as B. Bolzano (1781–1848) (Abel was among the first to pay attention to Bolzano) and Cauchy. That is why Abel always spoke favorably of Cauchy—this in spite of the bitterness engendered by his long, and ultimately fruitless, waiting for the (for him) extremely important expert opinion of his ambitious paper on elliptic functions that Cauchy was supposed to referee. Abel said of Cauchy that "at the moment he is the one who knows best how to do mathematics" (Cauchy sei "derjenige, der gegenwärtig am besten wisse, wie man Mathematik macht.") quoted after [545, p. 147].

74. In a letter to Holmboe, dated 24 October 1826, Abel referred to himself as the author. In that letter he says, among other things, that "an excerpt of my paper on the impossibility of the solution of algebraic equations was inserted in the bulletin of M. *Férussac*. I did that myself" ("Un extrait de mon mémoire sur l'impossibilité de résoudre équations algébriques a été inséré dans le bulletin de M. Férussac. Je l'ai fait moi-même.") [3, p. 260].

75. The weaknesses of Ruffini's proof—which are not group-theoretic in nature—have been analyzed in detail by L. Sylow and S. Lie in their notes to Abel's works [3, p. 281ff], and by H. Burkhardt in the paper [69], from which we quoted extensively. Since Ruffini did not yet know the field concept, he was not in a position to work out the concepts of the base field determined by the coefficients of the given equation and of the successive extension of a field by adjunction, or to determine the exact meaning of "reducible" or "irreducible" with reference to an equation or a resolvent, or of "algebraically solvable."

76. In this connection, see [3, pp. 219–220].

77. See [186].

78. As early as 16 January 1826, Abel writes to Holmboe: "Since my arrival in Berlin I have also been applying myself to the solution of the following general problem: *To find all the equations that can be solved algebraically*. I have not yet succeeded, but as far as I can judge, I will yet succeed. As long as the degree of the equation is a prime, it is not all that difficult, but when the degree is a composite number, things get devilishly complicated" ("Depuis mon arrivée à Berlin je me suis aussi occupé de la solution du problème général suivant. *Trouver toutes les équations qui sont résolubles algébriquement*. Je ne l'ai pas encore achevée, mais autant que j'en puis juger, j'y réussirai. Tant que le degré de l'équation est un nombre premier, la difficulté n'est pas si grande, mais lorsque ce nombre est composé, le diable s'en mêle.") [3, p. 256].

79. One of the direct references to Gauss runs as follows: "The method that we used before to solve the equation $\varphi x = 0$ is basically the same as that which *Gauss* had used in his *Disquisitiones arithmeticae*, articles 359 and beyond, to solve a certain class of equations that he had come across in his study of the equation $x^n - 1 = 0$. These equations have the same property as our equation $\varphi x = 0$, namely, all its roots can be represented by

$$x, \theta x, \theta^2 x, \ldots, \theta^{\mu-1} x,$$

where θx is a rational function" ("La méthode que nous avons suivie précédemment pour résoudre l'équation $\varphi x = 0$ est au fond la même que celle dont s'est servi M. *Gauss* dans ses 'Disquisitiones arithmeticae' art. 359 et suiv. pur résoudre une certaine classe d'équations, auxquelles il était parvenu dans ses recherches sur l'équation $x^n - 1 = 0$. Ces équations ont la même propriété que notre équation $\varphi x = 0$; savoir que toutes ses racines peuvent être représentées par

$$x, \theta x, \theta^2 x, \ldots, \theta^{\mu-1} x,$$

θx étant une fonction rationelle.") [2, p. 491].

80. This makes Galois's methodological remarks decisive for the analysis of his work. It is very surprising that this viewpoint has never been elaborated in the many works on Galois, including those of G. Sarton [604] and L. Infeld [301].

81. A similar document is furnished by a passage from the autobiography of the 20-year-old G. Eisenstein (written in connection with his request for admission to the *Abitur* examination), in which he describes a new kind of mathematics. The passage reads as follows: "In contrast to the older school, the fundamental principle of the new school, founded by Gauss, Jacobi and Dirichlet, is this. Where the old school sought to attain its ends through long and involved calculation and deductions (as is the case even in Gauss's *Disquisitiones*), the new avoids it by the use of a brilliant expedient; it comprehends a whole area in a single main idea, and in one stroke presents the final result with utmost elegance. Where the former, advancing from theorem to theorem, finally reaches fruitful ground after many steps, the latter sets down from the very beginning a formula containing in concentrated form the full circle of truths of a whole area, which need only be read off and spelled out. Following the old method, one could, in a pinch, prove the theorems; now, however, one can see the true nature of the whole theory, the essential inner machinery and wheel-work" ("Das wesentliche Princip der neueren mathematischen Schule, die durch Gauss, Jacobi und Dirichlet begründet ist, ist im Gegensatz mit der älteren, dass während jene ältere durch langwierige und verwickelte Rechnung (wie selbst noch in Gauss' Disquisitiones) und Deduktionen zum Zweck zu gelangen suchte, diese mit Vermeidung derselben durch Anwendung eines genialen Mittels in einer Hauptidee die Gesammtheit eines ganzen Gebietes umfasst und gleichsam durch einen einzigen Schlag das Endresultat in der höchsten Eleganz darstellt. Während jene, von Satz zu Satz fortschreitend, nach einer langen Reihe endlich zu einigem fruchtbaren Boden gelangt, stellt diese gleich von vorn herein eine Formel hin, in welcher der vollständige Kreis der Wahrheiten eines ganzen Gebietes konzentriert enthalten ist und nur herausgelesen und ausgesprochen zu werden darf. Auf die frühere Art konnte man die Sätze zwar auch

zur Not beweisen, aber jetzt sieht man erst das wahre Wesen der ganzen Theorie, das eigentliche innere Getriebe und Räderwerk.") [589, pp. 158–159].

82. This was essentially also Gauss's position. In this connection, see Gauss's opinion of the uses of new calculi and, in particular, of Möbius's barycentric calculus. See also note 19.

83. The *Discours préliminaire* was first published in 1908, in [222]. In this connection see pp. 21 and 16.

84. In a manuscript of 1797 or 1798, Gauss considered the congruence $f(x) = x^n + a_1x^{n-1} + \cdots + a_n \equiv 0 \pmod p$ and suggested that if the congruence has factors of degree higher than the first, then one should admit as roots not only entire rational numbers but also "as it were, imaginary" ("gleichsam imaginäre") roots. Gauss did not come back to this issue. In this connection see [238, pp. 90–91; 234, pp. 212–240].

85. Gauss writes $Fx, \varphi x, \psi x$, and χx, that is, he does not put the argument in parentheses.

86. See note 64.

87. In [221] we have b instead of β.

88. Thus the claim that Galois first presented the Galois theory in his letter to Auguste Chevalier, dated 29 May 1832, is superficial. (This claim is made, for example, in the mathematical dictionary published by J. Naas and H. L. Schmid, Leipzig 1961 [740].)

89. However, equipped with the perfected method, Galois returned once more to primitive equations; his results are contained in the paper "Des équations primitives qui sont solubles par radicaux" [231], based on the Galois *Nachlass* and published by Liouville in 1846.

90. On the other hand, Auguste Chevalier insisted emphatically that the following paper was not an abstract: "This manuscript is precisely the one that the author presented to the Academy" ("Ce manuscrit est précisement celui que l'auteur présenta à l'Académie.") [222, p. 7].

91. In spite of the fact that this theorem fitted the traditional pattern from the point of view of content and customary terminology, it was thought to be very deep. One indication of this is that the referee (Poisson) appointed by the Academy set down at this point the marginal note: "The proof of this lemma is inadequate but the lemma is true according to number 1000 of Lagrange's paper, Berlin 1775" ("La démonstration de ce lemme n'est pas suffisante; mais il est vrai, d'après le n^0 1000 du Mémoire de Lagrange, Berlin 1775."). Galois's retort was, "... We have transcribed word for word the proof of this lemma that we gave in a paper presented in 1830. We add to it, as an historical document, the note that M. Poisson felt obliged to append. One can then judge" ("Nous avons transcrit textuellement la démonstration que nous avons donnée de ce lemme dans un Mémoire présenté en 1830. Nous y joignons comme document historique la note suivante qu'a cru devoir y apposer M. Poisson. On jugera.") [222, p. 7].

92. Here Galois notes that "invariant" ("invariable") refers to the constancy of the form as well as to the numerical value of the function of the roots.

93. As the realization of Galois's significance grew, so did the number of mathematical-historical works and biographies devoted to Galois. The opinion of Galois became ever more favorable. In this connection see the Galois biographies [155, 499, 587, 490, 178, 558, 29, 704, 25, 604, 45, 418, 301, 161] and the bibliographies in [301, pp. 328–33] and [604, p. 259]. In addition, there are detailed studies devoted to the circumstances surrounding the court proceedings, the police files of the revolutions of 1830 and 1832, and so on. See once more the bibliography [301, pp. 328 to 331].

94. We find in the same issue a sketchy obituary, written by Chevalier, which contains next to no material of mathematical-historical interest. On the other hand, it contains a description of the circumstances of Galois's death as reflected by Chevalier. The publishers of the *Revue*

encyclopédique had added to Galois's letter to Chevalier the remark that the manuscripts left by Galois would soon be published. But this did not happen. In the meantime, Chevalier made every effort faithfully to fulfil Galois's legacy. Galois's brother Alfred had also greatly involved himself; among other things, a copy of a letter from him to Jacobi has been preserved. In turn—admittedly, only after 1846—Jacobi called T. Schönemann's attention to Galois.

95. J. Tannery, who in 1906 and 1907 edited the Galois manuscripts passed over by Liouville, writes in his introduction only that Liouville had obtained the manuscripts from Chevalier. As for Liouville, he left his whole library and his manuscripts to one of his sons-in-law. The latter's wife took on the troublesome task of searching out the Galois manuscripts. Eventually the manuscripts ended up in the French Academy and were edited by Tannery, the then assistant head (sous-directeur) if the *Paris Ecole Normale*—the very school that had once expelled Galois for his participation in revolutionary activities. The manuscripts, supplied with commentaries and descriptions, were published in the *Bulletin des Sciences mathématiques*, 1906, pp. 226–248, 255 to 263, and 1907, pp. 275–308. They were reprinted in [222].

96. Toward the end of the century there was a increase in interest in Galois due to the recognition, by that time, of his importance, and Liouville's earlier selection appeared in book form [221] in 1897 with an introduction by Emile Picard.

97. Also [224, 228, 225–227, 232, 229, 231].

98. Serret translated Kronecker's first 1853 paper on algebraically solvable equations [424] and included it, under the title *Sur les équations résolubles* algébriquement, as a supplement in his *Cours d'Algèbre supérieure*, second edition, Paris, 1854. This translation also appeared in the later editions. At this point, there is an error in the bibliography of Kronecker's works; the first edition of the *Cours d'Algèbre supérieure* had appeared already in 1849.

99. Thus H. Weber's claim, contained in his obituary of Kronecker [719, p. 6], to the effect that Kronecker first mentioned Galois's name in writing in 1856, is incorrect.

100. The reference is to [11].

101. At the time, this theorem caused something of a sensation. It has since been proved again by, among others, H. Weber (in [715–718]), D. Hilbert (in his "Zahlbericht" [279, chapter 23], F. Mertens (in [511]) and A. Speiser (in [625]). See also K. Hensel's addendum in [422, p. 497].

102. According to E. Schering [423, p. 411, footnote], it was presented for the first time in lectures in the winter of 1861–62. See [434].

103. E. Schering points out [423, p. 412] that Kronecker wrote him letters in which he made these "viewpoints" (Gesichtspunkte) precise in the sense that, at the time—that is, early in 1856—he (Kronecker) was occupied with the elaboration of the concept of a "domain of rationality" (Rationalitäts-Bereich).

104. Nothing in this assessment is changed either by the additional papers on the theory of equations, such as [425, 427, 429, 428, 431], and on "substitution theory" (Substitutionentheorie) [430] published by Kronecker during the period under discussion (that is, between 1846 and 1870) or by his new definition of the Galois group of an equation. Moreover, the latter belongs to the next period, in which the elaborated permutation-theoretic group concept had long since become the common property of mathematicians. As regards terminology, it should be noted that since 1856 [423, p. 413]—for example, in 1858 ([422], p. 42) and in 1861 ([422], p. 61)—Kronecker, influenced by an expression of Jacobi, used the phrase that two equations have "the same affect" ("denselben Affekt" haben). To be sure, by 1861 Kronecker himself indicated that these two expressions had the same meaning: "The general equations of the same degree which have *the same* affect, or, to use *Galois's* terminology, *the same* group of the equation ..." ("Diejenigen

allgemeinen Gleichungen desselben Grades, welche *denselben* Affect oder, nach der *Galois*schen Ausdrucksweise, *dieselbe* Gruppe der Gleichung haben ...") [422, p. 61]. Of course, Kronecker had used the word "group" in the sense of a mathematical group as early as 1856 in connection with the preparation of number-theoretic tools for the study of algebraically solvable equations. In his words: "First of all, the numbers k defined ... by the congruence form a group such that the product of any two numbers contained in the latter is again congruent to a number k modulo m" ("Zuvörderst bilden nämlich die Zahlen k, wie sie durch die Congruenz ... definirt worden sind, eine Gruppe von der Beschaffenheit, daß das Produkt von je zwei in derselben enthaltenen Zahlen wiederum einer Zahl k nach dem Modul m congruent ist." [422, p. 33].

105. R. Dedekind was one of those who became familiar with Galois theory rather early. He was probably the first at any German university to lecture (in the winter of 1857–58) on higher algebra and Galois theory, apparently without elaborating its group-theoretic core. Dedekind's extremely valuable contributions to group theory, concerning largely the study of commutator groups, belong to a later period and began only in the eighties. (III.4).

106. The papers involved are mainly [32–40]. For the spread of Galois's ideas to Italy see [43].

107. What is meant here is Galois's investigations of finite commutative fields.

108. Testimonies by Serret, Betti, Kirkman, and others to the effect that they consciously embarked on the process of amalgamation of the directions represented, respectively, by Cauchy and Galois would, of course, be extremely valuable. In view of the paucity of such materials, one must be satisfied with the mathematical papers and with contemporary descriptions.

109. The differences are superficial. In the second edition this material is labeled "Introduction" and is part of the first lesson. In the third edition, it is separated from the text proper.

110. And not as late as 1866, as has been claimed until now.

111. In this connection see also note 43.

112. This concept is absent from the second edition.

113. Recall the conventions concerning the use of the terms "permutation" and "substitution" (see note 43).

114. This may be why, in the course of his investigations, Galois abandoned the initially carefully explained logical difference between "permutation" and "substitution."

115. The three dots after 1, a, a_1 are missing in the original. Also, in the original the groups are separated by commas rather than, as here, by semicolons.

116. It is not until the *Traité* that we encounter the phrase "the subgroup contained" ("le groupe partiel contenue").

117. The concept of a complex, in connection with the theory of permutation groups, is due to Frobenius [209].

118. On a number of occasions I have pointed out the important role played by commentaries on Galois's work. Further "Commentaries on Galois" were published after 1870. Of these we mention those of J. König (1849–1913) [415], P. Bachmann (1837–1920) [17], and H. Weber (1842–1913) [720], published in 1879, 1881, and 1893, respectively. What distinguishes all of these papers from Jordan's view of Galois theory is their transition to the field-theoretic approach adopted in German-speaking lands following the precedent set by Kronecker and Dedekind in the seventies and eighties. We shall return to this development (III.4) in connection with the elaboration of the abstract group concept and will, in particular, discuss the relevant papers [714–718, 726–727, 729, 723] of H. Weber. At this point we shall only quote a passage from P. Bachmann's

paper [17], which reflects the conservative position still alive in the eighties: "In spite of the fact that new editors, chief among them Herr C. Jordan [Bachmann has in mind Jordan's second Galois commentary, that is, [316] (Wussing)], have brought Galois's investigations of the algebraic solvability of equations closer to the common understanding, many may be deterred from the study of this principal part of the subject of equations by their extensive use of the very abstract substitution theory. Such persons may well welcome the present exposition of the subject which avoids the said theory as much as the nature of things allows, and bases itself essentially on the two fundamental concepts of field and irreducibility. We believe that this treatment, due, incidentally, to Herr Dedekind, gives a far more concrete view of the phases of the solution process of equations than the usual mode of presentation" ("Obwohl neuere Bearbeiter, unter ihnen besonders Herr C. Jordan die Untersuchungen von GALOIS über die algebraische Auflösbarkeit der Gleichungen dem allgemeinen Verständnisse durch eine weitere Ausführung und systematischere Entwicklung näher gebracht haben, so dürfte doch Mancher vom Studium dieses principiellen Theiles der Lehre von den Gleichungen durch den ausgedehnten Gebrauch der sehr abstracten Substitutionentheorie abgehalten worden sein, welcher von Jenen gemacht wird. Solchen wird vielleicht nachfolgende neue Darstellung des Gegenstandes nicht unwillkommen sein, welche die genannte Theorie, soweit es der Natur der Sache nach möglich scheint, vermeidet, indem sie sich im Wesentlichen nur auf die beiden fundamentalen Begriffe des Zahlenkörpers und der Irreductibilität gründet. Bei dieser Behandlung, welche übrigens schon von Herrn Dedekind angegeben worden ist, wird zugleich – meinen wir – von den Phasen des Auflösungsprocesses der Gleichungen eine viel concretere Anschauung, als bei der gebräuchlichen Darstellungsweise, gewonnen.") [17, p. 450]. In this connection I must also mention four papers, [620–623], of J. T. Söderberg (1856–1927). As well, the reader should consult O. Hölder's paper [288] and three papers, [607–609], of the Russian mathematician D. F. Selivanov (1855–1932). On the other hand, Weber's work, which I shall discuss in III.4, is based entirely on the abstract group concept. In fact, Weber advances the view [720, p. 256] that a field should be regarded as a group with a double commutative law of composition. Thus this development flows directly into the axiomatization of algebra by E. Steinitz [687]. The emphasis on the field-theoretic position gave rise to—and this is the other side of the development just sketched—other views and definitions of the Galois group of an equation; in this connection we mention Kronecker's 1882 definition (in [435]) reworked in 1888 by A. Kneser [411], and Weber's 1886 definition in terms of the substitutions of the normal field (in [715–718]). Here too the development flowed into the contemporary formulation with the aid of the abstract group concept.

119. As given by Serret, the theorem is to the following effect: "*Let T be an arbitrary substitution of order μ, formed with n letters. If we form all the substitutions S such that product STS^{-1} reduces to a power of T whose exponent is congruent modulo μ to a power of some fixed number belonging to the exponent θ with respect to the modulus μ, then these substitutions S will constitute a conjugate system of order θM. When the number μ has primitive roots, θ can be an arbitrary divisor of $\varphi(\mu)$*" ("*Une substitution quelconque T, d'ordre μ, étant formée avec n lettres, si l'on forme toutes les substitutions S telles, que le produit STS^{-1} se réduise à une puissance de T, dont l'exposant, soit congru, suivant le module μ, à une puissance d'un nombre donné e appartenant à l'exposant θ, par rapport au module μ, les substitutions S constitueront un système conjugué d'ordre θM. Lorsque le nombre μ admet des racines primitives, θ peut être un diviseur quelconque de $\varphi(\mu)$.*") [618, Vol. II, p. 263].

120. See note 62. Another relevant and important theorem asserts that if for $n > 6$ the index is $> n$, then it is at least equal to $2n$: for $n > 9$, the theorem is due to J. Bertrand [28]; for $n > 6$, without proof, to Cauchy [111]; for $n > 6$, to E. Mathieu [505]. If the index is $> 2n$, then it is $\geq \frac{1}{2}n(n-1)$: Cauchy [111], Serret [613]. Other theorems for the determination of possible indices, that is, of possible group orders, are due to C. Jordan (1868, 1869, 1871, 1872, 1873), L. Sylow (1872, 1888), E. Netto (1877, 1878, 1888), G. Frobenius (1887), A. Borchert (1889, 1892,

1897), E. Maillet (1894), and G. A. Miller (1897). In this connection see the report on the literature in [187, p. 214].

121. See note 120.

122. See note 64.

123. There are $p(p-1)$ entire linear functions of the form $f + \Delta z$, f constant, and $p^2(p-1)$ fractional linear functions of the form $f - \Delta/(z+g)$, f, g constant.

124. This notation is due to Serret.

125. In this way, the results of Galois and Cauchy (see note 64) can be embedded in a general context.

126. Serret writes that the use of ∞ obeys the rules of ordinary algebra: If for some value of z the denominator of θz is congruent to 0 mod p, then, says Serret, the function takes on the value ∞. This convention, Serret continues, may be followed provided it does not conflict with one of the fundamental theorems of the theory of congruences.

127. The most important of the many papers on permutation groups and their applications are [305, 307–309, 311, 312, 314, 315].

128. This point of view is represented by a few more articles in the *Encyclopädie der mathematischen Wissenschaften* written some thirty years after the *Traité*, for example, in the volume *Geometrie*, first part, first half, Leipzig 1907–1910, section 5a: Konfigurationen der projectiven Geometrie [189, pp. 481–514], completed by E. Steinitz in April 1910. On the other hand, the article "Kontinuierliche geometrische Gruppen. Die Gruppentheorie als geometrisches Einteilungsprinzip" [189, pp. 291–386], completed by G. Fano in July 1907, represents fully the point of view of the theory of transformation groups.

129. See note 64, which describes the results of Galois and Cauchy relating to the group that Jordan called the "linear group." Using modern notation, we can describe the results obtained after Cauchy by Jordan as follows: The congruences with respect to a prime modulus given by

$$z_i \equiv \sum_{k=1}^{\mu} c_{ik} \pmod{p}, \qquad i = 1, \ldots, \mu; \quad |c_{ik}| \neq 0,$$

represent a permutation group that Jordan called the linear group. With regard to content, the elements of this group are identical with Cauchy's "geometric substitutions." Its order is $(p^\mu - 1)(p^\mu - p) \cdots (p^\mu - p^{\mu-1})$. This result was established by Gauss for $\mu = 1$ and $\mu = 2$. In 1879, Kronecker [434] referred to this group as the "metacyclic group" ("metazyklische Gruppe"). Some of the mathematicians who concerned themselves with this group are Betti [34], Mathieu [506–509], Jordan (as reported), Burnside [72], and Moore [535]. The linear group that leaves unchanged the element $1_{0,0,\ldots,0}$ (in Jordan's notation), that is, all z_i with $z_i \equiv 0$, partitions, for $\mu = 2$, the remaining $p^\mu - 1$ elements into $p + 1$ systems of $p - 1$ elements that coincide in the ratio of $z_1 : z_2 = z$. Here z can take on the values $0, 1, 2, \ldots, p-1, \infty$ (see note 126, which applies to Jordan as well). Hence for $\mu = 2$ we can represent the group by means of the congruences

$$z \equiv \frac{\alpha z + \beta}{\gamma z + \delta} \pmod{p}.$$

The permutations for which

$$\begin{vmatrix} \alpha & \beta \\ \gamma & \delta \end{vmatrix}$$

is a quadratic residue mod p form the "group of the modular equation." With this definition, Jordan built on the work of Betti [34, 35], Serret [617], Hermite [276], Mathieu [506], and Kirkman [344, 345].

130. Referring to his eulogy of Jordan on 4 June 1923, at the Paris Academy, Lebesgue writes, "One would be astonished to hear that Jordan had been a revolutionary or that he had tried to be different in order to attract attention; if, in the main body of his work, Jordan used almost exclusively synthetic reasoning, it was because *he was following in the footsteps of Galois.* ... The mathematician who hardly uses any calculations is substituting a mathematics of qualities for a mathematics of quantities ..." ("On s'étonnerait d'apprendre que Jordan a été révolutionnaire ou qu'il a cherché à se singulariser pour capter l'attention; si, dans la principale des ses oeuvres. Jordan a utilisé presque uniquement les raisonnements synthétiques, c'est qu'il *a suivi la voie et l'exemple de Galois* ... Le mathématicien qui n'use guère du calcul substitue à la mathématique des quantités une mathématique des qualités ..."). Reprinted as an abstract in [583], p. 56. Of course, this is a twentieth-century assessment, and it could hardly have been so formulated around 1870 without a sufficient historical distance from the *Traité*, which, by present standards, actually contains many calculations. Thus Lebesgue's assessment must definitely also be regarded as a document of its time showing how the concept of "an algebraic structure" cast off its shell only in the twentieth century.

131. This may be the reason why there has been no second edition of the *Traité*. Only in 1957 did there appear in Paris (Libraires Gauthier-Villars and A. Blanchard) a photographic reproduction [339]. The part comprising the permutation groups was quickly absorbed in the better expositions of group theory published at the end of the century, such as those of E. Netto [541], A. Capelli and G. Garbieri [87, Part I], H. Vogt [706], A. Capelli [88], L. Bianchi [43], W. Burnside [75], G. Chrystal [156], L. E. Dickson [171], H. Burkhardt [70], and A. Speiser [626]. The content of the rest, comprising the extension of the permutation-theoretic group concept and the corresponding methods to geometry and analysis, soon became obsolete. The effect on Jordan of the change in the group concept initiated by Klein can be seen in his many later papers [320, 321, 323, 325–328] devoted to his linear group and to transitive and primitive groups. Since 1877, especially after the publication of P. Gordan's important paper "Über endliche Gruppen linearer Substitutionen einer Veränderlichen" [249], Jordan viewed a group as a group of transformations; see [330–338].

132. Here, for example, it is certain that between 1860 and 1862 Cayley, independently of Plücker, had discovered line coordinates [146, 147]. But one cannot validly claim, as does A. R. Forsyth in his biographical notes on Cayley [195, p. XXXV], that Cayley was the first to construct and to apply line coordinates.

133. The facts are as follows. Every algebraic curve of degree higher than the second has a number of points where its tangents intersect the curve. All these inflection points lie on a curve that arises from the original curve in a unique manner. The inflection points are invariant under collineations [195, p. XXXIX].

134. The paper begins with these words: "THE following investigations were suggested to me by a very elegant paper on the same subject, published in the *Journal* by Mr. Boole" [129, p. 80]. This is followed by the exposition of Boole's results.

135. Volumes I–VII of Cayley's collected mathematical papers appeared in the author's lifetime. To each of these volumes Cayley added supplementary remarks whose character is partly historical and partly biographical. In Volume II, pp. 598–601, Cayley gives a bibliography of the decisive invariant-theoretic papers of Boole, Sylvester, Salmon, Hermite and Aronhold and refers, in addition, to the relevant papers of Eisenstein, Hesse and Schläfli, on which he had already reported in notes 13, 14, 15, 16 and 100 in Vol. I.

136. Cayley formulated this extremely comprehensive task as follows: "IN continuing my researches on the present subject [that is, on linear transformations (Wussing)], I have been led to a new manner of considering the question ... In fact, the question may be proposed, 'To find all the derivatives of any number of functions, which have the property of preserving their form unaltered

after any linear transformations of the variables.' By Derivative I understand a function deduced in any manner whatever from the given functions, and I give the name of Hyperdeterminant Derivative, or simply of Hyperdeterminant, to these derivatives which have the property just enunciated ... there remains a question to be resolved, which appears to present very great difficulties, that of determining the *independent* derivatives, and the relation between these and the remaining ones" [129, p. 95]. See also note 138.

137. In his book *The Development of Mathematics* [26, p. 351], E. T. Bell conveys the mistaken impression that Cayley had stipulated all the five requirements that we now associate with a distance function $D(x, y)$.

138. In his "Second Memoir on Quantics" [143] of 1856 Cayley found finite complete asyzygetic systems of invariants and covariants for binary forms of order 2, 3, 4, 5 and 6, and asyzygetic systems of covariants for binary forms of order 2, 3, and 4. Due to an error, Cayley concluded that there are no finite fundamental systems for binary forms of higher order. P. Gordan's 1869 paper "Beweis, dass jede Covariante und Invariante einer binären Form eine ganze Function mit numerischen Coefficienten einer endlichen Anzahl solcher Formen ist" [246] showed him that he was wrong. Thereupon Cayley returned to the problem, found his error, and published in 1871 his "Ninth Memoir on Quantics," dated 7 April 1870.

139. Already Eisenstein had found a complete system of invariants for a binary cubic form [403, p. 165]. Cayley was aware of this fact, and it served him as a model for the general question of complete invariant systems for arbitrary forms. See note 135.

140. See [513] for the development of the theory of invariants up to 1890.

141. Hilbert's 1890 proof of the basis theorem was a departure from the older invariant-theoretic approach. Typical of the opposition of that school to the abstract viewpoint was the opinion of Gordan, the "king of invariants," of Hilbert's method of proof: "This is not mathematics, this is theology" ("Das ist nicht Mathematik, das ist Theologie") (German quotation after [26, p. 429]). See also III.3 for Klein's attitude toward the abstract view of groups.

142. It is an open question whether Cayley knew of noneuclidean hyperbolic geometry already in the fifties. He constantly published in Crelles Journal, the journal that had published Lobachevski's "Pangeometrie."

143. The conception preceded the publication (in 1871) by at least a year. Thus early in 1870, Klein lectured on Cayley metrics in the seminar of Kummer and Weierstrass and concluded with the question whether this might not also be the way to establish a connection with noneuclidean geometry.

144. What is important here is not the effect of these papers of Klein on the technical development of noneuclidean geometry, but rather the fact that Klein destroyed essential reservations against noneuclidean geometry. In this connection see A. Schoenflies's paper "Klein und die nichteuklidische Geometrie" [652].

145. See the drastic assessment of this development in E. T. Bell's book *The Development of Mathematics* [26, p. 428].

146. These investigations, and to some extent also the theory of elementary divisors [731] that K. Weierstrass developed at that time in its full generality, supplied the basis for Klein's dissertation "Über die Transformation der allgemeinen Gleichung des zweiten Grades zwischen Linien-Coordinaten auf eine canonische Form." The defense of the dissertation took place in Bonn on 12 December 1868. It was reprinted, with minor changes and additions, in the *Mathematische Annalen* **23** (1884) and in [346, pp. 7–49]. In his dissertation, Klein corrects Battaglini's [22] canonical form of second degree complexes by giving their true general form. It was Clebsch who had called Klein's attention to Battaglini. In 1920, looking back at these matters,

Klein wrote that he had found it "not at all easy to switch from the more elementary methods of the Plücker presentation to the consistent method of projective coordinates applied by Battaglini. The study of the textbooks of Salmon-Fiedler and of some original papers helped me surmount these difficulties" ("nicht ganz leicht, von den mehr elementaren Methoden der Plückerschen Darstellung zu dem konsequenten Verfahren der projektiven Koordinaten überzugehen, wie es von Battaglini gehandhabt wurde. Das Studium der Lehrbücher von Salmon-Fiedler und mancher Originalabhandlung half mir über diese Schwierigkeit weg.") [346, p. 3].

147. On 1 October 1872, Klein moved to Erlangen. He was accompanied by Lie, who had been visiting him in Göttingen since 1 September: "Lie ... had in the meantime secured the foundations of his theory of first order partial differential equations and, in particular, of contact transformations, and these soon became the subject of our daily conversation (the first note on his new ideas published by Lie, see the *Göttinger Nachrichten* of 30 October 1872, was actually edited by me). On the other hand, Lie entered with the greatest enthusiasm into my ideas on the group-theoretic classification of the different ways of dealing with geometry. The external reason for the writing of my paper was that in Erlangen, in addition to the lecture with which he introduced himself to the circle of his colleagues, the newly appointed professor was expected to present a printed program" ("Lie ... hatte inzwischen die Grundzüge seiner Theorie der partiellen Differentialgleichungen erster Ordnung, wie insbesondere der Berührungstransformationen, gewonnen, die bald unser tägliches Gespräch bildeten (wie denn die erste Note, welche Lie über seine neuen Auffassungen veröffentlichte, siehe die Göttinger Nachrichten vom 30. Okt. 1872, von mir redigiert war). Andererseits ging Lie nun mit größtem Eifer auf meine Ideen über die gruppentheoretische Klassifikation der verschiedenen Behandlungsweisen der Geometrie ein. Die äußere Veranlassung für die Entstehung meiner Schrift aber war, daß in Erlangen der neuernannte Professor neben einem Vortrage, mit dem er sich in dem Kreis seiner Kollegen einführte, herkömmlicherweise ein gedrucktes Programm vorzulegen hatte.") [346, p. 411].

148. Klein's other papers from that period also bear the direct stamp of mechanical-physical interests. One relevant example is the line-geometric paper "Notiz, betreffend den Zusammenhang der Liniengeometrie mit der Mechanik starrer Körper" [354], which contains a presentation and elaboration of the "intimate connection" (der "intime Zusammenhang") of the different directions of development in geometry and in mechanics (Poinsot, Möbius, Chasles), but also the first written statement on the significance of Cayley metrics for noneuclidean geometry.

149. A footnote states that "this argument ... has been chiefly used in kinematic investigations, see Schell, *Theorie der Bewegung und der Kräfte*, part 1, chapter III" [346, p. 424]. The reference is to [635].

150. The review of this joint paper of Klein and Lie in *Bulletin des sciences mathématiques et astronomiques* [**3** (1872) 330–332] emphasizes the importance of this principle and pays virtually no attention to the actual content of the paper, thus recognizing the fact that the methodological advance was more important than the results.

151. Two months later, in May 1871, there appeared Klein's paper "Über eine geometrische Repräsentation der Resolventen algebraischer Gleichungen" [355], which refers to [351] and repeats this definition of a closed system: "By a closed system ... we mean here a system whose transformations, combined with each other, always yield transformations of the system" ("Unter einem geschlossenen System soll hier ... ein System verstanden werden, dessen Transformationen miteinander kombiniert immer wieder Transformationen des Systems ergeben.") [347, p. 263]. §7 of (B), which, as Klein emphasizes, is entirely due to Lie, contains the following theorem, which is important in itself as well as indicative of the future direction of Lie's research: "*Differential equations* $dy/dx = f(y, x)$, *which go over into themselves under infinitely many linear transformations can be solved by quadratures alone*" ("*Differentialgleichungen* $(dy)/(dx) = f(y, x)$ *welche durch unendlich viele lineare Transformationen in sich übergehen, verlangen zu ihrer Integration*

nur Quadraturen") [346, p. 458] and, generally, the integration of ordinary (first-order) differential equations which admit the same transformations presents the same difficulties. Thus (B) shows that between the conclusion of (A) and (B), Lie took essential steps toward a coherent elaboration of the theory of the transformation groups named for him.

152. Of course, Klein later took into account the gap that still existed in 1871, and in 1893, when the Erlangen Program was published in the *Mathematische Annalen*, he added the following footnote: "This definition [of a group (Wussing)] requires completion. In the case of the groups in the text we are making the unstated assumption that they contain the inverses of all of their operations. As had first been pointed out by Lie, for groups with infinitely many operations, this is in no way a consequence of the group concept as such; this assumption must therefore be explicitly added to the definition given in the text" ("Diese Definition bedarf noch der Ergänzung. Bei den Gruppen des Textes wird nämlich stillschweigend vorausgesetzt, daß dieselben neben jeder Operation, die sie enthalten mögen, immer auch deren inverse enthalten; dies ist aber, wie wohl zuerst Lie hervorhob, bei unendlicher Zahl der Operationen keineswegs eine Folge des Gruppenbegriffs als solchem; unsere Voraussetzung sollte also der im Texte gegebenen Definition dieses Begriffs ausdrücklich zugefügt werden.") [346, p. 462]. For the same reason, Klein added a similar remark to the footnote mentioned in my discussion of [359].

153. See note 154.

154. It is interesting that already in 1872 Klein regarded the antithesis between the synthetic and analytic directions in geometry as out of date. He emphasizes this explicitly in one of the seven notes to the Erlangen Program, entitled "Über den Gegensatz der synthetischen und analytischen Richtung in der neueren Geometrie." The note states that "at this time, one must no longer regard the difference between newer synthetic and newer analytic geometry as essential, for the content and the reasoning on both sides have gradually become very similar. That is why, in the text, we refer to both as 'projective geometry'" ("Den Unterschied zwischen neuerer Synthese und neuerer analytischer Geometrie hat man zurzeit nicht mehr als einen wesentlichen zu betrachten, da der gedankliche Inhalt sowohl als die Schlußweise sich auf beiden Seiten allmählich ganz ähnlich gestaltet haben. Daher wählen wir im Texte zur gemeinsamen Bezeichnung beider das Wort ‚projektivische Geometrie'.") [346, p. 490].

155. For one thing, there are geometries—such as the Riemannian geometries—that do not fit the framework of the Erlangen Program. In recent times, spaces consisting of a set of points and a given group of transformations—and thus fitting the Erlangen Program—are called kleinian spaces, and the theory of invariants of a kleinian space is called a kleinian geometry.

156. The Italian translation of the Erlangen Program appeared in 1890; the French, in 1891; the English, in 1893; the Polish, in 1895; and the Russian, in 1895–96.

157. L. Euler, *Introductio in analysin infinitorum*, Lausanne, 1748, Vol. II, chapter XVIII, article 442.

158. It should be noted that the editors (R. Courant and S. Cohn-Vossen) partly shortened and altered this section. To judge by a remark, the original text had been considerably blunted. It is likely that, for the most part, it was polemics against the formalist development in geometry and the further development of the vector calculus that were eliminated. The author had no access to the manuscript, but this is probably of little consequence in the case of Klein's own overall assessment of the genesis, influence, and significance of the Erlangen Program.

159. Klein quotes the sentence from the Erlangen Program: "One should develop the invariant theory related to a group" ("Man entwickele die auf eine Gruppe bezügliche Invariantentheorie"), and he continues: "If we write instead 'the theory of the relations which are invariant *relative to the group*,' then we are only a step away from the words *relativity theory* used by modern physi-

cists for the cases of the general goal [of invariant theory (Wussing)] that belong in their domain" ("Schreibt man statt dessen: , die Theorie der Beziehungen, welche *relativ zur Gruppe* invariant sind', so ist nur noch ein Schritt bis zu dem Worte *Relativitätstheorie*, welches die modernen Physiker für die in ihren Bereich gehörigen Fälle der allgemeinen Zielsetzung gebrauchen.") [404, p. 38].

160. Our judgment of Klein's faulty assessment of the significance of the Erlangen Program is not weakened by his further remarks, in which he describes the Erlangen Program as belonging to the class of writings "that would call forth the new by ordering what is on hand" ("welche zu Neuem" haben "anregen wollen, indem sie Vorhandenes ordnen") [404, p. 28], and says that it "may be regarded as a notch in the development of, especially, geometry, or, to use the general Grassmannian term, of the theory of extension" ("Als ein Einschnitt in der Entwicklung speziell der Geometrie oder, um den allgemeinen Graßmannschen Ausdruck anzuwenden, der Ausdehnungslehre angesehen werden kann") [404, p. 29].

161. In this connection see also the popular-scientific lecture "Über den Ursprung und die Bedeutung der geometrischen Axiome" ("On the Origin and Significance of Geometric Axioms"), presented to the *Dozentenverein* (Society of University Lecturers) in Heidelberg, in 1870.

162. Similar approaches are found already in the work of Lobachevski; I.1.

163. The relevant papers are [465, p. 437; 361, p. 116; 385, p. 544; 390; 575, p. 203; 281, supplement 4, p. 121]; for additional references to the literature see [189, pp. 107–112]. For the present state of the "space problem" see H. Freudenthal: Neuere Fassungen des Riemann-Helmholtz-Lieschen Raumproblems [198] in [540, pp. 92–97].

164. In the introduction, the problem formulated in group-theoretic terms is changed into a physical question; both are viewed as equally valid. "To form in all possible ways systems of molecules that can be superimposed upon themselves in different positions" ("Former de toutes les manières possibles des systèmes de molécules superposables à euxmêmes dans diverses positions.") [313, p. 168]. Here Jordan points to the origin of his paper, namely, to mineralogical questions: "It is from this second point of view that M. Bravais has studied this question: the special cases that he has treated, and which he has applied in a remarkable way to crystallography, are the most important ones. I believe, nevertheless, that it is still of some interest today to treat the problem in its full generality" ("C'est sous ce second point de vue que M[r]. Bravais a étudié cette question: les cas particuliers qu'il a traités et dont il a fait une remarquable application à la cristallographie sout les plus importants. Je crois néanmoins qu'il y a encore aujourd'hui quelque intérêt à traiter le problème dans toute sa généralité.") [331, p. 168]. In this connection see also P. Groth, *Entwicklungsgeschichte der mineralogischen Wissenschaften*, Berlin, 1926; for Bravais's work see, in particular, pp. 110–113. See also note 165.

165. At this point Jordan refers in a footnote to the results of Cauchy and Betti and to his independence: "This result is remarkably similar to a theorem, extremely useful in the theory of substitutions, that has been given by Cauchy and M. Betti and that I have since rediscovered at the beginning of my researches on the same subject" ("Cette proposition présente une analogie remarquable avec un théorème fort utile dans la théorie des substitutions, qui a été donne par Cauchy et par M. Betti, et que j'ai retrouvé depuis au début de mes recherches sur le même sujet.") [313, p. 171]. Jordan devoted a section of the preparatory remarks to this "Mémoire" to the question of when two such motions commute; term for this is "précédée ou suivie."

166. The relevant passage reads, „When studying a manifold with an underlying group we look first for objects invariant under all the transformations of the group. However, there may exist objects that admit some but not all the transformations of the group. In the sense of the group-based study, these objects are of special interest; they have distinguished properties. What it comes down to is to distinguish solids that are symmetric, regular, in the sense of ordinary

geometry, as well as surfaces of rotation and helicoids" ("Bei der Behandlung einer Mannigfaltigkeit unter Zugrundelegung einer Gruppe fragen wir entsprechend zunächst . . . nach den Gebilden, die durch alle Transformationen der Gruppe ungeändert bleiben. Aber es gibt Gebilde, welche nicht alle aber einige Transformationen der Gruppe zulassen, und diese sind dann im Sinne der auf die Gruppe gegründeten Behandlung besonders interessant, sie haben ausgezeichnete Eigenschaften. Es kommt das also darauf hinaus, im Sinne der gewöhnlichen Geometrie symmetrische, reguläre Körper, Rotations- und Schraubenflächen auszuzeichnen.") [346, pp. 489–490].

167. The subject was treated by Riemann in the winter term of 1858–59, as Klein learned by consulting the stenographic copy of Riemann's lecture transmitted to the Göttingen university library by the physicist W. v. Bezold. In this connection, see Klein's note in the *Göttinger Nachrichten* (1897) math. phys. class, 190ff. See also Klein's 1921 explanations of how the concept of the Riemann sphere and the corresponding approaches of Gauss became known [347, p. 256, footnote 1].

168. See also H. Stoltz, "Die geometrische Bedeutung der komplexen Elemente in der analytischen Geometrie" [630]. This paper inspired Klein to write [357] and [360].

169. In Kronecker's case, the relevant papers are [429] and [431]; in Brioschi's, [55–61, 63–64]. In 1878, Brioschi published a summary of his results in [65].

170. In this connection see also the very personal introduction of Lie to the third volume of the *Theorie der Transformationsgruppen* [466, pp. V–XXV]. Lie returns to the topic of his relation to Klein, and we sense in him a striking undertone of dejection.

171. We refer the reader to the detailed, comprehensive presentations in the *Encyklopädie der mathematischen Wissenschaften*, which contain many references to the literature: A. Wiman, "Endliche Gruppen linearer Substitutionen," in Vol. 2, part 2, Leipzig, 1901–1921, pp. 349–470 [738]; R. Fricke, "Automorphe Funktionen mit Einschluss der elliptischen Modulfunktionen"; R. Fricke (assisted by preliminary studies of J. Harness and W. Wirtinger), "Elliptische Funktionen" [203].

172. We mention the most important papers of G. Frobenius, T. Molien, and I. Schur: [205–206, 210–215, 531–532, 667–669, 672–674]. Also, W. Burnside independently rediscovered many of Frobenius's results and included them in the second edition of his textbook the *Theory of Groups of Finite Order* [84], published in 1911. This book includes the first account of representation theory.

173. The purpose of the *Vorlesungen über das Ikosaeder* is precisely the clarification of Klein's new viewpoint. In this connection see, in particular, [379, pp. 139–161]. In the introduction we read, "It is now fully 25 years since the combined papers of *Brioschi*, *Hermite* and *Kronecker* created the modern theory of equations of the fifth degree. In spite of the fact that their investigations have been mentioned very frequently, they have not yet been really understood in wide circles of the mathematical public. In what follows, I begin with the study of the icosahedron [the theory of groups of isometries of regular polyhedrons (Wussing)] and view it as the actual foundation of the solution process. This gives rise to a form of the theory that one could hardly expect to be simpler or more transparent" ("Es sind jetzt volle 25 Jahre her, dass die Herren *Brioschi*, *Hermite* und *Kronecker* in vereinten Arbeiten die moderne Theorie der Gleichungen fünften Grades geschaffen haben. Aber so oft auch ihre Untersuchungen genannt werden, ein eigentliches Verständniss in weiteren Kreisen des mathematischen Publikums haben dieselben bis jetzt nicht gefunden. Indem ich im Folgenden die Lehre vom Ikosaeder voranstelle und als die eigentliche Grundlage des Auflösungsprocesses betrachte, entsteht eine Ansicht der Theorie, wie sie einfacher und durchsichtiger wohl nicht mehr gewünscht werden kann.") [379, p. 111].

174. In his letter to Chevalier, Galois stated the theorem that the transformation of elliptic functions leads to quintic equations whose group has order 60. This implies that there exist quintic equations solvable by elliptic functions.

175. Galois's claim, stated in note 174, provided Hermite—as well as Betti—with a direct inducement to look for its proof. To complete the historical record we also mention the papers by Hermite, Brioschi, Joubert (1834–1910), Gordan, and Kiepert (1846–1934) devoted to the solution of the quintic equation.

176. In this connection see also Klein's 1888 paper [384].

177. See note 182.

178. In the case of Gauss, this became known essentially from the Nachlass; see Vol. 8 of Gauss's works [235]. See also [203, pp. 196–197], which contains additional bibliographical references as well as a brief historical survey.

179. C. G. J. Jacobi, *Fundamenta nova theoria functionum ellipticarum*, Königsberg, 1829 [303].

180. The term „elliptische Modulfunktion" (elliptic modular function) is due to R. Dedekind, *Erläuterungen zu Riemanns Fragmenten über die Grenzfälle der elliptischen Modulfunktionen* [163, p. 438].

181. Beyond the papers noted in the text, we mention a few of the studies on modular functions and a few of the studies that, before Klein's work, led to the investigation of automorphic functions: Gauss [641, especially p. 59ff]; Abel [10]; Jacobi [303]; Hermite [275]; Eisenstein [183]; Weierstrass [730] (lectures given in the winter of 1862–63: [445, 732]); Riemann [584–586].

182. Reflections on "functions invariant with respect to (finite) groups of linear substitutions" ("Funktionen, welche gegenüber (endlichen) Gruppen linearer Substitutionen invariant sind") also play an important role in various places in Klein's *Vorlesungen über das Ikosaeder* [379]; in fact, they directly motivate, as a mathematical aid [379, p. 126ff], the transition to the treatment of infinite transformation groups and the group-theoretic treatment of the elliptic modular functions: "We now proceed to a second generalization ... We shall increase not the number of variables but the number of substitutions by taking as a basis *infinite* rather than finite groups. Setting aside the form-theoretic viewpoint, I propose to discuss here in function-theoretic form the very simplest examples. Instead of the rational functions of z (which remain invariant under finite groups of substitutions) we then have transcendental but single-valued functions" ("Wir schreiten jetzt zur zweiten Verallgemeinerung ... Nicht die Zahl der Variabelen werden wir vermehren, aber die Zahl der Substitutionen, indem wir statt endlicher Gruppen *unendliche* Gruppen zu Grunde legen. Unter Beiseitelassung des formentheoretischen Standpunktes will ich hier in functionentheoretischer Form nur die allereinfachsten Beispiele zur Sprache bringen. An Stelle der rationalen Functionen von z (die bei den Gruppen endlichvieler Substitutionen ungeändert blieben) haben wir dann transcendente, aber eindeutige Functionen." [379, pp. 126–127]. There follow as examples the simply periodic functions, the trigonometric functions, the elliptic modular functions and the doubly periodic functions.

183. As early as 12 June 1881, that is, during his Leipzig period, Klein wrote to Poincaré outlining the problem of automorphic functions: "At that time [summer of 1879 (Wussing)], my train of thought, which coincides in many points with that now followed by you, was this: 1. Periodic and doubly periodic functions are just examples of single-valued functions that admit linear transformations. The task of modern analysis is to determine all of these functions. 2. The number of these transformations can be finite; then one obtains the equations of the icosahedron, the octahedron, ..., which I considered earlier ... and which were my starting point in building this whole circle of ideas. 3. Groups of infinitely many linear transformations that give rise to useful functions (you call such groups "groupes discontinus") are obtained, *for example*, if one starts

with a polygon with circular sides whose circles cut a fixed circle at right angles and whose angles are exact submultiples of π" ("Mein Gedankengang, der mit dem jetzt von Ihnen eingeschlagenen nun vielfach zusammentrifft, war damals dieser: 1. Periodische und doppeltperiodische Funktionen sind nur Beispiele für eindeutige Funktionen mit linearen Transformationen in sich. Es ist Aufgabe der modernen Analysis, alle diese Funktionen zu bestimmen. 2. Die Anzahl dieser Transformationen kann eine endliche sein; dies gibt die Gleichungen des Ikosaeders, Oktaeders, ..., die ich früher betrachtete ... und von denen ich bei Bildung dieses ganzen Ideenkreises ausging. 3. Gruppen von unendlich vielen linearen Transformationen, die zu brauchbaren Funktionen Anlaß geben (groupe discontinu nach Ihrer Bezeichnung) erhält man zum *Beispiel*, wenn man von einem Kreisbogenpolygon ausgeht, dessen Kreise einen festen Kreis rechtwinklig schneiden und dessen Winkel genaue Teile von π sind.") [348, pp. 588–589].

184. L. Fuchs's papers [217–220]. For Poincaré, it was especially the last of these papers which provided the direct starting point for his involvement with automorphic functions. Hence the terms "le groupe fuchsien" and "la fonction fuchsienne" used by Poincaré. See note 185.

185. Of Poincaré's papers on automorphic functions I mention [569–574]. Concerning the relation between Klein and Poincaré, and especially their "race" involving publication dates in the summer of 1882, see Klein's autobiographical data [348, pp. 577–586] as well as the exchange of letters between Klein and Poincaré, printed in [348, pp. 587–621].

186. Klein described the genesis of the two-volume work *Vorlesungen über die Theorie der automorphen Funktionen* in a detailed report [348, pp. 742–747]. This report and Fricke's many relevant publications show that Fricke had played an essential part in the writing of the *Vorlesungen* as well as in providing the group-theoretic basis for the theory of automorphic functions; see, in particular, his 1892 paper "Zur gruppentheoretischen Grundlegung der automorphen Funktionen" [200].

187. As personal testimony bearing on the history of the subject, we quote the passage that follows the quotation in the text: "The exposition that follows is in a great many ways based on *Vorlesungen über das Ikosaeder* and especially on *Vorlesungen über die Theorie der elliptischen Modulfunctionen*, and its study requires, for the most part, that the reader be familiar with the basic sections of these works. In historical terms, the theory of automorphic functions has grown out of the theories of regular solids and of modular functions. At least, this is the path that Klein had once followed under the influence of the well-known papers of Schwarz [that is, [675, 676] (Wussing)] on the one hand and of the early papers of Poincaré [see note 185 (Wussing)] on the other. As for other considerations introduced by Poincaré, such as the arithmetical methods of Hermite ... and the function-theoretic questions of Fuchs [that is, [217–220] (Wussing)] concerning unique invertibility of solutions of 2nd-order differential equations, they belong to the very circle of ideas that gave rise to the theories of regular solids and of modular functions" ("Die nachfolgende Darstellung ist durch die ,*Vorlesungen über das Ikosaeder*' sowie namentlich durch die ,*Vorlesungen über die Theorie der elliptischen Modulfunctionen*' aufs mannigfachste vorbereitet und setzt ihrerseits zumeist die Bekanntschaft mit den grundlegenden Teilen dieser Werke beim Leser voraus. Auch historisch hat sich die Theorie der automorphen Functionen aus derjenigen der regulären Körper und der Modulfunctionen entwickelt. Wenigstens ist dieses der Weg, den seinerzeit Klein unter Einfluß einmal der bekannten Arbeiten von Schwarz, andrerseits der beginnenden Publicationen Poincaré's eingeschlagen hat. Wenn Poincaré daneben auch andere Momente heranzieht, nämlich die arithmetischen Methoden von Hermite u. a. ... und die functionentheoretischen Fragestellungen von Fuchs betreffend eindeutige Umkehr der Lösungen linearer Differentialgleichungen 2ter Ordnung, so gehen eben diese Ansätze ihrerseits doch wieder genau auf dieselben Gedankenkreise zurück, aus denen die Theorien der regulären Körper und der elliptischen Modulfunctionen erwachsen sind.") [201, pp. 1–2].

188. Both papers appeared in German in slightly reworked form in *Mathematische Annalen* **5** (1872). The dissertation [470] appeared for the first time in German in the *Collected Works* [459, pp. 105–152].

189. Already in his dissertation Lie points out the close connection with Klein's work in terms of subject matter as well as ideas. In the introduction Lie writes, "While writing the present paper, I maintained a lively exchange of ideas with Plücker's student Dr. Felix Klein, to whom I am indebted for many ideas—undoubtedly more than I have acknowledged by quotation" ("Während ich mich mit der gegenwärtigen Abhandlung beschäftigte, habe ich in lebhaftem Gedankenaustausche mit Plückers Schüler Dr. Felix Klein gestanden, dem ich viele Ideen verdanke, ohne Zweifel mehr, als es mir durch Zitate auszudrücken gelungen ist.") [459, p. 106].

190. In contrast to the Erlangen Program and the joint papers of Klein and Lie, there is no reference in the whole paper to the fact that the group concept had been taken over from the theory of permutation groups.

191. Retrospectively (in 1893), Lie described the years between 1870 and 1874, of decisive importance for his program of work, in these words: "On the other hand, between 1870 and 1874 I developed the concept of the *finite continuous group* and recognized its far reaching significance for geometry and for the theory of differential equations. ... In this way, the concepts of a *transformation* and of a group of *transformations* have for me moved ever more to the foreground and, step by step, I developed a general *transformation theory*" ("Andrerseits entwickelte ich in den Jahren 1870–74 den Begriff der *endlichen continuirlichen Gruppe* und erkannte seine weitreichende Bedeutung für die Geometrie und für die Theorie der Differentialgleichungen ... So traten für mich die Begriffe *Transformation* und *Transformationsgruppe* immer mehr in den Vordergrund und ich entwickelte nach und nach eine allgemeine *Transformationstheorie*.") [466, p. XVI].

192. Thus, Lie's papers directed toward geometry, differential geometry and differential equations, in which group theory plays just the role of a tool, embody this tendency, which directly furthers group theory. See, for example, the 1879 paper "Klassifikation der Flächen nach der Transformationsgruppe ihrer geodätischen Kurven" [481]. The introduction reads, "In the study of ordinary or partial differential equations, the properties of such equations that are unchanged by arbitrary point transformations (or contact transformations) deserve special attention. ... If I am not mistaken, I have succeeded in carrying out the research program defined thereby—which I have consistently pursued since 1872—for first-order partial differential equations in a number of papers published in the *Mathematische Annalen*, volumes VIII, IX, XI [see [473, 476, and 477] (Wussing)] ... When it comes to partial as well as ordinary differential equations of higher order, I have thus far been obliged to limit myself, essentially, to suggestions. In fact, I found it necessary first to develop an extensive auxiliary theory, the theory of transformation groups. In the *Göttinger Nachrichten* 1874/359, number 22 [see [472] (Wussing)] I listed all the groups of a doubly extended manifold, and added that this could provide a basis of a reasonable integration theory of equations

$$f(x, y, y', \ldots, y^{(n)}) = 0$$

that have a transformation group to begin with. I then presented, in four papers [[474–475, 478–479] (Wussing)] in the *Archiv for Mathematik og Naturvidenskab*, volumes I and III, a detailed theory of transformation groups of a doubly extended manifold. I intend to develop the transformation theory of an n-tuply extended manifold in future papers. When it comes to the application of my transformation theory to ordinary differential equations, I think it expedient first to illustrate the scope and essential nature of my method of investigation by means of a special example. I think that the differential equation of the geodesic curves of a surface lends itself very well to this purpose. I shall therefore attempt, in the present paper, to study the differential equation just mentioned" ("Bei Untersuchungen über gewöhnliche oder partielle Differentialgleichungen verdienen solche Eigenschaften derselben eine besondere Aufmerksamkeit, die bei beliebigen Punkttransformationen (oder Berührungstransformationen) ungeändert bleiben ... Die hiermit definierte Untersuchungsrichtung, die ich seit 1872 konsequent verfolgt habe, führte ich für die partiellen Differentialgleichungen erster Ordnung, wenn ich nicht irre, glücklich durch in einigen Abhandlungen, die in den Mathematischen Annalen Bd. VIII, IX, XI gedruckt sind ... Was partielle Differentialgleichungen wie auch gewöhnliche Differentialgleichungen höherer Ordnung betrifft, so habe ich mich bis jetzt wesentlich auf Andeutungen beschränken müssen. Es war mir in der Tat notwendig, zuerst eine umfangreiche Hilfstheorie, die Theorie der Transformationsgruppen zu entwickeln. In den Göttinger Nachrichten 1874/359, Nr. 22 gab ich eine Aufzählung von allen Gruppen einer zweifach ausgedehnten Mannigfaltigkeit, indem ich zugleich angab, daß sich hierauf eine rationelle Integrationstheorie solcher Gleichungen:

$$f(x, y, y', \ldots, y^{(n)}) = 0,$$

die überhaupt eine Transformationsgruppe besitzen, begründen ließe. Sodann gab ich im Archiv for Mathematik og Naturvidenskab, Bd. I und III in vier Abhandlungen eine ausführliche Theorie der Transformationsgruppen einer zweifach ausgedehnten Mannigfaltigkeit. Und in weiteren Abhandlungen beabsichtige ich, die Transformationstheorie einer n-fach ausgedehnten Mannigfaltigkeit zu entwickeln.

Indem ich mich jetzt zur Verwertung meiner Transformationstheorie für gewöhnliche Differentialgleichungen wende, halte ich es für zweckmäßig, zuerst an einem speziellen Beispiele die Tragweite und überhaupt das Wesen meiner Untersuchungsmethode auseinanderzusetzen. Und hierzu scheint mir die Differentialgleichung der geodätischen Kurven einer Fläche sich sehr gut zu eignen. Ich werde daher versuchen, in der nachstehenden Abhandlung die Transformationsgruppe der soeben besprochenen Differentialgleichung zu untersuchen.") [459, pp. 358 to 359].

193. In a footnote [461, p. 5], Lie observes that one can think of contact transformations of two variables as point transformations of the variables x_1, x_2, p "that have the special property of taking the equation $dx_1 - p\,dx_2 = 0$ into itself" ("welche die besondere Eigenschaft haben, die Gleichung $dx_1 - p\,dx_2 = 0$ in sich überzuführen"). Later, in 1878, Lie discovered that there exist groups of contact transformations which can not be represented as groups of point transformations (see [479]).

194. At this point Lie just states the result [461, p. 6] that groups of contact transformations of two variables leave invariant a second-order differential equation or at least a third-order differential equation.

195. In what follows, Lie sheds light on the kinds of connections that exist between such results and the Erlangen Program: Since the (geometric) treatment of a manifold invariably depends on the underlying group of transformations, and since the theory of transformation groups of one variable always yields groups of linear transformations, "every conceivable treatment of the binary region is included in the ordinary linear invariant theory" ("jede Behandlungsweise des binären Gebietes, die man ersinnen könnte, in die gewohnliche lineare Invariantentheorie eingeschlossen ist"). As Lie had stated earlier, in the case of two variables one obtains, among other things, certain groups that resemble the circular transformations and the linear transformations. Lie is right in saying that this "assigns a distinguished position to metric and projective studies of the plane in the totality of all possible modes of study to which we can expose the plane" ("dies weist den metrischen und den projektivischen Untersuchungen der Ebene eine ausgezeichnete Stellung zu unter allen möglichen Untersuchungsarten, denen man die Ebene unterwerfen kann") [461, p. 7].

196. In a letter from Lie to Klein, dated December 1885, we read, "I am often amazed at the sureness with which my instinct had led me, already in the years 1872–74. For all things arrange themselves under the group concept and the theory of complete systems with known infinitesimal

transformations. It is strange that so many apparently completely different, and difficult, theories reduce to finding the invariant objects of a group" ("Ich wundere mich häufig, wie sicher mein Instinkt mich schon in den Jahren 1872–74 führte. Denn alles ordnet sich unter den Begriff Gruppe und unter die Theorie der vollständigen Systeme mit bekannten infinitesimalen Transformationen. Ganz kuriös ist es, wie viele, anscheinend absolut verschiedene, schwierige Theorien sich darauf reduzieren, die bei einer Gruppe invarianten Gebilde zu finden."). Quoted by F. Engel, in his introduction to Vol. IV of the collected works of Lie [460, p. X].

197. Lie's 1880 paper "Theorie der Transformationsgruppen I" [482] begins with almost the same words as those just quoted. Incidentally, this paper is the only one that actually appeared of a series of papers on transformation groups promised by Lie. It brings together many of Lie's earlier results. From the historical point of view the paper deserves special attention, for at the end (§20: "General Considerations") [462, pp. 90–94] Lie explicitly designates his theory of transformation groups as an extension of the theory of permutation groups, and takes this occasion to assess, from this viewpoint, the papers of Abel, Galois, Jordan, the algebraic invariant theory of Cayley and Sylvester, the papers of Klein, and so on. This, then, is an assessment of the decisive turning points of the theory of permutation groups. Admittedly, Lie's judgment is at times subjective and incomplete, inasmuch as he does not mention all the relevant papers and may well not have known them. In this connection see also the self-refereed three numbers of the paper "Untersuchungen über unendliche kontinuierliche Gruppen" [489], in which Lie retrospectively (1894) assesses his studies of infinite continuous groups between 1883 and 1886.

198. See note 205.

199. Lie gave the first definition of the fundamental concept of an infinitesimal contact transformation already in 1872.

200. I do not go into the question how concepts such as normal subgroup, simple group, transitive group and imprimitive group had to be modified upon transition from permutation groups to transformation groups.

201. At the suggestion [466, p. XXIV] of F. Klein and A. Mayer, F. Engel went in 1884 to Christiania to help Lie work out his comprehensive presentations. In this connection note the high praise and appreciation of Engel's contribution expressed by Lie in the introduction to the third volume of *Theorie der Transformationsgruppen* [466, p. XXIV].

202. Section V of the third volume of the *Theorie der Transformationsgruppen* [466, pp. 393–543], entitled *Untersuchungen über die Grundlagen der Geometrie* contains Lie's critique of Helmholtz's axioms of the theory of motions mentioned at the beginning of III. 3. See especially pp. 437–470.

203. I refer the reader to Cartan's own assessment of his scientific papers, an assessment undertaken in 1931 and published in the first volume of his collected papers [91, pp. 15–112]. This assessment is extremely valuable from the historical point of view.

204. To clarify the position reached by Cartan in the application of the concept of structure—not yet group as structure but structure of group—and to illustrate the clear program developed by Cartan I quote from the introduction to his fundamental 1894 paper "Sur la structure des groupes de transformations finis et continus" [96]. These passages are, at the same time, a remarkable document from the time of the recognition of the decisive significance of the group concept. It reads, "The concept of the *structure* of a finite and continuous group of transformations came to M. Lie at the very beginning of his studies [Lie actually spoke of the type and *not* of the structure of a transformation group (Wussing)] that led the great Norwegian to create the theory of groups, a theory that by its fruitfulness renewed, so to speak, several branches of the mathematical sciences. In his method of solving algebraic equations that admit a given group of substitutions, Galois reduced the problem to the solution of a series of auxiliary equations of which the number and nature depend only on the *structure* of the group under consideration. In the same way, Mr.

Lie reduced the integration of a system of partial differential equations which admits a finite continuous group of transformations to the integration of a series of auxiliary systems, and the number of these systems, their nature, and the manner in which they are interconnected, depends again only on the *structure* of the group involved. Again in a similar manner, the theory of linear differential equations created by Mr. Picard and Mr. Vessiot and, more generally, the theory of differential equations that admit a *fundamental system* of integrals, point out the importance of the *structure* of finite continuous groups. Properly speaking, the problem of the structure of finite continuous groups can be formulated thus: *To find all possible structures of groups with an arbitrary number of parameters*" ("La notion de la *structure* d'un groupe de transformations fini et continu s'est présentée à M. Lie dès le début des recherches qui ont amené le grand géomètre norvégien à fonder sa théorie des groupes, théorie qui, par sa fécondité, a renouvelé pour ainsi dire plusieurs branches de la science mathématique. Dans sa méthode de résolution des équations algébriques qui admettent un groupe de substitutions donné, Galois a ramené le problème à la résolution d'une série d'équations auxiliaires dont le nombre et la nature ne dépendent que de la *structure* du groupe de substitutions considéré. De la même façon, M. Lie a ramené l'intégration d'un système d'équations aux dérivés partielles qui admet un groupe de transformations *fini et continu* à l'intégration d'une série de systèmes auxiliaires, et le nombre de ces systèmes, leur nature, la manière dont ils se relient les uns aux autres ne dépendent encore que de la *structure* du groupe considéré. D'une manière analogue encore, la théorie des équations différentielles linéaires fondée par MM. Picard et Vessiot et plus généralement celle des équations différentielles qui admettent un *système fondamental* d'intégrales, mettent en évidence l'importance de la *structure* des groupes finis et continus.
Le problème proprement dit de la structure des groupes de transformations finis et continus pourrait s'énoncer ainsi: *Trouver toutes les structures possibles des groupes à un nombre quelconque des paramètres.*") [91, p. 137].

205. It is true that Lie occasionally considered infinite groups—for example, in 1872, the infinite group of all contact transformations of space that leave the group of transformations invariant could not develop for infinite groups a theory analogous, to his theory of finite continuous groups until (as he writes in retrospect, in 1891), he had "in the beginning of 1883 the lucky idea of selecting from among all groups those whose transformations could be defined by differential equations" ("Anfang des Jahres 1883 die glückliche Idee unter allen Gruppen diejenigen herauszugreifen, deren Transformationen durch Differentialgleichungen definiert werden können") [462, p. 396]. Of course, as Lie himself had pointed out in, for example, [472, p. 302], there exist infinite continuous groups that cannot be defined by means of differential equations.

206. Lie explains in the introduction that he could not turn down the invitation of the board of the Ecole Normale Supérieure to write a commemorative article on Galois: "Without doubt, I owe this honor, which I do not think I can possibly refuse, to the fact that for 25 years I have particularly striven solely to extend his [Galois's (Wussing)] ideas on algebraic equations, so original and so fertile, to other areas of the mathematical sciences" ("Je dois sans doute cet honneur, que je n'ai pas cru pouvoir refuser, à ce que, depuis vingt-cinq ans, je me suis tout particulièrement efforcé d'étendre à d'autres domaines de la science mathématique ses idées sur les équations algébriques, si originales et si fécondes.") [462, p. 592].

207. After a series of pertinent papers on transformation groups, H. Poincaré (1854–1912)—like Lie—wrote a comprehensive critical overview of the state of group theory, which had come to a temporary conclusion after the emergence of transformation groups. The paper, entitled "On the Foundations of Geometry" [576] (English translation by T. J. McCormack), appeared in 1889. Its content is primarily philosophical and epistemological. It makes an attempt, based on Poincaré's subjectively idealistic viewpoint, to provide foundations for geometry. Under the circum-

stances, the paper had to deal with the significance of group theory and make assessments of the group-theoretic contributions of the geometers. Seen abstractly, Poincaré's definition of the "group of displacements" contains only the requirement of closure of the set under successive displacements that are exclusively geometric in nature. Thus Poincaré is seen to be far behind Lie in spite of the fact that he values him as the person "who has contributed most towards making prominent the importance of the notion of group and laying the foundations of the theory that I have just expounded. It was he, in fact, who gave the present form of the mathematical theor yof continuous groups." [576, p. 37] In terms of content, Poincaré considers groups of displacements. Specifically, he deals with just three types of groups of geometric motions: discontinuous, continuous, and semicontinuous groups, all of which are subgroups of the group of all geometric motions that Poincaré just calls "the group." The group is partitioned as follows: "All the operations of the group can be divided into sheaves; for 'discontinuous' groups the different operations of the same sheaf are only a single operation repeated once, twice, three times etc; for 'continuous' groups properly so called the different operations of the same sheaf correspond to different whole numbers, commensurable or incommensurable; finally, for groups that may be called 'semicontinuous,' these operations correspond to different commensurable numbers" [576, p. 37].

208. This is shown in Lie's article "Influence de Galois sur le développement des Mathématiques" [490] by the fact that, in his discussion of the fruitfulness of the group concept he completely disregarded the development of number theory and, in particular, of field theory, and thus failed to consider a development that made an essential contribution to the elaboration of the abstract, axiomatic group concept (III.4).

209. Klein's reservations about the shift of mathematics toward the abstract and axiomatic, a shift that was strong by the end of the century, are already clearly present in his 1895 lecture "Über Arithmetisierung der Mathematik" [392]. In spite of his "recognition of the extraordinary importance of these developments" ("Anerkennung der außerordentlichen Wichtigkeit dieser Entwicklungen"), he speaks of the "rejection of the view that arithmetized science is a kind of extract that already completely contains the true substance of mathematics" ("Zurückweisung der Auffassung, als sei in der arithmetisierten Wissenschaft wie in einem Extrakt der eigentliche Inhalt der Mathematik bereits erschöpfend enthalten") [347, p. 233].

210. The scolia in the original have not been numbered. See [222, p. 8].

211. In a footnote to his first expressly group-theoretic paper, Cayley, who had evidently misunderstood the historical facts (II.1) writes that "the idea of a group as applied to permutations or substitutions is due to Galois, and the introduction of it may be considered as marking an epoch in the progress of the theory of algebraical equations" [130, p. 124].

212. See the footnote quoted in note 211.

213. It is well known that only after many failures did Hamilton become convinced that one could not retain for quaternions the commutativity of multiplication as well as the distributive law and eventually recognized that the latter was the more essential; see the introduction to his *Lectures of Quaternions* [260, pp. 8–30]. In this connection see also, as a later proof of the growing general recognition of noncommutative algebra, H. Hankel's historical remarks—marked by deep understanding—in his *Theorie der complexen Zahlensysteme, insbesondere der gemeinen imaginären Zahlen und der Hamiltonschen Quaternionen* [262], in particular pp. 104–106.

214. Hölder begins this paper with these words: "Herr Cayley has posed the problem of finding all possible groups of a given order. Also, a group is to be thought of in very general terms, its operations are no longer permutations of letters, rotations or collineations but simply symbols for which a law of composition has been defined with properties yet to be discussed. Holoedrically isomorphic groups [that is, isomorphic groups, as we call them today (Wussing)] are regarded

as not being different" ("Herr Cayley hat die Aufgabe gestellt, alle möglichen Gruppen irgend einer gegebenen Ordnungszahl aufzufinden. Die Gruppe soll dabei ganz allgemein gedacht werden, ihre Operationen sind keine Buchstabenvertauschungen oder Rotationen oder Collineationen mehr, sondern lediglich Symbole, für welche ein Gesetz der Verknüpfungen definiert ist mit Eigenschaften, die noch zur Sprache kommen werden. Holoedrisch isomorphe Gruppen gelten als nicht verschieden.") [291, p. 301].

215. One of the four papers, namely, [148] was even published in the *Mathematische Annalen* **12** (1878).

216. The word "basis" does not appear in the statement of the basis *theorem*, but is used in the text itself in its modern sense: "If $A, B, C, \ldots$ are elements of a group, then all elements of the form $A^x B^y C^z \cdots$ belong to the group. Certain elements of a group form its basis if all the elements of the group can be obtained from them by exponentiation and multiplication" ("Sind $A, B, C, \ldots$ mehrere Elemente einer Gruppe, so gehören auch alle Elemente der Form $A^x B^y C^z \ldots$ der Gruppe an. Mehrere Elemente einer Gruppe bilden eine *Basis* derselben, wenn sich aus ihnen durch Potenziren und Multipliciren alle Elemente der Gruppe zusammensetzen lassen.") [207, p. 219].

217. Namely, there are two entirely different types of infinite groups: "There also exist groups of uncountably many elements; for example, if they are not all roots of unity, then the units of an algebraic field form an infinite group of finite rank [that is, with finitely many generators (Wussing)]" ("Es giebt auch Gruppen von unzählig vielen Elementen, z. B. bilden die Einheiten eines algebraischen Körpers, falls sie nicht sämtlich Wurzeln aus 1 sind, eine unendliche Gruppe von endlichem Range.") [207, p. 218]. "There also exist systems of uncountably many elements from which we can construct finite groups, for example the roots of unity of all degrees" ("Es giebt auch Systeme von unzählig vielen Elementen, aus denen sich endliche Gruppen bilden lassen, z. B. die Einheitswurzeln aller Grade.") [207, p. 219].

218. The book consists of two sections, "Theorie der Substitutionen und der ganzen Funktionen" and "Anwendung der Substitutionentheorie auf die algebraischen Gleichungen." The first section contains, among others, the chapters "Symmetrische oder einwertige Funktionen. Alternierende und zweiwertige Funktionen," "Mehrwertige Funktionen und Substitutionsgruppen," "Transitivität und Primitivität. Einfache und zusammengesetzte Gruppen. Isomorphismus," "Die Anzahl der Werte ganzer Funktionen," "Untersuchung einiger besonderer Arten von Gruppen" and "Analytische Darstellung der Substitutionen. Die lineare Gruppe." The second contains the classical Galois theory and treats, in particular, the equations of degree up to the fourth, cyclotomic equations and abelian equations.

219. Netto explains in the introduction that "it was not only the new, fundamental ideas [in Jordan's *Traité* (Wussing)] that had to be included. At this point I wish explicitly to emphasize that in spite of their generally different modes, I make appropriate use of certain proofs and trains of thought .... Granted that certain details can be traced back to the *Traité* and to these investigations, the author is indebted to his venerable teacher Herr Kronecker for the views underlying the whole of this work. The author has tried to utilize the fruits derived from the lectures and the study of the papers of this scientist as well as from the stimulating personal contacts with him, and hopes that traces of this will show somewhere in this work. The one thing he regrets is that Herr L. Kronecker's newest and distinguished publication, 'Grundzüge einer arithmetischen Theorie der algebraischen Grössen,' has appeared too late for him to derive the kind of benefit that he and his readers might otherwise have derived" ("es waren nicht nur die neuen, grundlegenden Begriffe, welche aufgenommen werden mussten; auch manche Beweise und Gedankenfolgen konnten, wie hier ausdrücklich hervorgehoben werden mag, trotz der Verschiedenheit des Ganges im allgemeinen, passend verwendet werden ... Wenn aber auch manche Einzelheiten auf jenen 'Traité' und auf diese Untersuchungen zurückgeführt werden müssen, so verdankt doch der Ver-

fasser seinem verehrten Lehrer Herrn L. Kronecker die Anschauungen, welche seinem gesamten Werke zu Grunde liegen. Er hat sich bemüht, die Früchte, die ihm aus den Vorlesungen und aus dem Studium der Abhandlungen dieses Gelehrten, die ihm aus dem anregenden persönlichen Verkehre mit diesem Manne geworden sind, zu verwerten; und er hofft, dass die Spuren hiervon an manchen Stellen seiner Arbeit hervortreten mögen. Eines bedauert er: dass die neueste, bedeutende Publikation des Herrn L. Kronecker: ‚Grundzüge einer arithmetischen Theorie der algebraischen Grössen' zu spät erschien, als dass er von denselben den Nutzen hätte ziehen können, den zu ziehen er sich und seinen Lesern gewünscht hätte.") [541, p. IV].

220. At this point Netto refers to Kronecker's paper [432] and adds that "what follows is, for the most part, a quotation from that paper" ("das Nachfolgende ist dieser Abhandlung grossenteils wörtlich entnommen") [541, p. 144].

221. For this Jordan uses the terms "holoedric" and "meriedric" isomorphism, and Netto, "simple and complex isomorphism" (einstufiger und mehrstufiger Isomorphismus).

222. Weber gives a group-theoretic proof, and to this end he collects in §1 of his paper "Lemmas on Groups." He gives the following definition of a group: "*Definition:* A system G of any h elements $\Theta_1, \Theta_2, \ldots, \Theta_h$ is called a *group of degree h* if it satisfies the following conditions: I. By means of some prescription, called the composition or the multiplication, we obtain from two elements of the system a new element of that system. In symbols, $\Theta_r\Theta_s = \Theta_t$. II. We always have $(\Theta_r\Theta_s)\Theta_t = \Theta_r(\Theta_s\Theta_t) = \Theta_r\Theta_s\Theta_t$. III. $\Theta\Theta_r = \Theta\Theta_s$ and $\Theta_r\Theta = \Theta_s\Theta$ always imply $\Theta_r = \Theta_s$" ("*Definition:* Ein System G von h Elementen irgend welcher Art, $\Theta_1, \Theta_2, \ldots, \Theta_h$ heißt eine *Gruppe vom Grade h*, wenn es den folgenden Bedingungen genügt:
I. Durch irgend eine Vorschrift, welche als Composition oder Multiplication bezeichnet wird, leitet man aus zwei Elementen des Systems ein neues Element desselben Systems her. In Zeichen

$$\Theta_r\Theta_s = \Theta_t$$

II. Es ist stets

$$(\Theta_r\Theta_s)\Theta_t = \Theta_r(\Theta_s\Theta_t) = \Theta_r\Theta_s\Theta_t$$

III. Aus $\Theta\Theta_r = \Theta\Theta_s$ und aus $\Theta_r\Theta = \Theta_s\Theta$ folgt stets $\Theta_r = \Theta_s$.) [713, p. 302].
Then Weber proves the existence of a "principal element" (Hauptelement) [identity (Wussing)] and of the "reciprocal" (das reziproke Element) of each given element. After defining the "period" (die Periode) and "degree" (der Grad) of en element, he goes on to define abelian groups. Finally, using a group-theoretic interpretation of the theorem given by Schering [627] in 1868 (I.3) and referring to Frobenius and Stickelberger (but using the term "basis"), Weber states the basis theorem for (finite) abelian groups: "*In an abelian group G of degree h one can always find elements* $\Theta_1, \Theta_2, \ldots, \Theta_\nu$ *of degrees* $n_1, n_2, \ldots, n_\nu$ *respectively such that every element of G can be written just once in the form* $\Theta_1^{s_1}\Theta_2^{s_2} \ldots \Theta_\nu^{s_\nu}$ *provided that each of the numbers* $s_1, s_2, \ldots, s_\nu$ *belongs to a complete residue system modulo* $n_1, n_2, \ldots, n_\nu$ *respectively.* Also, $h = n_1 n_2 \cdots n_\nu$" ("*In einer Abel'schen Gruppe G vom Grade h kann man stets die Elemente* $\Theta_1, \Theta_2, \ldots, \Theta_\nu$ *von den Graden* $n_1, n_2, \ldots, n_\nu$ *so auswählen, dass in der From*

$$\Theta_1^{s_1}\Theta_2^{s_2} \ldots \Theta_\nu^{s_\nu}$$

jedes Element Θ *von G und jedes nur einmal enthalten ist, wenn* $s_1, s_2, \ldots, s_\nu$ *je einem vollständigen Restsystem nach den Modulen* $n_1, n_2, \ldots, n_\nu$ *entnommen werden.* Zugleich ergiebt sich

$$h = n_1 n_2 \ldots n_\nu.\text{") [713, pp. 306–307].}$$

In much the same way, by a mere group-theoretic formulation of well known facts, Weber, following Lejeune-Dirichlet's "Vorlesungen über Zahlentheorie" [455], goes on to develop, among other things, a theory of characters.

223. In this connection see the introduction to G. Peano, *Calcolo geometrico secondo l'Ausdehnungslehre di Grassmann, preceduto dalle operazioni della logica deduttiva* [546] dated 1888 and

L. Schendel, *Grundzüge der Algebra nach Grassmann'schen Prinzipien* [636] dated 1885. It was not until 1889 that Peano obtained his postulates for arithmetic. Peano's *I principii di Geometria* [547] also shows Grassmann's influence.

224. It should be emphasized that one of Hankel's aims in his *Theorie der complexen Zahlensysteme* [262] was to transplant Hamilton's calculus of quaternions to Germany in a "transparent" ("durchsichtig gemachten") manner, that is, in the formalistic manner. This also provided the occasion for the adoption of the technical terms "distributive," "commutative," and associative" used as technical terms in England since 1840. In spite of the fact that it does not include the word "group," the second section, "Allgemeine Formenlehre," is especially important for its relation to group theory. The same is true if §4, "Algorithmus associativer Rechnungsoperationen ohne Commutation" [262, pp. 18–24].

225. While he used a terminology that is completely foreign to us, Schröder was fully aware of the connection between his "absolute" (absolute) algebra and the theory of permutation groups. With reference to [660, p. 7] Booles *An investigation of the Laws of Thought on which are Founded the Mathematical Theories of Logic and Probability*, dated 1854, and in connection with [660, p. 30] Hankel's [262], Jordan's *Traité* and Serret's *Cours d'Algèbre supérieure*, he introduces the "multiplication with substitution symbols" ("Multiplication mit den Substitutionssymbolen") as a special case of "an associative but noncommutative operation" (einer "associativen aber nichtcommutativen Operation") [660, p. 30], and asserts, from the context, its logical possibility. Schröder indicates the generality of his approach—his language is reminiscent of the later formulations of G. Cantor—as follows: "We suppose that there is given an unlimited manifold of objects (of any kind) that are conceptually different from each other, the difference being indicated by sign or boundary. The various elements of this thought manifold are denoted by $a, b, c, \ldots$.

As examples of objects constituting such manifolds, some of which will occasionally be quoted in the sequel, I mention: names, concepts, opinions, algorithms, numbers, symbols denoting magnitudes and operations, points and point systems, or any geometric objects, quantities of substances, and so on. The given manifold may be called a *domain of numbers* (ein *Zahlengebiet*) in the broadest sense of the word" ("Es wird angenommen, dass eine unbegrenzte Mannigfaltigkeit von Objecten (irgend einer Art) gegeben sei, welche begrifflich — durch ein Merkmal oder eine Grenze — von einander unterschieden sind. Beliebige Elemente dieser gedachten Mannigfaltigkeit werden mit Buchstaben $a, b, c, \ldots$ bezeichnet.

Als Beispiele von solchen eine Mannigfaltigkeit constituierenden Objecten, welche zum Theil selbst im Verlaufe dieser Mittheilung gelegentlich citirt werden, führe ich an: Eigennamen, Begriffe, Urtheile, Algorithmen, Zahlen, Grössen- und Operationssymbole, Punkte und Punktsysteme, oder irgend welche geometrische Gebilde, Quantitäten von Substanzen, u. a. m. Die gegebene Mannigfaltigkeit kann ein *Zahlengebiet* — im weitesten Sinne des Wortes — genannt werden.") [660, p. 3].

226. v. Dyck's formulation is the following: "*All these infinitely many operations of the group which thus correspond to the identity of $\bar{G}$, form a group H and, as follows directly from the origin of its operations, it commutes with all the operations S, S′, ... of the group G or, to use the term introduced by Herr Lie, this group H is distinguished in G*" ("*Alle diese unendlich vielen Operationen der Gruppe G, welche sonach der Identität in $\bar{G}$ entsprechen, bilden eine Gruppe H und diese ist, wie aus der Entstehungsweise ihrer Operationen unmittelbar hervorgeht, mit allen Substitutionen S, S′ ... der Gruppe G vertauschbar, oder um die von Herrn Lie eingeführte Bezeichnungsweise zu gebrauchen, diese Gruppe H ist in G ausgezeichnet enthalten.*") [180, p. 12].

227. This paper is dated March 1884.

228. Cauchy had proved [108, p. 250] that every system of conjugate permutations of n elements, where n is divisible by a prime p, always contains a permutation of order p.

229. See also note 214. When it comes to the abstract group concept, the 1892 Hölder paper [290] refers explicitly also to Frobenius [208].

230. See note 219.

231. See, for example, P. du Bois-Reymond, Allgemeine Functionstheorie, Tübingen, 1882 [46]. Here Du Bois-Reymond rejects "arithmetization" (Arithmetisierung)—by that time this term, coined by Kronecker, had become the watchword of his followers—as formalistic: It sees analysis as mere play with symbols.

232. See also A. Kneser, "Arithmetische Begründung einiger algebraischer Fundamentalsätze" [411]. Here similar views are developed, but this paper's historical influence has been minor compared to that of Weber's paper quoted here.

233. In connection with commutative fields, Weber adds that "one could try to dispense with the commutativity of the elements under the operation, or to introduce a third composition. I have not further pursued the logical possibility or fruitfulness of such an extension and therefore stay, for the time being, with the simple earlier assumption" ("man könnte auch versuchen, die Vertauschbarkeit der Elemente in den Operationen aufzugeben, oder noch eine dritte Art der Composition einfügen. Ob solche Erweiterungen logisch möglich oder fruchtbar sind, habe ich nicht weiter untersucht und bleibe also für jetzt bei der gemachten einfachen Annahme stehen.") [720, p. 528]. The same page of Weber's paper even contains a hint of the concept of a zero divisor.

234. That is why H. Hasse and R. Baer published in 1930 a new edition of E. Steinitz's *Algebraische Theorie der Körper* (see [687]). See also the assessment of this paper of Steinitz by N. Bourbaki [51, p. 77].

235. The second volume of Weber's *Lehrbuch der Algebra* [722] consists of four books. The first three are of special interest from the viewpoint of group theory. Book 1—"Groups"—contains among other things the sections "General Group Theory," "Abelian Groups," and "Constitution of General Groups." Book 2—"Linear Groups"—treats the theory of finite transformation groups, including the "Group of Linear Substitutions" (in section 6) and the "Polyhedral Groups" (in section 8). Book 3 is devoted to "Applications of Group Theory," some of whose topics are metacyclic groups, applications of group theory to geometry (inflection points, double tangents), the theory of equations of degree 5 and 7, and, finally, special cases of the relation between group theory and the theory of forms, for example, the case of the relation between groups of linear ternary substitutions and forms.

236. The second volume (on the theory of permutation groups) announced in 1904 was published by J.-A. de Séguier in 1912 under the title *Groupes des Substitutions* [606].

237. A second edition appeared in Moscow in 1933. In this connection see [631, 492, 244, 458, 444].

238. We mention the following works on group axiomatics from the period up to 1910: L. E. Dickson, "Definition of a Group and a Field by Independent Postulates" [172]; and E. V. Huntington, "Note on the Definition of Abstract Groups and Fields by Sets of Independent Postulates" [299]. Both date from 1905. The following works on group axiomatics belong to a different period: R. Baer and F. Levi, Vollständige irreduzible Systeme von Gruppenaxiomen" (1932) [20]; P. Lorenzen: "Ein Beitrag zur Gruppenaxiomatik (1944) [495]; B. Stolt, *Über Axiomensysteme, die eine abstrakte Gruppe bestimmen* (1953) [629].

239. Generalizations of the group concept (groupoids, semigroups, loops) are due to, among others, H. Brandt, "Über eine Verallgemeinerung des Gruppenbegriffes" (1926) [52]; A. A. Albert: "Quasigroups" (1943–44) [14, 15]; and R. H. Bruck, "Contributions to the Theory of Loops" (1946) [67].

240. One of the earliest papers devoted explicitly to finiteness conditions is due to A. Loewy, "Zur Theorie der Gruppen linearer Substitutionen" [493], dated 1900. It reads, "In the sequel, we shall deduce a few theorems and give excellent characteristic criteria for groups of the type of a finite group [a group of linear substitutions with certain additional requirements but not necessarily finite (Wussing)] that distinguish the finite groups in this more general variety of groups, and therefore also in the class of all groups of linear substitutions" ("Ueber Gruppen vom Typus einer endlichen Gruppe wollen wir im Folgenden einige Sätze herleiten und vorzüglich charakteristische Kennzeichen angeben, durch welche die endlichen Gruppen innerhalb dieser allgemeineren Gruppengattung und daher auch unter allen Gruppen linearer Substitutionen ausgezeichnet erscheinen.") [493, pp. 225–226].

241. In this connection see, for example, S. N. Tschernikow, O. J. Schmidt, and P. S. Nowikow, *Endlichkeitsbedingungen in der Gruppentheorie*, Berlin, 1963 [544], with many additional relevant references to the literature [544, pp. 62–65, 75].

Bibliography

Abh. Gesch. Math.	Abhandlungen zur Geschichte der Mathematik
Abh. Ges. Wiss. Leipzig	Abhandlungen der mathematisch-physikalischen Klasse der Königlich-Sächsischen Gesellschaft der Wissenschaften
Abh. Göttingen	Abhandlungen der Königlichen Gesellschaft der Wissenschaften zu Göttingen. Mathematische Klasse
Abh. Göttingen	Abhandlungen der Königlichen Gesellschaft der Wissenschaften zu Göttingen
Abh. Math. Sem. Hamburg	Abhandlungen aus dem mathematischen Seminar der Hamburgischen Universität
Acta math.	Acta mathematica
Amer. Journ. Math.	American Journal of Mathematics
Ann. Ecole Norm.	Annales de l'Ecole Normale
Ann. math. Gergonne	Annales de mathématiques pures et appliquées
Ann. mat. pur. appl.	Annali di Matematica pura ed applicata
Ann. Sc. mat. fis.	Annali di Scienze matematiche e fisiche
Arch. Hist. Ex. Sc.	Archive for History of Exact Sciences
Arch. Math. Naturv.	Archiv for Mathematik og Naturvidenskab
Atti Acc. Neapel	Atti della Reale Accademia di Napoli
Atti Ist. Lombardo	Atti dell' Istituto Reale Lombardo di scienze, lettere ed arti
Ber. Verh. Sächs. Ges. Wiss.	Berichte über die Verhandlungen der Sächsischen Gesellschaft der Wissenschaften, Mathematisch-physikalische Klasse
Bull. Amer. Math. Soc.	Bulletin of the American Mathematical Society
Bull. Sc. math. Darboux	Bulletin des Sciences mathématiques et astronomiques
Bull. Sc. math. Férussac	Bulletin des Sciences mathématiques de M. Férussac
Bull. Soc. math. France	Bulletin de la Société mathématique de France
Chr. Forh. Aar	Christiania Forhandlinger i Videnskabs — Selskabet i Christiania. Aar
Denkschr. Akad. Wien	Denkschriften der kaiserlichen Akademie der Wissenschaften zu Wien. Mathematisch-naturwissenschaftliche Klasse
Giorn. Matem.	Giornale di Matematiche
Hist. Acad. Sc. Paris	Histoire de l'Académie des Sciences Paris
Jahrb. Fortschr. Math.	Jahrbuch über die Fortschritte der Mathematik
JB. dt. Math. Ver.	Jahresbericht der deutschen Mathematikervereinigung
Journ. Crelle	Journal für die reine und angewandte Mathematik
Journ. Ecole polyt.	Journal de l'Ecole polytechnique
Journ. math. Liouville	Journal de mathématiques pures et appliquées

Math. Ann.	Mathematische Annalen
Math. Journ. Cambridge Dublin	Cambridge and Dublin Mathematical Journal
Math. Zeitschr.	Mathematische Zeitschrift
MB. Akad. Berlin	Monatsberichte der Berliner Akademie
Mém. Acad. Bruxelles	Mémoires couronnés par l'Académie Royale des sciences et belles lettres de Bruxelles
Mém. Acad. Paris	Mémoires de l'Académie de Paris
Mem. Ist. Lombardo-Veneto	Memoire dell' Imperiale Regio Istituto del Regno Lombardo-Veneto
Mem. Ist. Naz. Ital.	Memoire dell' Istituto Nazionale Italiano, Classe di Fisica e Matematica
Mem. Mat. Fis. Soc. Ital.	Memoire di Matematica e di Fisici della Società Italiana delle Scienze
Mem. Phil. Soc. Manchester	Memoirs of the Literary and Philosophical Society of Manchester
Monatsh. Math. Phys.	Monatshefte für Mathematik und Physik
Nachr. Göttingen	Nachrichten der Königlichen Gesellschaft der Wissenschaften zu Göttingen
Nouv. Ann. Math.	Nouvelles Annales de Mathématiques
Nouv. Mém. Acad. Berlin	Nouvelles Mémoires de l'Académie Berlin
Nova acta Upsala	Nova acta regiae societatis scientiarum Upsaliensis
Nov. com. acad. Petropol.	Novi commentarii academiae Petropolitanae
Phil. Mag.	Philosophical Magazine
Phil. Trans. London	Philosophical Transactions of the Royal Society London
Proc. Amer. Acad.	Proceedings of the American Academy of Arts and Sciences
Proc. Math. Soc. London	Proceedings of the London Mathematical Society
Quart. Journ. Math.	Quaterly Journal of Pure and Applied Mathematics
Rendiconti Acc. Lincei	Rendiconti della Reale Accademia dei Lincei
SB. Akad. Berlin	Sitzungsberichte der königlich preußischen Akademie der Wissenschaften zu Berlin
SB. Akad. Heidelberg	Sitzungsberichte der Heidelberger Akademie der Wissenschaften
SB. Akad. Wiss. München	Sitzungsberichte der Akademie der Wissenschaften zu München
SB. Erlangen	Sitzungsberichte der physikalisch-medizinischen Sozietät zu Erlangen
Scient. Monthly	Scientific Monthly
Trans. Amer. Math. Soc.	Transactions of the American Mathematical Society
Verh. Ver. Heidelberg	Verhandlungen des naturhistorisch-medizinischen Vereins zu Heidelberg
WA	Reprint

[1] Abbati, P.: Lettera di Pietro Abbati Modenese al socio Paolo Ruffini da questo presentata il di 16 dicembre 1802. Mem. Mat. Fis. Soc. Ital. **10** (Part 2) (1803), 385–409.

[2] Abel, N. H.: Oeuvres complètes. Publiée aux frais de l'état norvegien par MM. L. Sylow et S. Lie, Bd. 1. Christiania 1881.

[3] Abel, N. H.: Oeuvres complètes. Publiée aux frais de l'état norvegien par MM. L. Sylow et S. Lie, Bd. 2, Christiania 1881.

[4] Abel, N. H.; Galois, E.: Abhandlungen über die Algebraische Auflösung von Gleichungen. German ed. H. Maser. Berlin 1889.

[5] Abel, N. H.: Mémoire sur les équations algébriques, ou l'on démontre l'impossibilité de la résolution de l'équation générale du cinquième degré. Christiania 1824. WA: [2], pp. 28–33.

[6] Abel, N. H.: Beweis der Unmöglichkeit der algebraischen Auflösung der allgemeinen Gleichungen, welche den vierten Grad übersteigen. Journ. Crelle **1** (1826). WA: [4], pp. 8–28.

[7] Abel, N. H.: Démonstration de l'impossibilité de la résolution algébrique des équations générales qui passent le quatrième degré. Journ. Crelle **1** (1826). WA: [2], pp. 66–87.

[8] Abel, N. H.: Recherches sur la série

$$1 + \frac{m}{1} x + \frac{m(m-1)}{1 \cdot 2} x^2 + \frac{m(m-1)(m-2)}{1 \cdot 2 \cdot 3} x^3 + \cdots$$

Journ. Crelle **1** (1826). WA: [2], pp. 219–250.

[9] Abel, N. H.: Analyse du mémoire précédent. ([7].) Bull. sc. math. Férussac **6** (1826). WA: [2], pp. 87–94.

[10] Abel, N. H.: Recherches sur les fonctions elliptiques. Journ. Crelle **2** (1827). WA: [2], pp. 263–388.

[11] Abel, N. H.: Mémoire sur une classe particulière d'équations résolubles algébriquement. Journ. Crelle **4** (1829). WA: [2], pp. 478–507.

[12] Abel, N. H.: Sur la résolution algébrique des équations. (*Nachlaß.*) In: [3], pp. 217–243.

[13] Ahrens, W.: Ueber discrete Schaaren von continuirlichen Transformationen. Math. Ann. **50** (1898), pp. 518–524.

[14] Albert, A. A.: Quasigroups, I. Trans. Amer. Math. Soc. **54** (1943), 507–519.

[15] Albert, A. A.: Quasigroups, II. Trans. Amer. Math. Soc. **55** (1944), 401–419.

[16] Astronomen, Die, auf der Pleissenburg. Ed. C. Bruhns. Leipzig 1877.

[17] Bachmann, P.: Ueber Galois' Theorie der algebraischen Gleichungen. Math. Ann. **18** (1881), 449–468.

[18] Bachmann, P.: Grundlehren der neueren Zahlentheorie. Leipzig 1907.

[19] Bachmann, P.: Materialien für eine wissenschaftliche Biographie von Gauß. Ed. F. Klein and M. Brendel. Issue **1**. Über Gauß' zahlentheoretische Arbeiten. Leipzig 1911.

[20] Baer, R.; Levi, F.: Vollständige irreduzibele Systeme von Gruppenaxiomen. SB. Akad. Heidelberg, **2** (1932), 1–12.

[21] Baltzer, R.: Analytische Geometrie. Leipzig 1882.

[22] Battaglini, G.: Intorno ai sistemi di rette di primo ordine. Atti Acc. Neapel **3** (1866). WA: Giorn. matem. **6** (1868), 24–36.

[23] Battaglini, G.: Intorno ai sistemi di rette di secondo grado. Atti Acc. Neapel **3** (1866). WA: Giorn. matem. **7** (1869), 55–71.

[24] Behnke, H.: Felix Klein und die heutige Mathematik. Mathematisch-Physikalische Semesterberichte **2** (1961), 129–150.

[25] Bell, E. T.: Men of Mathematics. New York 1937.

[26] Bell, E. T.: The Development of Mathematics. 2nd ed., New York/London 1945.

[27] Bernays, P.: Die Bedeutung Hilberts für die Philosophie der Mathematik. Die Naturwissenschaften **4** (1922) 93–99.

[28] Bertrand, J.: Mémoire sur le nombre de valeurs que peut prendre une fonction quand on y permute les lettres qu'elle renferme. Journ. Ecole polyt. **18** (1845), 123–140.

[29] Bertrand, J.: La vie d'Evariste Galois par P. Dupuy. Eloges académiques, Paris, **1902**, 329–345.

[30] Bessel, A.: Ueber die Invarianten der einfachsten Systeme simultaner binärer Formen. Math. Ann. **1** (1869), 173–194.

[31] Betti, E.: Opere matematiche. Pubblicate per cura della R. Accademia de' Lincei. Vol. 1. Milano 1903.

[32] Betti, E.: Sopra la risolubilità per radicali delle equazioni algebriche irriduttibili di grado primo. Ann. Sc. mat. fis. **2** (1851). WA: [31], pp. 17–27.

[33] Betti, E.: Un teorema sulle risolventi delle equazioni risolubili per radicali. Ann. Sc. mat. fis. **2** (1851). WA: [31], pp. 28–29.

[34] Betti, E.: Sulla risoluzione delle equazioni algebriche. Ann. Sc. mat. fis. **3** (1852). WA: [31], pp. 31–80.

[35] Betti, E.: Sopra l'abbassamento delle equazioni modulari delle funzioni ellittiche. Ann. Sc. mat. fis. **4** (1853). WA: [31], pp. 81–95.

[36] Betti, E.: Un teorema sulla risoluzione analitica delle equazioni algebriche. Ann. Sc. mat. fis. **5** (1854). WA: [31], pp. 96–101.

[37] Betti, E.: Sopra la teorica delle sostituzioni. In: Ann. Sc. mat. fis. **6** (1855). WA: [31], pp. 102–123.

[38] Betti, E.: Sopra la più generale funzione algebrica che può soddisfare una equazione il grado della quale è potenza di un numero primo. Ann. Sc. mat. fis. **6** (1855). WA: [31], pp. 126–135.

[39] Betti, E.: Sur la résolution par radicaux des équations dont le degré est un puissance d'un nombre premier. Comptes rendus Paris **48** (1859). WA: [31], pp. 183–187.

[40] Betti, E.: Sur les substitutions de six lettres. Comptes rendus Paris **63** (1866), 878.

[41] Bezout, E.: Mémoire sur la résolution générale des équations de tous les degrés. Mém. Acad. Paris **1765**, 533–545.

[42] Bianchi, L.: Sopra alcune classi di gruppi di sostituzioni lineari a coefficienti complessi. Math. Ann. **43** (1893), 101–135.

[43] Bianchi, L.: Teoria dei gruppi di sostituzioni e delle equazioni algebriche secondo Galois. Pisa 1900.

[44] Birembaut, A.: Précisions sur la biographie du mathématicien Vandermonde et de sa famille. Actes du Congrès de Luxembourg, July **1953**, 530–533.

[45] Birkhoff, G.: Galois and Group Theory. Osiris **3** (1937), 260–268.

[46] Bois-Reymond, P. du: Allgemeine Functionstheorie. Tübingen 1882.

[47] Bonola, R.: Die nichteuklidische Geometrie. Historisch-kritische Darstellung ihrer Entwicklung. Authorized German translation, ed. H. Liebmann. Leipzig, Berlin 1919.

[48] Boole, G.: Researches on the Theory of Analytical Transformations with a Special Application to the Reduction of the General Equation of the Second Order. Math. Journ. Cambridge Dublin **2** (1841), 64–73.

[49] Boole, G.: Exposition of a General Theory of Linear Transformations. Math. Journ. Cambridge Dublin **3** (1843), 1–20, 106–119.

[50] Boole, G.: An Investigation of the Laws of Thought, on which are Founded the Mathematical Theories of Logic and Probabilities. London 1854.

[51] Bourbaki, N.: Eléments d'histoire des mathématiques. Paris 1960.

[52] Brandt, H.: Über eine Verallgemeinerung des Gruppenbegriffes. Math. Ann. **96** (1927), 360–366.

[53] Breuer, S.: Das Abelsche Gleichungsproblem bei Euler. JB. dt. Math. Ver. **30** (1921), 158–169.

[54] Briefwechsel zwischen Gustav Lejeune-Dirichlet und Herrn Leopold Kronecker. Ed. E. Schering. Nachr. Göttingen **1885**. WA: [423], pp. 407–431.

[55] Brioschi, F.: Sulle equazioni del moltiplicatore per la trasformazione delle funzioni ellitiche. Ann. mat. pur. appl. **1** (1858), 175–177.

[56] BRIOSCHI, F.: Sulla risoluzioni delle equazioni del quinto grado. Ann. mat. pur. appl. **1** (1858), 256–259, 326–328.
[57] BRIOSCHI, F.: Sur diverses équations analogues aux équations modulaires dans la théorie des fonctions elliptiques. Comptes rendus Paris **47** (1858), 337–341.
[58] BRIOSCHI, F.: Sul metodo di Kronecker per la risoluzioni delle equazioni di quinto grado. Atti Ist. Lombardo **1** (1859), 275–282.
[59] BRIOSCHI, F.: Sulla risolvente di Malfatti per le equazioni del quinto grado. Ann. mat. pur. appl. **5** (1863), 233–250.
[60] BRIOSCHI, F.: Sopra alcune nuove relazioni modulari. Ann. Sc. mat. fis. **3** (1866–68), 1–16.
[61] BRIOSCHI, F.: La soluzione più generale della equazioni del quinto grado. Ann. mat. pur. appl., 2nd Series, **1** (1867), 222–231.
[62] BRIOSCHI, F.: Des substitutions de la forme $\Theta(r) \equiv \varepsilon(r^{n-2} + ar^{(n-3)/2})$ pour un nombre n premier de lettres. Math. Ann. **2** (1870), 467–470.
[63] BRIOSCHI, F.: Sur l'équation du cinquième degré. Comptes rendus Paris **80** (1875), 753–757, 815–819.
[64] BRIOSCHI, F.: Sopra una classe di forme binaire. Ann. mat. pur. appl., 2nd Series, **8** (1877), 24–42.
[65] BRIOSCHI, F.: Ueber die Auflösung der Gleichungen vom fünften Grade. Math. Ann. **13** (1878), 109–160.
[66] BROWN, E. W.: The History of Mathematics. Scient. Monthly **12** (1921), 385–413.
[67] BRUCK, R. H.: Contributions to the Theory of Loops. Trans. Amer. Math. Soc. **60.** (1946), 245–354.
[68] Bulletin des sciences mathématiques et astronomiques, rédigé par MM. G. DARBOUX et J. HOUËL. Vol. 2, Paris 1871.
[69] BURKHARDT, H.: Die Anfänge der Gruppentheorie und Paolo Ruffini. Abh. Gesch. Math. **6** (1892), 119–159.
[70] BURKHARDT, H.: Endliche discrete Gruppen. Encyklopädie der mathematischen Wissenschaften mit Einschluß ihrer Anwendungen. Vol. 1, Part 1. Leipzig 1898–1904, pp. 208–226.
[71] BURNSIDE, W.: Notes on the Theory of Groups of Finite Order. Proc. Math. Soc. London **25** (1894), 9–18.
[72] BURNSIDE, W.: On a Class of Groups Defined by Congruences. Proc. Math. Soc. London **25** (1894), 113–139; **26** (1895), 58–106.
[73] BURNSIDE, W.: On the Isomorphism of a Group with Itself. Proc. Math. Soc. London **27** (1896), 354–367.
[74] BURNSIDE, W.: Note on the Symmetric Group. Proc. Math. Soc. London **28** (1897), 119–129.
[75] BURNSIDE, W.: Theory of Groups of Finite Order. Cambridge 1897.
[76] BURNSIDE, W.: On Linear Homogeneous Continuous Groups whose Operations are Permutable. Proc. Math. Soc. London **29** (1898), 325–352.
[77] BURNSIDE, W.: On the Continuous Group that is Defined by Any Given Group of Finite Order. Proc. Math. Soc. London **29** (1898), 546–565.
[78] BURNSIDE, W.: On the Reduction of a Linear Substitution to its Canonical Form. Proc. Math. Soc. London **30** (1899), 180–194.
[79] BURNSIDE, W.: Note on the Simple Group of Order 504. Math. Ann. **52** (1899), 174–176.
[80] BURNSIDE, W.: On Transitive Groups of Degree n and Class $n-1$. Proc. Math. Soc. London **32** (1901), 240–246.
[81] BURNSIDE, W.: On Group-Characteristics. Proc. Math. Soc. London **33** (1901), 146–162.
[82] BURNSIDE, W.: On some Properties of Groups of Odd Order. Proc. Math. Soc. London **33** (1901), 162–185, 257–268.
[83] BURNSIDE, W.: On the Condition of Reducibility of Any Group of Linear Substitutions. Proc. Math. Soc. London, 2nd Series, **3** (1905), 430–434.

[84] Burnside, W.: Theory of Groups of Finite Order. 2nd ed., Cambridge 1911.
[85] Cajori, F.: Horner's Method of Approximation Anticipated by Ruffini. Bull. Amer. Math. Soc. 2nd Series, **17**, 409–414.
[86] Capelli, A.: Sopra l'isomorfismo dei gruppi di sostituzioni. Giorn. Matem. **16** (1878), 32–68.
[87] Capelli, A.; Garbieri, G.: Corso di analisi algebraica. Part 1: Teorie introduttiorie. Padova 1886.
[88] Capelli, A.: Lezioni di algebra completamente. Napoli 1895.
[89] Carathéodory, C.: Die Bedeutung des Erlanger Programms. Die Naturwissenschaften, Jg. 7 (1919), 297–300.
[90] Carnot, L. N. M.: Geometrie der Stellung. German. ed. H. C. Schumacher. Part 1. Altona 1808.
[91] Cartan, E.: Oeuvres complètes. Partie I: Groupes de Lie. Vol. 1. Paris 1952.
[92] Cartan, E.: Sur la structure des groupes simples finis et continus. Comptes rendus Paris **116** (1893). WA: [91], pp. 99–101.
[93] Cartan, E.: Sur la structure des groupes finis et continus. Comptes rendus Paris **116** (1893). WA: [91], pp. 103–105.
[94] Cartan, E.: Ueber die einfachen Transformationsgruppen. Abh. Ges. Wiss. Leipzig 1893. WA: [91], pp. 107–132.
[95] Cartan, E.: Sur la réduction de la structure d'un groupe à sa forme canonique. Comptes rendus Paris **119** (1894). WA: [91], pp. 133–135.
[96] Cartan, E.: Sur la structure des groupes de transformations finis et continus. Paris 1894. 2nd. ed., Paris 1933. WA: [91], pp. 137–286.
[97] Cassina, U.: Sur l'histoire des concepts fondamentaux de la géometrie projective. Université de Paris, Palais de la Découverte, Série D, Nr. 50, 1957.
[98] Cauchy, A.-L.: Méthode pour déterminer à priori le nombre des racines réelles négatives d'une équation d'un degré quelconque. Paris 1812.
[99] Cauchy, A.-L.: Sur le nombre des valeurs qu'une fonction peut acquérir, lorsqu'on y permute de toutes les manières possibles les quantités qu'élle renferme. Journ. Ecole polyt. **10** (1815), 1–28.
[100] Cauchy, A.-L.: Sur les Fonctions qui ne peuvent obtenir que deux valeurs égales et de signes contraires par suite des transpositions opérées entre les variables qu'elles renferment. Journ. Ecole polyt. **10** (1815), 29–112.
[101] Cauchy, A.-L.: Sur la détermination du nombre des racines dans les équations algébriques. Journ. Ecole polyt. **10** (1815), 547–548.
[102] Cauchy, A.-L.: Sur les racines imaginaires des équations. Journ. Ecole polyt. **11** (1820), 411–416.
[103] Cauchy, A.-L.: Sur la résolution des équations. (Extrait d'une lettre adressée à M. Libri.) Comptes rendus Paris **4** (1837), 362–365.
[104] Cauchy, A.-L.: Note sur la résolution des équations de degré quelconque. Comptes rendus Paris **5** (1837), 301–304.
[105] Cauchy, A.-L.: Sur la théorie des nombres, et en particulier sur les formes quadratiques des nombres premiers. Comptes rendus Paris **9** (1839), 473–474.
[106] Cauchy, A.-L.: Méthode simple et nouvelle pour la détermination complète des sommes alternées, formées avec les racines primitives des équations binomes. Journ. math. Liouville **5** (1840), 154–168.
[107] Cauchy, A.-L.: Sur la synthèse algébrique. Comptes rendus Paris **16** (1843), 867–881, 967–976, 1039–1052.
[108] Cauchy, A.-L.: Mémoire sur les arrangements que l'on peut former avec des lettres données, et sur les permutations ou substitutions à l'aide desquelles on passe d'un arrangement à un autre. In: Exercises d'analyse et de physique mathématique, 3, Paris 1844, pp. 151–252.

[109] Cauchy, A.-L.: Sur le nombre des valeurs égales ou inégales que peut acquérir une fonction de n variables indépendantes, quand on permute ces variables entre elles d'une manière quelconque. Comptes rendus Paris **21** (1845), 593–607, 668–679, 727–742, 779–797.
[110] Cauchy, A.-L.: Mémoire sur diverses propriétés remarquables des substitutions régulières ou irrégulières, et des systèmes de substitutions conjuguées. Comptes rendus Paris **21** (1845), 835–852, 895–902, 931–933, 972–987, 1025–1041.
[111] Cauchy, A.-L.: Mémoire sur les premiers termes de la série des quantités qui sont propres à représenter le nombre des valeurs distinctes d'une fonction des n variables indépendantes. Comptes rendus Paris **21** (1845), 1093–1101.
[112] Cauchy, A.-L.: Mémoire sur la résolution des équations linéaires symboliques, et sur les conséquences remarquables que cette résolution entraîne après elle dans la théorie des permutations. Comptes rendus Paris **21** (1845), 1123–1134.
[113] Cauchy, A.-L.: Mémoire sur les substitutions permutables entre elles. Comptes rendus Paris **21** (1845), 1188–1199.
[114] Cauchy, A.-L.: Note sur la réduction des fonctions transitives aux fonctions intransitives, et sur quelques propriétés remarquables des substitutions qui n'altèrent pas la valeur d'une fonction transitive. Comptes rendus Paris **21** (1845), 1199–1201.
[115] Cauchy, A.-L.: Note sur les substitutions qui n'altèrent pas la valeur d'une fonction, et sur la forme régulière que prennent toujours celles d'entre elles qui renferment un moindre nombre de variables. Comptes rendus Paris **21** (1845), 1234–1238.
[116] Cauchy, A.-L.: Mémoire sur diverses propriétés des systèmes de substitutions, et particulièrement de ceux qui sont permutables entre eux. Comptes rendus Paris **21** (1845), 1238–1254.
[117] Cauchy, A.-L.: Note sur les fonctions caractéristiques des substitutions. Comptes rendus Paris **21** (1845), 1254–1255.
[118] Cauchy, A.-L.: Mémoire sur le nombre et la forme des substitutions qui n'altèrent pas la valeur d'une fonction de plusieurs variables indépendantes. Comptes rendus Paris **21** (1845), 1287–1300.
[119] Cauchy, A.-L.: Applications diverses des principes établis dans les précédents Mémoires aux fonctions qui ne renferment pas plus de six variables. Comptes rendus Paris **21** (1845), 1356–1369.
[120] Cauchy, A.-L.: Mémoire sur les fonctions de cinq ou six variables, et spécialement sur celles qui sont doublement transitives. Comptes rendus Paris **21** (1845), 1401–1409.
[121] Cauchy, A.-L.: Mémoire sur les fonctions de cinq ou six variables, et spécialement sur celles qui sont doublement transitives. Comptes rendus Paris **22** (1846), 2–31.
[122] Cauchy, A.-L.: Mémoire sur un nouveau calcul qui permet de simplifier et d'étendre la théorie des permutations. Comptes rendus Paris **22** (1846), 53–63.
[123] Cauchy, A.-L.: Applications diverses de ce nouveau calcul. Comptes rendus Paris **22** (1846), 99–107.
[124] Cauchy, A.-L.: Recherches sur un système d'équations simultanées, dont les unes se déduisent des autres à l'aide d'une ou de plusieurs substitutions. Comptes rendus Paris **22** (1846), 159–160.
[125] Cauchy, A.-L.: Note sur diverses propriétés de certaines fonctions algébriques. Comptes rendus Paris **22** (1846), 160.
[126] Cauchy, A.-L.: Sur la résolution directe d'un système d'équations simultanées, dont les unes se déduisent des autres à l'aide d'une ou de plusieurs substitutions. Comptes rendus Paris **22** (1846), 193–196.
[127] Cauchy, A.-L.: Sur la résolution des équations symboliques non linéaires. Comptes rendus Paris **22** (1846), 235–238.
[128] Cauchy, A.-L.: Note sur un théorème fondamental relatif à deux systèmes de substitutions conjuguées. Comptes rendus Paris **22** (1846), 630–632.
[129] Cayley, A.: The Collected Mathematical Papers. Vol. 1, Cambridge 1889.

[130] CAYLEY, A.: The Collected Mathematical Papers. Vol. 2, Cambridge 1889.
[131] CAYLEY, A.: The Collected Mathematical Papers. Vol. 4, Cambridge 1891.
[132] CAYLEY, A.: The Collected Mathematical Papers. Vol. 8, Cambridge 1895.
[133] CAYLEY, A.: The Collected Mathematical Papers. Vol. 10, Cambridge 1896.
[134] CAYLEY, A.: On Certain Results Relating to Quaternions. Phil. Mag. **26** (1845). WA: [129], pp. 123–126.
[135] CAYLEY, A.: On the Theory of Linear Transformations. Math. Journ. Cambridge Dublin **4** (1845). WA: [129], pp. 80–94.
[136] CAYLEY, A.: On Linear Transformations. Math. Journ. Cambridge Dublin **1** (1846). WA: [129], pp. 95–112.
[137] CAYLEY, A.: On Homogenous Functions of the Third Order with Three Variables. Math. Journ. Cambridge Dublin **1** (1846). WA: [129], pp. 230–233.
[138] CAYLEY, A.: Note on the Theory of Permutations. Phil. Mag. **34** (1849), 527–529. WA: [129], pp. 423–424.
[139] CAYLEY, A.: Note sur la théorie des hyperdéterminants. Journ. Crelle **42** (1851). WA: [129], pp. 577–579.
[140] CAYLEY, A.: On the Theory of Groups, as Depending on the Symbolic Equation $\theta^n = 1$. Phil. Mag. **7** (1854). WA: [130], pp. 123–130.
[141] CAYLEY, A.: On the Theory of Groups, as Depending on the Symbolic Equation $\theta^n = 1$. — Second Part. Phil. Mag. **7** (1854). WA: [130], pp. 131–132.
[142] CAYLEY, A.: An Introductory Memoir upon Quantics. Phil. Trans. London **144** (1854). WA: [130], pp. 221–234.
[143] CAYLEY, A.: A Second Memoir upon Quantics. Phil. Trans. London **146** (1856). WA: [130], pp. 250–275.
[144] CAYLEY, A.: A Memoir on the Conditions for the Existence of Given Systems of Equalities among the Roots of an Equation. In: Phil. Trans. London **147** (1857). WA: [130], pp. 465–470.
[145] CAYLEY, A.: A Sixth Memoir upon Quantics. Phil. Trans. London **149** (1859). WA: [130], pp. 561–592.
[146] CAYLEY, A.: On a New Analytical Representation of Curves in Space. Quart. Journ. Math. **3** (1860). WA: [131], pp. 446–455.
[147] CAYLEY, A.: On a New Analytical Representation of Curves in Space. Quart. Journ. Math. **5** (1862). WA: [131], pp. 490–494.
[148] CAYLEY, A.: A Theorem on Groups. Math. Ann. **13** (1878). WA: [133], pp. 149–152.
[149] CAYLEY, A.: On the Theory of Groups. Proc. Math. Soc. London **9** (1878). WA: [133], pp. 324–330.
[150] CAYLEY, A.: The Theory of Groups. Amer. Journ. Math. **1** (1878). WA: [133], pp. 401–403.
[151] CAYLEY, A.: The Theory of Groups; Graphical Representation. Amer. Journ. Math. **1** (1878). WA: [133], pp. 403–405.
[152] Centenaire, Le, de l'Ecole normale 1795–1895. Paris 1895.
[153] CHASLES, M.: Aperçu historique sur l'origine et le developpement des méthodes en géométrie, particulièrement de celles qui se rapportent à la géométrie moderne, suivi d'une Mémoire de géométrie sur deux principes généraux de la science, la dualité et l'homographie. Mém. Acad. Bruxelles, XI (1837), 1–851.
[154] CHASLES, M.: Rapport sur les progrès de la géometrie. Paris 1870.
[155] CHEVALIER, A.: Nécrologie. (First article on GALOIS.) In: Revue encyclopédique, Paris 1832, pp. 744–754.
[156] CHRYSTAL, G.: Textbook of Algebra. Vol. 1, Edinburgh/London 1886. Vol. 2, Edinburgh/London 1889.
[157] CLAIRAUT, A.-C.: Elémens d'algèbre. 5th. ed., Paris 1797.
[158] CLEBSCH, A.: Ueber die Abbildung algebraischer Flächen, insbesondere der vierten und fünften Ordnung. Math. Ann. **1** (1869), 253–316.

[159] Clebsch, A.: Ueber die Anwendung der quadratischen Substitution auf die Gleichungen 5ten Grades und die geometrische Theorie des ebenen Fünfseits. Math. Ann. **4** (1871), 284–345.
[160] Courant, R.: Felix Klein. Die Naturwissenschaften, Jg. 13 (1925), 765–772.
[161] Dalmas, A.: Evariste Galois, révolutionnaire et géomètre. Paris 1956.
[162] Dedekind, R.: Sur la théorie des nombres entiers algébriques. Bull. Sc. math. Darboux **11** (1876), 278–288; **12** (1877), 17–41, 69–92, 144–164, 207–248.
[163] Dedekind, R.: Erläuterungen zu Riemanns Fragmenten über die Grenzfälle der elliptischen Modulfunctionen. In: [582], pp. 466–478.
[164] Dedekind, R.; Weber, H.: Theorie der algebraischen Functionen einer Veränderlichen. Journ. Crelle **92** (1882), 181–290.
[165] Dedekind, R.: Was sind und was sollen die Zahlen? Braunschweig 1888. WA: 10. Reprinted, Braunschweig and Berlin 1965.
[166] Dedekind, R.: Stetigkeit und irrationale Zahlen. 2nd ed. Braunschweig 1892. WA: 8. Reprinted, Braunschweig and Berlin 1965.
[167] Dedekind, R.: Ueber Gruppen, deren sämmtliche Theiler Normaltheiler sind. Math. Ann. **48** (1897), 548–561.
[168] Dehn, M.: Hilberts geometrisches Werk. Die Naturwissenschaften. Jg. **10** (1922), 77–80.
[169] Dickson, L. E.: Systems of Simple Groups Derived from the Orthogonal Group. Bull. Amer. Math. Soc. **4** (1898), 382–389.
[170] Dickson, L. E.: The Structure of the Hypoabelian Groups. Bull. Amer. Math. Soc. **4** (1898), 495–510.
[171] Dickson, L. E.: Linear Groups with an Exposition of the Galois Field Theory. Leipzig 1901.
[172] Dickson, L. E.: Definitions of a Group and a Field by Independent Postulates. Trans. Amer. Math. Soc. **6** (1905), 198–204.
[173] Dickson, L. E.: History of the Theory of Numbers. Vol. 3. Quadratic and Higher Forms. Washington 1923.
[174] Dickson, L. E.: Modern algebraic Theories. Chicago 1926.
[175] Dieudonné, J.: La Géométrie des groupes classiques. Ergebnisse der Mathematik, New Series, Issue 5. Berlin/Göttingen/Heidelberg 1955.
[176] Dieudonné, J.: L'oeuvre mathématique de C. F. Gauss. Université de Paris, Palais de la Découverte, Série D, N^0 79, 1961.
[177] Druckenmüller, N.: Die Übertragungsprincipien der analytischen Geometrie. Vol. 1. Trier 1842.
[178] Dupuy, P.: La vie d'Evariste Galois. Ann. Ecole Norm. **13** (1896), 197–266.
[179] Dyck, W. v.: Über regulär verzweigte Riemann'sche Flächen und durch sie definirte Irrationalitäten. Munich 1879.
[180] Dyck, W. v.: Gruppentheoretische Studien. Math. Ann. **20** (1882), 1–44.
[181] Dyck, W. v.: Gruppentheoretische Studien. II. Ueber die Zusammensetzung einer Gruppe discreter Operationen, über ihre Primitivität und Transitivität. Math. Ann. **22** (1883), 70–108.
[182] Eisenstein, G.: Beiträge zur Theorie der elliptischen Functionen. III. Fernere Bemerkungen zu den Transformationsformeln. Journ. Crelle **32** (1846), 59–70.
[183] Eisenstein, G.: Beiträge zur Theorie der elliptischen Functionen. VI. Genaue Untersuchung der unendlichen Doppelproducte, aus welchen die elliptischen Functionen als Quotienten zusammengesetzt sind, und der mit ihnen zusammenhängenden Doppelreihen. Journ. Crelle **35** (1847), 153–184, 185–274.
[184] Engel, F.; Stäckel, P.: Die Theorie der Parallellinien von Euklid bis auf Gauss. Leipzig 1895.
[185] Engel, F.: Sophus Lie. (Nachruf). Ber. Verh. Sächs. Ges. Wiss. **51** (1899), XI–LXI.
[186] Enneper, A.: Elliptische Funktionen. Theorie und Geschichte. Halle 1876.

[187] Encyklopädie der mathematischen Wissenschaften mit Einschluß ihrer Anwendungen. Vol. 1, Part 1. Leipzig 1898–1904.
[188] Encyklopädie der mathematischen Wissenschaften und Einschluß ihrer Anwendungen. Vol. 1, Part 1. Leipzig 1900–1904.
[189] Encyklopädie der mathematischen Wissenschaften mit Einschluß ihrer Anwendungen. Vol. 3, Part 1, 1st Half. Leipzig 1907–1910.
[190] Encyklopädie der mathematischen Wissenschaften mit Einschluß ihrer Anwendungen. Vol. 3, Part 1, 2nd Half. Leipzig 1914–1931.
[191] ERDMANN, B.: Die Axiome der Geometrie. Leipzig 1877.
[192] EULER, L.: De resolutione aequationum cuiusuis gradus. Nov. com. acad. Petropol. **9** (for 1762/63) (1764), 70–98.
[193] EULER, L.: Theoremata circa residua ex divisione potestatum relicta. Nov. com. acad. Petropol (1758/59), Petersburg, 1761. WA: [194], pp. 493–518.
[194] EULER, L.: Opera omnia. Ser. I, Bd. II. Leipzig/Berlin 1915.
[195] FORSYTH, A. R.: Biographical Notice of Arthur Cayley. In: [132], pp. IX–XLIV.
[196] FREGE, G.: Begriffsschrift. Eine der arithmetischen nachgebildete Formelsprache des reinen Denkens. Halle 1879.
[197] FREGE, G.: Die Grundlagen der Arithmetik. Eine logisch mathematische Untersuchung über den Begriff der Zahl. Breslau 1884.
[198] FREUDENTHAL, H.: Neuere Fassungen des Riemann-Helmholtz-Lieschen Raumproblems. In: [540], pp. 92–97.
[199] FREUDENTHAL, H.: Die Grundlagen der Geometrie um die Wende des 19. Jahrhunderts. Mathematisch-Physikalische Semesterberichte **8** (1961), 1–25.
[200] FRICKE, R.: Zur gruppentheoretischen Grundlegung der automorphen Functionen. Math. Ann. **42** (1893), 564–594.
[201] FRICKE, R.; KLEIN, F.: Vorlesungen über die Theorie der automorphen Functionen. Vol. 1. Die gruppentheoretischen Grundlagen. Leipzig 1897.
[202] FRICKE, R.; KLEIN, F.: Vorlesungen über die Theorie der automorphen Functionen. Vol. 2 in 3 Parts. Leipzig 1901, 1911, 1912.
[203] FRICKE, R.: Elliptische Funktionen. (With preliminary studies by J. HARKNESS and W. WIRTINGER.) In: Encyklopädie der mathematischen Wissenschaften mit Einschluß ihrer Anwendungen. Vol. 2, Part 2. Leipzig 1901–1921, pp. 177–348.
[204] FRICKE, R.: Automorphe Funktionen mit Einschluß der elliptischen Modulfunktionen. In: Encyklopädie der mathematischen Wissenschaften mit Einschluß ihrer Anwendungen. Vol. 2: Analysis. Part 2. Leipzig 1901–1921, pp. 349–470.
[205] FROBENIUS, G.: Ueber lineare Substitutionen und bilineare Formen. Journ. Crelle **84** (1878), 1–63.
[206] FROBENIUS, G.: Theorie der linearen Formen mit ganzen Coefficienten. Journ. Crelle **86** (1879), 146–208.
[207] FROBENIUS, G.; STICKELBERGER, L.: Ueber Gruppen von vertauschbaren Elementen. Journ. Crelle **86** (1879), 217–262.
[208] FROBENIUS, G.: Neuer Beweis des Sylowschen Satzes. Journ. Crelle **100** (1887), 179–181.
[209] FROBENIUS, G.: Über endliche Gruppen. SB. Akad. Berlin **1895**, 81–112.
[210] FROBENIUS, G.: Über Gruppencharaktere. MB. Akad. Berlin **1896**, 985–1021.
[211] FROBENIUS, G.: Über die Primfactoren der Gruppendeterminante. MB. Akad. Berlin **1896**, 1343–1382.
[212] FROBENIUS, G.: Über die Darstellung der endlichen Gruppen durch lineare Substitutionen. I. MB. Akad. Berlin **1897**, 994–1015.
[213] FROBENIUS, G.: Über die Darstellung der endlichen Gruppen durch lineare Substitutionen. II. MB. Akad. Berlin **1899**, 482–500.
[214] FROBENIUS, G.; SCHUR, I.: Über die reellen Darstellungen der endlichen Gruppen. MB. Akad. Berlin **1906**, 186–208.

[215] FROBENIUS, G.; SCHUR, I.: Über die Äquivalenz der Gruppen linearer Substitutionen. MB. Akad. Berlin, **1906**, 209–217.
[216] FUBINI, G.: Introduzione alla teoria dei gruppi discontinui e delle funzioni automorfe. Pisa 1908.
[217] FUCHS, L.: Ueber die linearen Differenzialgleichungen zweiter Ordnung, welche algebraische Integrale besitzen, und eine neue Anwendung der Invariantentheorie. Nachr. Göttingen **1875**, 568–581.
[218] FUCHS, L.: Ueber die linearen Differentialgleichungen zweiter Ordnung, welche algebraische Integrale besitzen, und eine neue Anwendung der Invariantentheorie. Journ. Crelle **81** (1876), 97–142.
[219] FUCHS, L.: Ueber die linearen Differentialgleichungen zweiter Ordnung, welche algebraische Integrale besitzen. Journ. Crelle **85** (1878), 1–25.
[220] FUCHS, L.: Ueber eine Klasse von Funktionen mehrerer Variabeln, welche durch Umkehrung der Integrale von Lösungen der linearen Differenzialgleichungen mit rationalen Coefficienten entstehen. Nachr. Göttingen 1880, 170–176.
[221] GALOIS, E.: Oeuvres mathématiques d'Evariste Galois, publiées sous les auspices de la Société Mathématique de France. Avec une introduction par M. E. PICARD. Paris 1897.
[222] GALOIS, E.: Manuscripts de Evariste Galois, publiés par J. Tannery. Paris 1908.
[223] GALOIS, E.; ABEL, N. H.: Abhandlungen über die Algebraische Auflösung von Gleichungen. German ed. H. MASER, Berlin 1889.
[224] GALOIS, E.: Démonstration d'un théorème sur les fractions continues périodiques. Ann. math. Gergonne **19** (1828/29). WA: [221], pp. 1–8.
[225] GALOIS, E.: Analyse d'un Mémoire sur la résolution algébrique des équations. Bull. Sc. math. Férussac **13** (April 1830). WA: [221], pp. 11–12.
[226] GALOIS, E.: Note sur la résolution des équations numériques. Bull. Sc. math. Férussac **13** (June 1830). WA: [221], pp. 13–14.
[227] GALOIS, E.: Sur la théorie des nombres. Bull. Sc. math. Férussac **13** (June 1830). WA: [221], pp. 15–23.
[228] GALOIS, E.: Notes sur quelques points d'analyse. Ann. math. Gergonne **21** (1830/31). WA: [221], pp. 9–10.
[229] GALOIS, E.: Mémoire sur les conditions de résolubilité des équations par radicaux. (Dated 6 January 1831.) Ed. J. LIOUVILLE, 1846. WA: [221], pp. 33–50.
[230] GALOIS, E.: Discours préliminaire. In: [222], pp. 21–22.
[231] GALOIS, E.: Des équations primitives qui sont solubles par radicaux. Ed. J. LIOUVILLE, 1846. WA: [221], pp. 51–61.
[232] GALOIS, E.: Lettre à Auguste Chevalier. (Dated 29 May 1832.) In: Revue encyclopédique, Sept. 1832. WA: [221], pp. 25–32.
[233] GARBIERI, G.; CAPELLI, A.: Corso di analisi algebraica. Part 1: Teorie introduttiorie. Padova, 1886.
[234] GAUSS, C. F.: Werke. Vol. 2. Göttingen 1876.
[235] GAUSS, C. F.: Werke. Vol. 8. Göttingen 1900.
[236] GAUSS, C. F.: Disquisitiones arithmeticae. Leipzig 1801.
[237] GAUSS, C. F.: Untersuchungen über höhere Arithmetik. German ed. H. MASER, Berlin 1889.
[238] Gedenkband anläßlich des 100. Todestages von C. F. Gauß am 23. Febr. 1955. Ed. H. REICHARDT. Leipzig 1957.
[239] GEISER, C. F.: Zur Erinnerung an Jakob Steiner. Schaffhausen 1874.
[240] GERGONNE, J.-D.: Considérations philosophiques sur les élémens de la science de l'étendue. Ann. math. Gergonne **16** (1825/26), 209–232.
[241] GERHARDT, C. J.: Geschichte der Mathematik in Deutschland. Munich 1877.
[242] GIERSTER, J.: Die Untergruppen der Galoisschen Gruppe der Modulargleichungen für den Fall eines primzahligen Transformationsgrades. In: Math. Ann. **18** (1881), 319–365.

[243] Gnedenko, B. V. (Гнеденко, Б. В.): M. V. Ostorogradski: His Life and His Scientific and Pedagogical Activity. Moscow 1952. (Russian.)
[244] Gnedenko, B. V. (Гнеденко, Б. В.): On the Development of Mathematics in the Ukraine. Istoriko-Matematičeskie Issledovanija **IX** (1956), 403–426. (Russian.)
[245] Gonseth, F.: La géométrie et le problème de l'espace. Neuchâtel 1955.
[246] Gordan, P.: Beweis, dass jede Covariante und Invariante einer binären Form eine ganze Function mit numerischen Coefficienten einer endlichen Anzahl solcher Formen ist. In: Journ. Crelle **69** (1868), 323–354.
[247] Gordan, P.: Ueber ternäre Formen dritten Grades. Math. Ann. **1** (1869), 90–128.
[248] Gordan, P.: Die simultanen Systeme binärer Formen. Math. Ann. **2** (1870), 227–280.
[249] Gordan, P.: Ueber endliche Gruppen linearer Transformationen einer Veränderlichen. Math. Ann. **12** (1877), 23–46.
[250] Gordan, P.: Ueber Gleichungen siebenten Grades mit einer Gruppe von 168 Substitutionen. Math. Ann. **20** (1882), 515–530.
[251] Gordan, P.: Vorlesungen über Invariantentheorie. Ed. G. Kerschensteiner. Vol. 1. Leipzig 1885.
[252] Gordan, P.: Vorlesungen über Invariantentheorie. Ed. G. Kerschensteiner. Vol. 2. Leipzig 1887.
[253] Grassmann, H.: Ausdehnungslehre, vollständig und in strenger Form. Berlin 1862.
[254] Grave, D. A. (Граве, Д. А.): Theory of Finite Groups. Kiev 1908.
[255] Grave, D. A. (Граве, Д. А.): Elements of Higher Algebra. Kiev 1914.
[256] Groth, P.: Entwicklungsgeschichte der mineralogischen Wissenschaften. Berlin 1926.
[257] Gundelfinger, S.: Zur Theorie der ternären cubischen Formen. Math. Ann. **4** (1871), 144–163.
[258] Gundelfinger, S.: Ueber einige allgemeine Theoreme aus der neueren Algebra. Math. Ann. **4** (1871), 164–168.
[259] Halphen, G. H.: Traité des fonctions elliptiques. Vol. 3. Paris 1891.
[260] Hamilton, W. R.: Lectures of Quaternions. Dublin 1853.
[261] Hamilton, W. R.: Elements of Quaternions. London 1866.
[262] Hankel, H.: Theorie der complexen Zahlensysteme, insbesondere der gemeinen imaginären Zahlen und der Hamilton'schen Quaternionen nebst ihrer geometrischen Darstellung. Leipzig 1867.
[263] Hankel, H.: Die Elemente der projectivischen Geometrie. Leipzig 1875.
[264] Hasse, H.: Die moderne algebraische Methode. JB. dt. Math. Ver. **39** (1930), 22–34.
[265] Hecker, J.: Über Ruffini's Beweis für die Unmöglichkeit der algebraischen Auflösung der allgemeinen Gleichung von einem höheren als dem vierten Grade. Bonn 1886.
[266] Heffter, L.: Ueber metacyklische Gruppen und Nachbarconfigurationen. Math. Ann. **50** (1898), 261–268.
[267] Heffter, L.; Köhler, C.: Lehrbuch der analytischen Geometrie. Vol. 1. Leipzig/Berlin 1905.
[268] Helmholtz, H.: Wissenschaftliche Abhandlungen von H. Helmholtz. Vol. 2. Leipzig 1883.
[269] Helmholtz, H.: Vorträge und Reden von H. Helmholtz. Vol. 2. 5th ed. Braunschweig 1903.
[270] Helmholtz, H.: Ueber die thatsächlichen Grundlagen der Geometrie. Verh. Ver. Heidelberg **4** (1866). WA: [268], pp. 610–617.
[271] Helmholtz, H.: Ueber die Thatsachen, die der Geometrie zum Grunde liegen. Nachr. Göttingen **1868**. WA: [268], pp. 618–639.
[272] Helmholtz, H.: Ueber den Ursprung und die Bedeutung der geometrischen Axiome. In: Populärwissenschaftliche Vorträge, Issue 3; Braunschweig, 1876. WA: [269], pp. 1–31.
[273] Hensel, K.: Theorie der algebraischen Zahlen. Leipzig/Berlin 1908.
[274] Hermite, Ch.: Sur les fonctions algébriques. Comptes rendus Paris **32** (1851), 458–461.

[275] HERMITE, CH.: Sur la résolution de l'équation du cinquième degré. Comptes rendus Paris **46** (1858), 508–515.
[276] HERMITE, CH.: Sur la théorie des équations modulaires. Comptes rendus Paris **48** (1859), 940–947, 1079–1084, 1095–1102; **49** (1859), 16–24, 110–118, 141–144.
[277] HILBERT, D.: Ueber die Theorie der algebraischen Formen. Math. Ann. **36** (1890), 473–534.
[278] HILBERT, D.: Ueber die Theorie der algebraischen Invarianten. Nachr. Göttingen **1891**, 232–242; **1892**, 6–16, 439–449.
[279] HILBERT, D.: Die Theorie der algebraischen Zahlkörper. JB. dt. Math. Ver. **4** (1894–95), 175–546.
[280] HILBERT, D.: Mathematische Probleme. (Lecture delivered at the International Mathematical Congress at Paris, 1900.) Nachr. Göttingen **1900**, 253–297.
[281] HILBERT, D.: Grundlagen der Geometrie. Leipzig 1900.
[282] HILBERT, D.: Grundlagen der Geometrie. 2nd ed., Leipzig 1903.
[283] HILBERT, D.: Über das Unendliche. Math. Ann. **95** (1926), 161–190.
[284] HINDENBURG, C. F.: Novi systematis permutationum, combinationum ac variationum primae lineae et logisticae serierum formulis analytico-combinatoriis per tabulas exhibendae conspectus et specimina. Leipzig 1781.
[285] Histoire de la Science. Ed. M. DAUMAS. Paris 1957.
[286] Histoire générale des sciences. Publiée sous la direction de R. TATON. Book III, Vol. I. Paris 1961.
[287] HÖLDER, O.: Zurückführung einer beliebigen algebraischen Gleichung auf eine Kette von Gleichungen. Math. Ann. **34** (1889), 26–56.
[288] HÖLDER, O.: Ueber den Söderberg'schen Beweis des Galois'schen Fundamentalsatzes. Math. Ann. **34** (1889), 454–462.
[289] HÖLDER, O.: Ueber den Casus Irreducibilis bei der Gleichung dritten Grades. Math. Ann. **38** (1891), 307–312.
[290] HÖLDER, O.: Die einfachen Gruppen im ersten und zweiten Hundert der Ordnungszahlen. Math. Ann. **40** (1892), 55–88.
[291] HÖLDER, O.: Die Gruppen der Ordnungen p^3, pq^2, pqr, p^4. Math. Ann. **43** (1893), 301–412.
[292] HÖLDER, O.: Die Gruppen mit quadratfreier Ordnungszahl. Nachr. Göttingen **1895**, 211–229.
[293] HÖLDER, O.: Bildung zusammengesetzter Gruppen. Math. Ann. **46** (1895), 321–422.
[294] HÖLDER, O.: Anschauung und Denken in der Geometrie. Leipzig 1900.
[295] HÖLDER, O.: Die Arithmetik in strenger Begründung. Leipzig 1914.
[296] HÖLDER, O.: Die mathematische Methode. Berlin 1924.
[297] HOFMANN, J. E.: Geschichte der Mathematik. Vol. 3. (From "Auseinandersetzungen um den Calculus bis zur Französischen Revolution.") Berlin 1957.
[298] HOFMANN, J. E.: Über zahlentheoretische Methoden Fermats und Eulers, ihre Zusammenhänge und ihre Bedeutung. Arch. Hist. Ex. Sc. **1**, Issue 2 (1961), 122–159.
[299] HUNTINGTON, E. V.: Note on the Definitions of Abstract Groups and Fields by Sets of Independent Postulates. Trans. Amer. Math. Soc. **6** (1905), 181–197.
[300] HURWITZ, A.: Grundlagen einer independenten Theorie der elliptischen Modulfunctionen und Theorie der Multiplicatorgleichungen erster Stufe. Math. Ann. **18** (1881), 528–592.
[301] INFELD, L.: Wen die Götter lieben. (Nach der engl. Ausgabe.) Vienna 1954.
[302] Исследования, Историко-математические. Ed. G. F. RYBKIN and A. P. JUŠKEVIČ. Vol. 3 (1950).
[303] JACOBI, C. G. J.: Fundamenta nova theoriae functionum ellipticarum. Königsberg 1829.
[304] Jahrbuch über die Fortschritte der Mathematik. Vol. 3 (1871), Berlin 1874.
[305] JORDAN, C.: Mémoire sur le nombre des valeurs des fonctions. Journ. Ecole polyt. **22** (1861), 113–194.
[306] JORDAN, C.: Commentaire sur le Mémoire de Galois. Comptes rendus Paris **60** (1865), 770–774.

[307] Jordan, C.: Note sur les équations modulaires. Comptes rendus Paris **66** (1866), 308–312.
[308] Jordan, C.: Recherches sur les polyèdres. Journ. Crelle **66** (1866), 22–85.
[309] Jordan, C.: Mémoire sur la résolution algébrique des équations. Journ. math. Liouville, 2nd Series, **12** (1867), 109–157.
[310] Jordan, C.: Sur les groupes de mouvements. Comptes rendus Paris **65** (1867), 229–232.
[311] Jordan, C.: Sur deux nouvelles séries de groupes. Comptes rendus Paris **67** (1868), 229–233.
[312] Jordan, C.: Sur la résolution algébrique des équations primitives de degré p^2 (p étant premier impair). Journ. math. Liouville, 2nd Series, **13** (1868), 111–135.
[313] Jordan, C.: Mémoire sur les groupes de mouvements. Ann. mat. pur. appl., 2nd Series, **2** (1868/69), 167–215, 322–345.
[314] Jordan, C.: Théorèmes sur les équations algébriques. Journ. math. Liouville, 2nd Series, **14** (1869), 139–146.
[315] Jordan, C.: Sur une équation du 16ème degré. Journ. Crelle **70** (1869), 182–184.
[316] Jordan, C.: Commentaire sur Galois. Math. Ann. **1** (1869), 141–160.
[317] Jordan, C.: Traité des substitutions et des équations algébriques. Paris 1870.
[318] Jordan, C.: Mémoire sur la résolution des équations algébriques les unes par les autres. Journ. math. Liouville, 2nd Series, **16** (1871), 1–20.
[319] Jordan, C.: Sur la résolution des équations les unes par les autres. Comptes rendus Paris **72** (1871), 283–290.
[320] Jordan, C.: Sur la classification des groupes primitifs. Comptes rendus Paris **73** (1871), 853–857.
[321] Jordan, C.: Théorèmes sur les groupes primitifs. Journ. math. Liouville, 2nd Series, **16** (1871), 383–408.
[322] Jordan, C.: Recherches sur les substitutions. Comptes rendus Paris **74** (1872), 975–977.
[323] Jordan, C.: Recherches sur les substitutions. Journ. math. Liouville, 2nd Series, **17** (1872), 351–367.
[324] Jordan, C.: Sur les formes réduites des congruences du second degré. Comptes rendus Paris **74** (1872), 1093–1095.
[325] Jordan, C.: Sur l'énumération des groupes primitifs pour les dix-sept premiers degrés. Comptes rendus Paris **75** (1872), 1754–1757.
[326] Jordan, C.: Sur la limite de transitivité des groupes non altérnés. Bull. Soc. math. France **1** (1873), 40–71.
[327] Jordan, C.: Mémoire sur une application de la théorie des substitutions à l'étude des équations différentielles linéaires. Bull. Soc. math. France **2** (1874), 100–127.
[328] Jordan, C.: Sur deux points de la théorie des substitutions. Comptes rendus Paris **79** (1874), 1149–1151.
[329] Jordan, C.: Sur la limite du degré des groupes primitifs qui contiennent une substitution donnée. Journ. Crelle **79** (1875), 248–258.
[330] Jordan, C.: Sur une classe de groupes d'ordre fini contenus dans les groupes linéaires. Bull. Soc. math. France **5** (1877), 175–177.
[331] Jordan, C.: Mémoire sur les équations différentielles linéaires à intégrale algébrique. Journ. Crelle **84** (1878), 89–215.
[332] Jordan, C.: Sur la réduction des substitutions linéaires. Comptes rendus Paris **90** (1880), 598–601.
[333] Jordan, C.: Sur la réduction des substitutions linéaires. Journ. Ecole polyt. **29** (1880), 151–161.
[334] Jordan, C.: Nouvelles recherches sur la limite de transitivité des groupes qui ne contiennent pas le groupe alterné. Journ. math. Liouville, 5th Series, **1** (1895), 35–60.
[335] Jordan, C.: Sur les groupes d'ordre $p^m q^2$. Journ. math. Liouville, 5th Series, **4** (1898), 21–26.
[336] Jordan, C.: Sur les groupes hypoabéliens. Comptes rendus Paris **138** (1904), 725–728.
[337] Jordan, C.: Groupes abéliens généraux contenus dans les groupes linéaires à moins de sept variables. Journ. math. Liouville, 6th Series, **3** (1907), 213–266.

[338] JORDAN, C.: Mémoire sur les groupes résolubles. Journ. math. Liouville, 7th Series, **3** (1917), 263–374.
[339] JORDAN, C.: Traité des substitutions et des équations algébriques. 2nd ed., Paris 1957.
[340] KAGAN, V. F. (Каган, В. Ф.): LOBACHEVSKI. Moscow/Leningrad 1944. (Russian.)
[341] KANT, I.: Werke. Vol. 1. Ed. A. BUCHENAU. Berlin 1922.
[342] KANT, I.: Kritik der reinen Vernunft. After the 2n ded. (1787) in I. KANT's Gesammelte Werke. Ed. E. CASSIRER. Vol. 3. Berlin 1922.
[343] KANTOR, S.: Theorie der endlichen Gruppen von eindeutigen Transformationen in der Ebene. Berlin 1895.
[344] KIRKMAN T. P.: Theory of Groups and Many-Valued Functions. Mem. Phil. Soc. Manchester, 3rd Series, **1** (1862), 274–397.
[345] KIRKMAN, T. P.: On Non-modular Groups. Mem. Phil. Soc. Manchester, 3rd Series, **2** (1865), 204–227.
[346] KLEIN, F.: Gesammelte mathematische Abhandlungen. Vol. 1. Ed. R. FRICKE and A. OSTROWSKI. Berlin 1921.
[347] KLEIN, F.: Gesammelte mathematische Abhandlungen. Vol. 2. Ed. R. FRICKE and H. VERMEIL. Berlin 1922.
[348] KLEIN, F.: Gesammelte mathematische Abhandlungen. Vol. 3. Ed. R. FRICKE, H. VERMEIL and E. BESSEL-HAGEN. Berlin 1923.
[349] KLEIN, F.: Zur Theorie der Liniencomplexe des ersten und zweiten Grades. Math. Ann. **2** (1870). WA: [346], pp. 53–80.
[350] KLEIN, F.; LIE, S.: Deux notes sur une certaine famille de courbes et de surfaces. Comptes rendus Paris **70** (1870). WA: [346], pp. 415–423.
[351] KLEIN, F.; LIE, S.: Über diejenigen ebenen Kurven, welche durch ein geschlossenes System von einfach unendlich vielen vertauschbaren linearen Transformationen in sich übergehen. Math. Ann. **4** (1871). WA: [346], pp. 424–459.
[352] KLEIN, F.: Über die sogenannte Nicht-Euklidische Geometrie. Nachr. Göttingen **1871**. WA: [346], pp. 244–253.
[353] KLEIN, F.: Über die sogenannte Nicht-Euklidische Geometrie. (1. Aufsatz) Math. Ann. **4** (1871). WA: [346], pp. 254–305.
[354] KLEIN, F.: Notiz, betreffend den Zusammenhang der Liniengeometrie mit der Mechanik starrer Körper. Math. Ann. **4** (1871). WA: [346], pp. 226–238.
[355] KLEIN, F.: Über eine geometrische Repräsentation der Resolventen algebraischer Gleichungen. Math. Ann. **4** (1871), WA: [347], pp. 262–274.
[356] KLEIN, F.: Über Liniengeometrie und metrische Geometrie. Math. Ann. **5** (1872). WA: [346], pp. 106–126.
[357] KLEIN, F.: Zur Interpretation der komplexen Elemente in der Geometrie. Nachr. Göttingen **1872**. WA: [346], pp. 402–405.
[358] KLEIN, F.: Vergleichende Betrachtungen über neuere geometrische Forschungen. Erlangen 1872. WA: Math. Ann. **43** (1893). WA: [346], pp. 460–497.
[359] KLEIN, F.: Über die sogenannte Nicht-Euklidische Geometrie. (2nd article.) Math. Ann. **6** (1873). WA: [346], pp. 311–343.
[360] KLEIN, F.: Eine Übertragung des Pascalschen Satzes auf Raumgeometrie. SB. Erlangen **1873**. WA: [346], pp. 406–408.
[361] KLEIN, F.: Nachtrag zu dem "Zweiten Aufsatz über Nicht-Euklidische Geometrie." Math. Ann. **7** (1874). WA: [346], pp. 344–350.
[362] KLEIN, F.: Über binäre Formen mit linearen Transformationen in sich selbst. Math. Ann. **9** (1875/76). WA: [347], pp. 275–301.
[363] KLEIN, F.: Über algebraisch integrierbare lineare Differentialgleichungen. (1st article.) SB. Erlangen **1876**. WA: [347], pp. 302–306.
[364] KLEIN, F.: Über algebraisch integrierbare lineare Differentialgleichungen. (2nd article.) Math. Ann. **12** (1877). WA: [347], pp. 307–320.

[365] KLEIN, F.: Weitere Untersuchungen über das Ikosaeder. Math. Ann. **12** (1877). WA: [347], pp. 321–380.
[366] KLEIN, F.: Ergänzende Bemerkungen über Gordans Arbeit über Gleichungen fünften Grades. Math. Ann. **13** (1878). WA: [347], pp. 380–384.
[367] KLEIN, F.: Über Gleichungen siebenten Grades. SB. Erlangen **1877/78**. WA: [347], pp. 388–389.
[368] KLEIN, F.: Über die Transformation der elliptischen Funktionen und die Auflösung der Gleichungen fünften Grades. Math. Ann. **14** (1878/79). WA: [348], pp. 13–75.
[369] KLEIN, F.: Über die Erniedrigung der Modulargleichungen. Math. Ann. **14** (1878/79). WA: [348], pp. 76–89.
[370] KLEIN, F.: Über die Transformation siebenter Ordnung der elliptischen Funktionen. Math. Ann. **14** (1878/79). WA: [348], pp. 90–135.
[371] KLEIN, F.: Über Multiplikatorgleichungen erster Stufe. Math. Ann. **15** (1878/79). WA: [348], pp. 137–139.
[372] KLEIN, F.: Über die Transformation elfter Ordnung der elliptischen Funktionen. Math. Ann. **15** (1879). WA: [348], pp. 140–165.
[373] KLEIN, F.: Über die Auflösung gewisser Gleichungen vom siebenten und achten Grade. Math. Ann. **15** (1879). WA: [347], pp. 390–425.
[374] KLEIN, F.: Zur Systematik der Theorie der elliptischen Modulfunktionen. SB. Akad. Wiss. Munich **1879**. WA: [348], pp. 169–178.
[375] KLEIN, F.: Über unendlich viele Normalformen des elliptischen Integrals erster Gattung. SB. Akad. Wiss. Munich **1880/81**. WA: [348], pp. 179–185.
[376] KLEIN, F.: Über eindeutige Funktionen mit linearen Transformationen in sich. Math. Ann. **19** (1882). WA: [348], pp. 622–626.
[377] KLEIN, F.: Über eindeutige Funktionen mit linearen Transformationen in sich. Math. Ann. **20** (1882). WA: [348], pp. 627–629.
[378] KLEIN, F.: Neue Beiträge zur Riemannschen Funktionentheorie. Math. Ann. **21** (1882/83). WA: [348], pp. 630–710.
[379] KLEIN, F.: Vorlesungen über das Ikosaeder und die Auflösung der Gleichungen vom fünften Grade. Leipzig 1884.
[380] KLEIN, F.: Über die elliptischen Normalkurven der n-ten-Ordnung. Abh. Ges. Wiss. Leipzig **1885**. WA: [348], pp. 198–254.
[381] KLEIN, F.: Neue Untersuchungen über elliptische Funktionen und Modulfunktionen. Erster Bericht. Abh. Ges. Wiss. Leipzig **1885**. WA: [348], pp. 255–273.
[382] KLEIN, F.: Neue Untersuchungen über elliptische Funktionen und Modulfunktionen. Zweiter Bericht. Math. Ann. **26** (1885/86). WA: [348], pp. 274–282.
[383] KLEIN, F.: Zur Theorie der allgemeinen Gleichungen sechsten und siebenten Grades. Math. Ann. **28** (1886/87). WA: [347], pp. 439–472.
[384] KLEIN, F.: Sur la résolution, par les fonctions hyperelliptiques, de l'équation du vingt-septième degré, de laquelle dépend la détermination des vingt-sept droites d'une surface cubique. Journ. math. Liouville, 4th Series, **4** (1888). WA: [347], pp. 473–478.
[385] KLEIN, F.: Zur Nicht-Euklidischen Geometrie. Math. Ann. **37** (1890). WA: [346], pp. 353–383.
[386] KLEIN, F.: Zur Theorie der allgemeinen Laméschen Funktionen. Nachr. Göttingen **1890**. WA: [347], pp. 540–549.
[387] KLEIN, F.: Vorlesungen über die Theorie der elliptischen Modulfunctionen. Prepared and completed by R. FRICKE. Vol. 1. Leipzig 1890.
[388] KLEIN, F.: Über den Begriff des funktionentheoretischen Fundamentalbereichs. Math. Ann. **40** (1891/92). WA: [348], pp. 711–720.
[389] KLEIN, F.: Vorlesungen über die Theorie der elliptischen Modulfunctionen. Prepared and completed by R. FRICKE. Vol. 2. Leipzig 1892.
[390] KLEIN, F.: Höhere Geometrie (Lecture notes.) Leipzig 1893.

[391] KLEIN, F.: Über die Komposition der binären quadratischen Formen. Nachr. Göttingen **1893**. WA: [348], pp. 283–286.

[392] KLEIN, F.: Über Arithmetisierung der Mathematik. (Lecture, Göttingen, 2 November 1895) Nachr. Göttingen **1895**. WA: [347], pp. 232–240.

[393] KLEIN, F.; FRICKE, R.: Vorlesungen über die Theorie der automorphen Functionen. Vol. 1. Die gruppentheoretischen Grundlagen. Leipzig 1897.

[394] KLEIN, F.: Nachruf auf E. Schering. JB. dt. Math. Ver. **6** (1898), 25–27.

[395] KLEIN, F.: Sulla risoluzione delle equazioni di sesto grado. Rendiconti Acc. Lincei, Series 5a, **8** (1899). WA: [347]. p. 480.

[396] KLEIN, F.; FRICKE, R.: Vorlesungen über die Theorie der automorphen Functionen. Vol. 2 in 3 Parts. Leipzig 1901, 1911, 1912.

[397] KLEIN, F.: Über die Auflösung der allgemeinen Gleichungen fünften und sechsten Grades. Journ. Crelle **129** (1905); Math. Ann. **61** (1905). WA: [347], pp. 481–502.

[398] KLEIN, F.: Beweis für die Nichtauflösbarkeit der Ikosaedergleichung durch Wurzelzeichen. Math. Ann. **61** (1905/06). WA: [347], pp. 385–387.

[399] KLEIN, F.: Über die geometrischen Grundlagen der Lorentzgruppe. JB. dt. Math. Ver. **19** (1910). WA: [346], pp. 533–552.

[400] KLEIN, F.: Zu Hilberts erster Note über die Grundlagen der Physik. Nachr. Göttingen **1917**. WA: [346], pp. 553–567.

[401] KLEIN, F.: Über die Differentialgesetze für die Erhaltung von Impuls und Energie in der Einsteinschen Gravitationstheorie. Nachr. Göttingen **1918**. WA: [346], pp. 568–585.

[402] KLEIN, F.: Über die Integralform der Erhaltungssätze und die Theorie der räumlich-geschlossenen Welt. Nachr. Göttingen **1918**. WA: [346], pp. 586–612.

[403] KLEIN, F.: Vorlesungen über die Entwicklung der Mathematik im 19. Jahrhundert. Vol. 1. Ed. R. COURANT and O. NEUGEBAUER, Berlin 1926.

[404] KLEIN, F.: Vorlesungen über die Entwicklung der Mathematik im 19. Jahrhundert. Vol. 2. Ed. R. COURANT and S. COHN-VOSSEN. Berlin 1927.

[405] KLEIN, F.: Vorlesungen über nicht-euklidische Geometrie. Ed. W. ROSEMANN. Berlin 1928.

[406] KLÜGEL, G. S.: Mathematisches Wörterbuch. Vol. 1. Leipzig 1803.

[407] KNESER, A.: Irreductibilität und Monodromiegruppe algebraischer Gleichungen. Berlin 1884.

[408] KNESER, A.: Die Monodromiegruppe einer algebraischen Gleichung bei linearen Transformationen der Variabeln. Math. Ann. **28** (1887), 125–132.

[409] KNESER, A.: Zur Theorie der algebraischen Functionen. Math. Ann. **29** (1887), 171–186.

[410] KNESER, A.: Ueber die Gattung niedrigster Ordnung, unter welcher gegebene Gattungen algebraischer Grössen enthalten sind. Math. Ann. **30** (1887), 179–202.

[411] KNESER, A.: Arithmetische Begründung einiger algebraischer Fundamentalsätze. Journ. Crelle **102** (1888), 20–55.

[412] KNESER, A.: Ein neuer Beweis der Unmöglichkeit, allgemeine Gleichungen höheren Grades algebraisch aufzulösen. Journ. Crelle **106** (1890), 48–64.

[413] KNESER, A.: Bemerkungen über den sogenannten casus irreducibilis bei cubischen Gleichungen. Math. Ann. **41** (1893), 344–348.

[414] KÖHLER, C.; HEFFTER, L.: Lehrbuch der analytischen Geometrie. Vol. 1. Leipzig/Berlin 1905.

[415] KÖNIG, J.: Ueber rationale Functionen von *n* Elementen und die allgemeine Theorie der algebraischen Gleichungen. Math. Ann. **14** (1879), 212–230.

[416] KÖNIG, J.: Einleitung in die allgemeine Theorie der algebraischen Grössen (from Hungarian). Leipzig 1903.

[417] KÖTTER, E.: Die Entwickelung der synthetischen Geometrie von Monge bis auf Staudt (1847). Erster Band eines Berichtes, erstattet an die Deutsche Mathematikervereinigung. JB. dt. Math. Ver. **5**, Issue 2 (1901), 1–484.

[418] KOLLROS, L.: Evariste Galois (Elemente der Mathematik, Beiheft 7). Basel 1949.
[419] KOWALEWSKY, G.: Große Mathematiker. Berlin 1938.
[420] KRONECKER, L.: Werke. Ed. under the auspices d. preuss. Akad. d. Wiss. by K. HENSEL. Vol. 1. Leipzig 1895.
[421] KRONECKER, L.: Werke. Ed. under the auspices d. preuss. Akad. d. Wiss. by K. HENSEL. Vol. 2. Leipzig 1897.
[422] KRONECKER, L.: Werke. Ed. under the auspices d. preuss. Akad. d. Wiss. by K. HENSEL. Vol. 4. Berlin/Leipzig 1929.
[423] KRONECKER, L.: Werke. Ed. under the auspices d. preuss. Akad. d. Wiss. by K. HENSEL. Vol. 5. Berlin/Leipzig 1930.
[424] KRONECKER, L.: Über die algebraisch auflösbaren Gleichungen (I. Abhandlung). MB. Akad. Berlin **1853**. WA: [422], pp. 1–11.
[425] KRONECKER, L.: Note sur les fonctions semblables des racines d'une équation. Journ. math. Liouville **19** (1854). WA: [422], pp. 13–16.
[426] KRONECKER, L.: Über die algebraisch auflösbaren Gleichungen (II. Abhandlung). MB. Akad. Berlin **1856**. WA: [422], pp. 25–37.
[427] KRONECKER, L.: Zwei Sätze über Gleichungen mit ganzzahligen Coefficienten. Journ. Crelle **53** (1857). WA: [420], pp. 103–108.
[428] KRONECKER, L.: Über Gleichungen des siebenten Grades. MB. Akad. Berlin **1858**. WA: [422], pp. 39–42.
[429] KRONECKER, L.: Sur la résolution de l'équation du cinquième degré (Extrait d'une lettre adressée à M. HERMITE). Comptes rendus Paris **46** (1858.) WA: [422], pp. 43–47.
[430] KRONECKER, L.: Extrait d'une lettre de M. Kronecker à M. Brioschi (sur la théorie des substitutions). Ann. mat. pur. appl., 2nd Series, **2** (1859). WA: [422], pp. 49–52.
[431] KRONECKER, L.: Mitteilung über algebraische Arbeiten (über Gleichungen fünften Grades). MB. Akad. Berlin **1861**; Journ. Crelle **59** (1861). WA: [422], pp. 53–62.
[432] KRONECKER, L.: Auseinandersetzung einiger Eigenschaften der Klassenanzahl idealer complexer Zahlen. MB. Akad. Berlin **1870**. WA: [420], pp. 271–282.
[433] KRONECKER, L.: Zur Geschichte des Reciprocitätsgesetzes. MB. Akad. Berlin **1875**. WA: [421], pp. 1–10.
[434] KRONECKER, L.: Einige Entwickelungen aus der Theorie der algebraischen Gleichungen. MB. Akad. Berlin **1879**. WA: [422], pp. 73–96.
[435] KRONECKER, L.: Grundzüge einer arithmetischen Theorie der algebraischen Grössen. Berlin **1882**, WA: [421], pp. 237–387.
[436] KRONECKER, L.: Vorlesungen über die Theorie der Determinanten. Leipzig 1903.
[437] KRULL, W.: Über verallgemeinerte endliche Abelsche Gruppen. Math. Zeitschr. **23** (1925), 161–196.
[438] KRULL, W.: Zur Theorie der allgemeinen Zahlringe. Math. Ann. **99** (1928), 51–70.
[439] KRULL, W.: Galoissche Theorie der unendlichen algebraischen Erweiterungen. Math. Ann. **100** (1928), 687–698.
[440] KUMMER, E. E.: Zur Theorie der complexen Zahlen. MB. Akad. Berlin **1845**. WA: Journ. Crelle **35** (1847), 319–326.
[441] KUMMER, E. E.: Über die Zerlegung der aus Wurzeln der Einheit gebildeten complexen Zahlen in ihre Primfactoren. Journ. Crelle **35** (1847), 327–367.
[442] KUMMER, E. E.: Bestimmung der Anzahl nicht äquivalenter Classen für die aus λ ten Wurzeln der Einheit gebildeten complexen Zahlen und die idealen Factoren derselben. Journ. Crelle **40** (1850), 93–116.
[443] KUMMER, E. E.: Über eine Eigenschaft der Einheiten der aus den Wurzeln der Gleichung $\alpha^\lambda = 1$ gebildeten complexen Zahlen und über den zweiten Faktor der Klassenzahl. MB. Akad. Berlin **1870**, 855–880.
[444] KUROSCH, A. G.: Gruppentheorie. (With an Appendix by B. H. NEUMANN.) Berlin 1953.
[445] LAMPE, E.: Gedächtnisrede auf K. Weierstraß. JB. dt. Math. Ver. **6** (1897), 27–44.

[446] LAGRANGE, J.-L.: Oeuvres de Lagrange. Publiées par les soins de M. J.-A. SERRET. Vol. 3. Paris 1869.
[447] LAGRANGE, J.-L.: Réflexions sur la résolution algébrique des équations. Nouv. Mém. Acad. Berlin, pour les années 1770/71. Berlin 1772/1773. WA: [446], pp. 203–421.
[448] LAGRANGE, J.-L.: Recherches d'Arithmétique. Nouv. Mém. Acad. Berlin, années 1773 et 1775. WA: [446], pp. 693–795.
[449] LAGRANGE, J.-L.: Traité de la résolution des équations numériques de tous les degrés. Nouv. ed., revue et augmentée par l'auteur. Paris 1808.
[450] LEBESGUE, H.: Notices d'histoire des mathématiques. Avec une introduction de M L. FÉLIX. Genf 1958.
[451] LEBESGUE, H.: Notice sur la vie et les travaux de Camille Jordan. In: [450], pp. 40–65.
[452] LEGENDRE, A.-M.: Recherches d'analyse indéterminée. In: Hist. Acad. Sc. Paris, année 1785, Paris 1788.
[453] LEGENDRE, A.-M.: Essai sur la théorie des nombres. Paris 1798.
[454] LEJEUNE-DIRICHLET, P. G.: Untersuchungen über verschiedene Anwendungen der Infinitesimalrechnung auf die Zahlentheorie. Part 1. Journ. Crelle **19** (1839). WA: Ostwald's Klassiker, Nr. 91. Leipzig 1897, pp. 3–36.
[455] LEJEUNE-DIRICHLET, P. G.: Vorlesungen über Zahlentheorie. Ed. and supplemented by R. DEDEKIND. 3rd rev. and enlarged ed., Braunschweig 1879.
[456] LEVI BEN GERSON: Die Praxis des Rechners. Ed. G. LANGE. Frankfurt/M. 1909.
[457] LEVI, F.; BAER, R.: Vollständige irreduzibele Systeme von Gruppenaxiomen. SB. Akad. Heidelberg, **2** (1932), 1–12.
[458] LEVIN, B. JA. (Левин, Б. Я.): O. YU. SCHMIDT. In: [492], pp. 400–411. (Russian.)
[459] LIE, S.: Gesammelte Abhandlungen. Vol. 1. Leipzig/Oslo 1934.
[460] LIE, S.: Gesammelte Abhandlungen. Vol. 4. Leipzig/Oslo 1929.
[461] LIE, S.: Gesammelte Abhandlungen. Vol. 5. Leipzig/Kristiania 1924.
[462] LIE, S.: Gesammelte Abhandlungen. Vol. 6. Leipzig/Oslo 1927.
[463] LIE, S.: Theorie der Transformationsgruppen. 1st Part. Leipzig/Berlin 1888. Reprint: Leipzig/Berlin 1930.
[464] LIE, S.: Theorie der Transformationsgruppen. 2nd Part. Leipzig/Berlin 1930. Reprint: Leipzig/Berlin 1930.
[465] LIE, S.: Theorie der Transformationsgruppen. 3rd and last Part. Leipzig 1893.
[466] LIE, S.: Theorie der Transformationsgruppen. 3rd and last Part. Leipzig/Berlin 1893. Reprint: Leipzig/Berlin 1930.
[467] LIE, S.: Geometrie der Berührungstransformationen. Exposition by S. Lie and G. Scheffers. Vol. 1. Leipzig 1896.
[468] LIE, S.; KLEIN, F.: Deux notes sur une certaine famille de courbes et de surfaces. Comptes rendus Paris **70** (1870). WA: [346], pp. 415–423.
[469] LIE, S.; KLEIN, F.: Über diejenigen ebenen Kurven, welche durch eine geschlossenes System von einfach unendlich vielen vertauschbaren linearen Transformationen in sich übergehen. In: Math. Ann. **4** (1871), WA: [346], pp. 424–459.
[470] LIE, S.: Over en Classe geometriske Transformationer. Chr. Forh. Aar **1871** (1872). WA in German: [459], pp. 105–152.
[471] LIE, S.: Über eine Klasse geometrischer Transformationen (Fortsetzung). Chr. Forh. Aar **1871** (1872). WA: [459], pp. 153–210.
[472] LIE, S.: Über Gruppen von Transformationen. Nachr. Göttingen **1874**. WA: [461], pp. 1–8.
[473] LIE, S.: Begründung einer Invariantentheorie der Berührungstransformationen. Math. Ann. **8** (1875). WA: [460], pp. 1–96.
[474] LIE, S.: Theorie der Transformationsgruppen. Erste Abhandlung. Arch. Math. Naturv. **1** (1876). WA: [461], pp. 9–41.
[475] LIE, S.: Theorie der Transformationsgruppen. Zweite Abhandlung. Arch. Math. Naturv. **1** (1876). WA: [461], pp. 42–75.

[476] Lie, S.: Allgemeine Theorie der partiellen Differentialgleichungen erster Ordnung. Math. Ann. **9** (1876). WA: [460], pp. 97–151.

[477] Lie, S.: Allgemeine Theorie der partiellen Differentialgleichungen erster Ordnung. Math. Ann. **11** (1877). WA: [460], pp. 163–262.

[478] Lie, S.: Theorie der Transformationsgruppen. Dritte Abhandlung. Bestimmung aller Gruppen einer zweifach ausgedehnten Punktmannigfaltigkeit. Arch. Math. Naturv. **3** (1878). WA: [461], pp. 78–133.

[479] Lie, S.: Theorie der Transformationsgruppen. Vierte Abhandlung. Arch. Math. Naturv. **3** (1878). WA: [461], pp. 136–197.

[480] Lie, S.: Theorie der Transformationsgruppen. Fünfte Abhandlung. Arch. Math. Naturv. **4** (1879). WA: [461], pp. 199–222.

[481] Lie, S.: Klassifikation der Flächen nach der Transformationsgruppe ihrer geodätischen Kurven. Kristiania 1879. WA: [459], pp. 358–408.

[482] Lie, S.: Theorie der Transformationsgruppen. I. Math. Ann. **16** (1880). WA: [462], pp. 1–94.

[483] Lie, S.: Über Flächen, die infinitesimale und lineare Transformationen gestatten. Arch. Math. Naturv. **7** (1882). WA: [461], pp. 224–234.

[484] Lie, S.: Selbstanzeige zu: Über Flächen, die infinitesimale und lineare Transformationen gestatten. Jahrb. Fortschr. Math. **14** (1882). Berlin 1885. pp. 641–642.

[485] Lie, S.: Über unendliche kontinuierliche Gruppen. Chr. Forh. Aar **1883** (1883). WA: [461], pp. 314–360.

[486] Lie, S.: Zur Theorie der Transformationsgruppen. Chr. Forh. Aar **1888** (1888). WA: [461], pp. 553–557.

[487] Lie, S.: Ein Fundamentalsatz in der Theorie der unendlichen Gruppen. Chr. Forh. Aar **1889** (1889). WA: [461], pp. 558–560.

[488] Lie, S.: Die Grundlagen für die Theorie der unendlichen kontinuierlichen Transformationsgruppen. 1. Abhandlung. Abh. Ges. Wiss. Leipzig **1891**. WA: [462], pp. 300–330.

[489] Lie, S.: Untersuchungen über unendliche kontinuierliche Gruppen. Abh. Ges. Wiss. Leipzig **1895**. WA: [462], pp. 396–493.

[490] Lie, S.: Influence de Galois sur le développement des Mathématiques. In: [152], pp. 481–489. WA: [462], pp. 592–601.

[491] Lie, S.: Aus Briefen an A. Mayer. In: [461], pp. 583–614.

[492] Men of Russian Science. Ed. I. V. Kuznecov. Moscow 1961. (Russian.)

[493] Loewy, A.: Zur Theorie der Gruppen linearer Substitutionen. Math. Ann. **53** (1900), 225–242.

[494] Loewy, A.: Inwieweit kann Vandermonde als Vorgänger von Gauß bezüglich der algebraischen Auflösung der Kreisteilungsgleichungen $x^n = 1$ angesehen werden! JB. dt. Math. Ver. **27** (1918), 189–195.

[495] Lorenzen, P.: Ein Beitrag zur Gruppenaxiomatik. Math. Zeitschr. **49** (1944), 313–327.

[496] Lorey, W.: Abhandlung über das Studium der Mathematik an den deutschen Universitäten seit Anfang des 19. Jahrhunderts. Leipzig 1916.

[497] Lüroth, J.: Ernst Schröder. JB. dt. Math. Ver., **12** (1903), 249–265.

[498] Maennchen, P.: Materialien für eine wissenschaftliche Biographie von Gauß. Ed. by F. Klein, M. Brendel, and L. Schlesinger. Issue 6. Die Wechselwirkung zwischen Zahlenrechnen und Zahlentheorie bei C. F. Gauß. Leipzig 1918.

[499] Magasin pittoresque. Vol. 16 (1848), Paris. pp. 227–228. (Anonymous article on E. Galois.)

[500] Malfatti, G.: Tentativo per la risoluzione delle equazioni di quinto grado. Pavia 1772.

[501] Malfatti, G.: Dubbi proposti al socio Paolo Ruffini sulla sua dimostrazione della impossibilità di risolvere le equazioni superiori al quarto grado. Mem. Fis. Soc. Ital. **11** (1804), 579–607.

[502] Manuel Moschopulos: Über magische Quadrate. Greek-German ed. S. Günther. In: S. Günther: Vermischte Untersuchungen. Leipzig 1876.

[503] MARX, K.: Nachwort zur zweiten Auflage vom 1. Bande des Kapitals. (1873) In: K. MARX, FR. ENGELS, Werke, Vol. 23. Berlin 1962.
[504] MASCHKE, H.: Ueber den arithmetischen Charakter der Coefficienten der Substitutionen endlicher linearer Substitutionsgruppen. Math. Ann. **50** (1898), 492–498.
[505] MATHIEU, E.: Nombre de valeurs d'une fonction. Paris 1859.
[506] MATHIEU, E.: Mémoire sur le nombre de valeurs que peut acquérir une fonction. Comptes rendus Paris **48** (1859), 840–841.
[507] MATHIEU, E.: Nombre de valeurs que peut acquérir une fonction quand on y permute ses variables de toutes les manières possibles Journ. math. Liouville, 2nd Series, **5** (1860), 9–42.
[508] MATHIEU, E.: Mémoire sur la résolution des équations dont le degré est une puissance d'un nombre premier. Ann. mat. pur. appl. **4** (1861), 113–132.
[509] MATHIEU, E.: Mémoire sur l'étude des fonctions de plusieurs quantités, sur la manière de les former et sur les substitutions qui les laissent invariables. Journ. math. Liouville, 2nd Series, **6** (1861), 241–323.
[510] MATHIEU, E.: Sur la fonction cinq fois transitive de 24 quantités. Journ. Liouville, 2nd Series, **18** (1873), 25–46.
[511] MERTENS, F.: Über zyclische Gleichungen. Journ. Crelle **131** (1906), 87–112.
[512] MESCHKOWSKI, H.: Wandlungen des mathematischen Denkens. Eine Einführung in die Grundlagenprobleme der Mathematik. Braunschweig 1956.
[513] MEYER, F.: Bericht über den gegenwärtigen Stand der Invariantentheorie. JB. dt. Math. Ver. **1** (1890/91), 79–292.
[514] MILLER, G. A.: A Simple Proof of a Fundamental Theorem of Substitution Groups, and Several Applications of the Theorem. Bull. Amer. Math. Soc. **2** (1896), 75–77.
[515] MILLER, G. A.: On the Lists of All the Substitution Groups that Can Be Formed with a Given Number of Elements. Bull. Amer. Math. Soc. **2** (1896), 138–145.
[516] MILLER, G. A.: The Substitution Groups whose Order is the Product of Two Unequal Prime Numbers. Bull. Amer. Math. Soc. **2** (1896), 332–336.
[517] MILLER, G. A.: Sur les groupes de substitutions. Comptes rendus Paris **122** (1896), 370–372.
[518] MILLER, G. A.: On the Commutator Groups. Bull. Amer. Math. Soc. **4** (1898), 135–139.
[519] MILLER, G. A.: On an Extension of Sylow's Theorem. Bull. Amer. Math. Soc. **4** (1898), 323–327.
[520] MILLER, G. A.: On the Hamilton Groups. Bull. Amer. Math. Soc. **4** (1898), 510–515.
[521] MILLER, G. A.: History of the Theory of Groups to 1900. The Collected Works, Vol. 1, pp. 427–467.
[522] MILLER, G. A.: The Group-theory Element of the History of Mathematics. Scient. Monthly **12** (1921), 75–82.
[523] MÖBIUS, A. F.: Gesammelte Werke. Vol. 1. Ed. R. BALTZER. Leipzig 1885.
[524] MÖBIUS, A. F.: Gesammelte Werke. Vol. 2. Ed. F. KLEIN. Leipzig 1886.
[525] MÖBIUS, A. F.: Gesammelte Werke. Vol. 3. Ed. F. KLEIN. Leipzig 1886.
[526] MÖBIUS, A. F.: Gesammelte Werke. Vol. 4. Ed. W. SCHEIBNER. Leipzig 1887.
[527] MÖBIUS, A. F.: Der barycentrische Calcul, ein neues Hülfsmittel zur analytischen Behandlung der Geometrie dargestellt und insbesondere auf die Bildung neuer Classen von Aufgaben und die Entwickelung mehrerer Eigenschaften der Kegelschnitte angewendet. Leipzig 1827.
[528] MÖBIUS, A. F.: Von den metrischen Relationen im Gebiete der Lineal-Geometrie. Journ. Crelle **4** (1829), 101–130.
[529] MÖBIUS, A. F.: Ueber eine allgemeinere Art der Affinität geometrischer Figuren. Journ. Crelle **12** (1834). WA: [523], pp. 517–544.
[530] MÖBIUS, A. F.: Theorie der elementaren Verwandtschaft. Ber. Verh. Sächs. Ges. Wiss. **15** (1863). WA: [524], pp. 433–471.
[531] MOLIEN, T.: Ueber Systeme höherer complexer Zahlen. Math. Ann. **41** (1893), 83–156.

[532] Molien, T.: Über die Invarianten der linearen Substitutionsgruppen. MB. Akad. Berlin **1897**, 1152–1156.
[533] Monge, G.: Géométrie descriptive. Paris 1794.
[534] Moore, E. H.: The Group of Holoedric Transformation into Itself of a Given Group. Bull. Amer. Math. Soc. **1** (1895), 61–66.
[535] Moore, E. H.: Concerning Jordan's Linear Groups. Bull. Amer. Math. Soc. **2** (1896), 33–43.
[536] Moore, E. H.: A Two-Fold Generalization of Fermat's Theorem. Bull. Amer. Math. Soc. **2** (1896), 189–199.
[537] Moore, E. H.: An Universal Invariant for Finite Groups of Linear Substitutions: With Application in the Theory of the Canonical Form of a Linear Substitution of Finite Period. Math. Ann. **50** (1898), 213–224.
[538] Müller, F.: Führer durch die mathematische Literatur mit besonderer Berücksichtigung der historisch wichtigen Schriften. Leipzig/Berlin 1909.
[539] Muth, P.: Grundlagen für die geometrische Anwendung der Invariantentheorie. Leipzig 1895.
[540] Naas, J.; Schröder, K.: Der Begriff des Raumes in der Geometrie. Berlin 1957.
[541] Netto, E.: Die Substitutionentheorie und ihre Anwendung auf die Algebra. Leipzig 1882.
[542] Noether, M.: Sophus Lie (Nachruf). Math. Ann. **53** (1900), 1–41.
[543] Nový, L.: Sur l'origine des études systématiques des problèmes de la théorie des groupes. In: Sommaires de XI-e Congrès Internationale d'Histoire des Sciences. Warszawa 1965, pp. 170–171.
[544] Nowikow, P. S.; Tschernikow, S. N.; Schmidt, O. J.: Endlichkeitsbedingungen in der Gruppentheorie. Ed. H. Grell. Berlin 1963. (Translated from Russian.)
[545] Ore, O.: Niels Henrik Abel. Mathematician Extraordinary. Minneapolis 1957.
[546] Peano, G.: Calcolo geometrico secondo l'Ausdehnungslehre di Grassmann, preceduto dalle operazioni della logica deduttiva. Torino 1888.
[547] Peano, G.: I principii di geometria, logicamente expositi. Torino 1889.
[548] Peano, G.: Arithmeticas principia, novo methodo exposita. Torino 1889.
[549] Peirce, B.: Linear Associative Algebra. Amer. Journ. Math. **4** (1881), 97–215.
[550] Peirce, C. S.: Upon the Logic of Mathematics. Proc. Amer. Acad. **7** (1865–68), 402–412.
[551] Peirce, C. S.: On the Algebra of Logic. Amer. Journ. Math. **3** (1880), 49–57.
[552] Peirce, C. S.: On the Relative Forms of the Algebras. Amer. Journ. Math. **4** (1881), 221–225.
[553] Peirce, C. S.: On the Algebras in Which Division is Unambiguous. Amer. Journ. Math, **4** (1881), 225–229.
[554] Peirce, C. S.: On the Algebra of Logic. Amer. Journ. Math. **7** (1885), 180–202.
[555] Peters, W. S.: Zum Begriff der Konstruierbarkeit bei I. Kant. Arch. Hist. Ex. Sc. **2** (1964), 153–167.
[556] Picard, E.: Sur les groupes de transformations des équations différentielles linéaires. Comptes rendus Paris **119** (1894), 584–589.
[557] Piccard, S.: Lobatchevsky, grand mathématicien russe. Sa vie, son oeuvre. Université de Paris, Palais de la Découverte, Série D, Nr. 47, 1957.
[558] Pierpont, J.: Early History of Galois's Theory of Equations. Bull. Amer. Math. Soc. **4** (1898), 332–340.
[559] Plücker, J.: Gesammelte wissenschaftliche Abhandlungen in zwei Bänden. Vol. 1. Ed. A. Schoenflies. Leipzig 1895.
[560] Plücker, J.: Ueber ein neues Coordinatensystem. Journ. Crelle **5** (1829). WA: [559], pp. 124–158.
[561] Plücker, J.: Analytisch-geometrische Entwicklungen. Vol. 1. Essen 1828.
[562] Plücker, J.: Über ein neues Coordinatensystem. Journ. Crelle **5** (1829), 1-36.
[563] Plücker, J.: Analytisch-geometrische Entwicklungen. Vol. 2. Essen 1831.

[564] PLÜCKER, J.: System der analytischen Geometrie, auf neue Betrachtungsweisen gegründet, und insbesondere eine ausführliche Theorie der Curven dritter Ordnung enthaltend. Berlin 1835.

[565] PLÜCKER, J.: Theorie der algebraischen Curven, gegründet auf eine neue Behandlungsweise der analytischen Geometrie. Bonn 1839.

[566] PLÜCKER, J.: System der Geometrie des Raumes in neuer analytischer Behandlungsweise, insbesondere die Theorie der Flächen zweiter Ordnung und Classe enthaltend. Düsseldorf 1846.

[567] PLÜCKER, J.: Neue Geometrie des Raumes, gegründet auf die Betrachtung der geraden Linie als Raumelement. Part 1. Leipzig 1868; Part 2. Leipzig 1869.

[568] POGGENDORFF, J. C.: Biographisch-literarisches Handwörterbuch der exakten Naturwissenschaften. Vols. I, II, III, IV, V, VI, VIIa.

[569] POINCARÉ, H.: Sur les fonctions uniformes qui se reproduisent par des substitutions linéaires. Math. Ann. **19** (1882), 553–564.

[570] POINCARÉ, H.: Théorie des groupes fuchsiens. Acta math. **1** (1882), 1–62.

[571] POINCARÉ, H.: Mémoire sur les fonctions fuchsiennes. Acta math. **1** (1882), 193–294.

[572] POINCARÉ, H.: Mémoire sur les groupes kleinéens. Acta math. **3** (1883), 49–92.

[573] POINCARÉ, H.: Sur les groupes des équations linéaires. Acta math. **4** (1884), 201–312.

[574] POINCARÉ, H.: Mémoire sur les fonctions zétafuchsiennes. Acta math. **5** (1884), 209–278.

[575] POINCARÉ, H.: Sur les hypothèses fondamentales de la Géométrie. Bull. Soc. math. France **15** (1887), 203–216.

[576] POINCARÉ, H.: On the Foundations of Geometry. (Translated from a manuscript of POINCARÉ by T. J. MCCORMACK.) The Monist **9** (1898), 1–43.

[577] PUISEUX, V.: Recherches sur les fonctions algébriques. Journ. math. Liouville **15** (1850), 365–480.

[578] REMAK, R.: Über die Zerlegung der endlichen Gruppen in direkte unzerlegbare Faktoren, Journ. Crelle **139** (1911), 293–308.

[579] REYE, T.: Das Problem der Configurationen. Acta math. **1** (1882), 93–96.

[580] REYE, T.: Die Hexaëder- und Octaëder-Configurationen. Acta math. **1** (1882), 97–108.

[581] REYE, T.: Die Geometrie der Lage. Part 1. 5th ed., Leipzig 1909.

[582] RIEMANN, B.: Gesammelte mathematische Werke. Leipzig 1876.

[583] RIEMANN, B.: Gesammelte Werke. 2nd ed., Leipzig 1892.

[584] RIEMANN, B.: Grundlagen für eine allgemeine Theorie der Functionen einer veränderlichen complexen Grösse. In: [583], pp. 3–45.

[585] RIEMANN, B.: Beiträge zur Theorie der durch die Gauss'sche Reihe $F(\alpha, \beta, \gamma; x)$ darstellbaren Functionen. In: Abh. Göttingen 1857. WA: [583], pp. 67–83.

[586] RIEMANN, B.: Gleichgewicht der Elektrizität auf Cylindern mit kreisförmigem Querschnitt und parallelen Axen. Aus dem Nachlaß. In: [582], pp. 440–444.

[587] RICHARD, M.: Notice sur Evariste Galois. Nouv. Ann. Math. **3** (1849), 448–452.

[588] ROSENTHAL, F.: Über die logische und erkenntnistheoretische Bedeutung des mathematischen Gruppenbegriffs (Inauguraldissertation). Jena 1928.

[589] RUDIO, F.: Eine Autobiographie von Gotthold Eisenstein. Mit ergänzenden biographischen Notizen. Abh. Gesch. Math. **7** (1895), 143–203.

[590] RUFFINI, P.: Opere matematiche. Ed. E. BORTOLOTTI. Vol. 1. Palermo 1915.

[591] RUFFINI, P.: Opere matematiche. Ed. E. BORTOLOTTI, Vol. 2. Roma 1953.

[592] RUFFINI, P.: Teoria generale delle Equazioni, in cui si dimostrata impossibile la soluzione algebraica delle equazioni generali di grado superiore al quarto. 2 Vols. Bologna 1799. WA: [590], pp. 1–324.

[593] RUFFINI, P.: Della soluzione delle equazioni algebraiche determinata particolari di grado superiore al quarto. Mem. Mat. Fis. Soc. Ital. **9** (1802). WA: [590], pp. 343–406.

[594] RUFFINI, P.: Della insolubilità delle equazioni algebraiche generali di grado superiore al quarto. Mem. Mat. Fis. Soc. Ital. **10**, Part 2 (1803). WA: [591], pp. 1–50.

[595] RUFFINI, P.: Sopra la determinazione delle radici nelle Equazioni numeriche di qualunque grado. Modena 1804. WA: [591], pp. 281–404.

[596] RUFFINI, P.: Risposta di Paolo Ruffini ai dubbi propostigli dal socio Gian-Francesco Malfatti sopra la insolubilità algebraica dell'equazioni di grado superiore al quarto. Mem. Mat. Fis. Soc. Ital. **12**, Part 1 (1804). WA: [591], pp. 53–90.

[597] RUFFINI, P.: Della insolubilità delle equazioni algebraiche generali di grado superiore al 4°, qualunque metodo si adoperi, algebraico esso siasi o trascendentale. Mem. Ist. Naz. Ital. **1**, Part 1 (1806). WA: [591], pp. 143–154.

[598] RUFFINI, P.: Alcune proprietà generali delle funzioni. Mem. Mat. Fis. Soc. Ital. **16**, Part 1 (1807). WA: [591], pp. 107–141.

[599] RUFFINI, P.: Riflessioni intorno alla soluzione delle equazioni algebraiche generali. Modena 1813. WA: [591], pp. 155–268.

[600] RUFFINI, P.: Di un nuovo metodo generale di estrarre le radici numeriche. Verona 1813. WA: [591], pp. 405–450.

[601] RUFFINI, P.: Intorno al metodo generale proposta dal Signor Hoëné Wronski onde risolvere le equaziono di tutti i gradi. Mem. Mat. Fis. Soc. Ital. **18** (1820). WA: [591], pp. 269–279.

[602] RUFFINI, P.: Alcune proprietà delle radici dell'unità. Mem. Ist. Lombardo-Veneto **3** (1824). WA: [591], pp. 451–466.

[603] SARTON, G.: The Study of the History of Mathematics. Cambridge (MA), 1936.

[604] SARTON, G.: Evariste Galois. Osiris **3** (1937), 241–259.

[605] SÉGUIER, J.-A. DE: Théorie des groupes finis. Eléments de la théorie des groupes abstraits. Paris 1904.

[606] SÉGUIER, J.-A. DE: Groupes des substitutions. Paris 1912.

[607] SELIVANOV, D. F.: Theory of the Algebraic Solution of Equations. (Russian.) St. Petersburg 1885.

[608] SELIVANOV, D. F.: Equations of Degree Five with Integer Coefficients. (Russian.) St. Petersburg 1889.

[609] SELIVANOV, D. F.: Expressions algébriques. Acta math. **19** (1895), 73–91.

[610] SERGESCU, P.: Histoire du nombre. Université de Paris, Palais de la Découverts, Série D, Nr. 23, 1953.

[611] SERRET, J.-A.: Remarque sur un Mémoire de M. Bertrand. Journ. Ecole polyt. **19** (1848), 147–148.

[612] SERRET, J.-A.: Remarque sur un Mémoire de M. Bertrand. Journ. math. Liouville, 1st Series, **14** (1849), 135–136.

[613] SERRET, J.-A.: Mémoire sur le nombre de valeurs que peut prendre une fonction quand on y permute les lettres qu'elle renferme. Journ. math. Liouville, 1st Series, **15** (1850), 1–44.

[614] SERRET, J.-A.: Mémoire sur les fonctions de quatre, cinq et six lettres. Journ. math. Liouville, 1st Series, **15** (1850), 45–70.

[615] SERRET, J.-A.: Développements sur une classe d'équations. Journ. math. Liouville. 1st Series, **15** (1850), 152–168.

[616] SERRET, J.-A.: Cours d'Algèbre supérieure. Paris 1854.

[617] SERRET, J.-A.: Sur les fonctions rationnelles linéaires prises suivant un module premier, et sur les substitutions auxquelles conduit la considération de ces fonctions. Comptes rendus Paris **48** (1859), 112–117, 178–182, 237–240.

[618] SERRET, J-.A.: Cours d'Algèbre supérieure. 3rd ed. 2 Vols. Paris 1866.

[619] SERRET, J -A.: Cours d'Algèbre supérieure. 4th ed. 2 Vols. Paris 1877.

[620] SÖDERBERG, J. T.: Deduktion af nödvändiga och tillräckliga vilkoret för möjligheten af algebraiska eqvationers solution med radikaler. In: Upsala Universitets Arsskrift. Matematik och naturvetenskap, 1886. Upsala 1886.

[621] SÖDERBERG, J. T.: Démonstration du théorème fondamental de Galois dans la théorie de la résolution algébrique des équations. Acta math. **11** (1888), 297–302.

[622] SÖDERBERG, J. T.: Einige Untersuchungen in der Substitutionentheorie und der Algebra. Nova acta Upsala, 3rd Series, **15** (1892), 1–38.

[623] SÖDERBERG, J. T.: Theorie der imprimitiven und der decomposablen auflösbaren Gruppen. Nova acta Upsala, 3rd Series, **20** (1899), pp. 1–26.

[624] SPECHT, W.: Gruppentheorie. Berlin/Göttingen/Heidelberg 1956.

[625] SPEISER, A.: Die Zerlegungsgruppe. Journ. Crelle **149** (1919), 174–188.

[626] SPEISER, A.: Theorie der Gruppen von endlicher Ordnung. Berlin 1923.

[627] SPEISER, A.: Theorie der Gruppen von endlicher Ordnung. 2nd ed., Berlin 1927.

[628] SPEISER, A.: Klassische Stücke der Mathematik. Zürich 1925.

[629] STOLT, B.: Über Axiomensysteme, die eine abstrakte Gruppe bestimmen. Uppsala 1953.

[630] STOLZ, O.: Die geometrische Bedeutung der complexen Elemente in der analytischen Geometrie. Math. Ann. **4** (1871), 416–441.

[631] SUŠKEVIČ, A. K. (Сушкевич, A. K.): Source Materials on the History of Algebra in Russia in the 19th and in the Beginning of the 20th Centuries. Istoriko-Matematičeskie Issledovanija **IV** (1951), 235–451. (Russian.)

[632] SYLOW, L.: Théorèmes sur les groupes des substitutions. Math. Ann. **5** (1872), 584–594.

[633] SYLVESTER, J. J.: Mathematical Papers. Vol. 1. Cambridge 1904.

[634] SYLVESTER, J. J.: On the General Theory of Associated Algebraical Forms. Math. Journ. Cambridge Dublin **6** (1851). WA: [633], pp. 198–202.

[635] SCHELL, W.: Theorie der Bewegung und der Kräfte. Leipzig 1870.

[636] SCHENDEL, L.: Grundzüge der Algebra, nach Grassmann'schen Prinzipien. Halle 1885.

[637] SCHERING, E.: Die Fundamental-Classen der zusammensetzbaren arithmetischen Formen. Abh. Göttingen **14** (1869), 3–13.

[638] SCHLEGEL, V.: Hermann Grassmann. Sein Leben und seine Werke. Leipzig 1878.

[639] SCHLESINGER, L.: Bericht über die Entwickelung der Theorie der linearen Differentialgleichungen seit 1865. JB. dt. Math. Ver. **18** (1909), 133–192.

[640] SCHLESINGER, L.: Materialien für eine wissenschaftliche Biographie von Gauß. Ed. by F. KLEIN and M. BRENDEL. Issue 2. Fragmente zur Theorie des arithmetisch-geometrischen Mittels aus den Jahren 1797–1799. Leipzig 1912.

[641] SCHLESINGER, L.: Materialien für eine wissenschaftliche Biographie von Gauß. Ed. by F. KLEIN and M. BRENDEL. Issue 3. Über Gauß' Arbeiten zur Funktionentheorie. Leipzig 1912.

[642] SCHMIDT, O. JU.: Über die Zerlegung endlicher Gruppen in direkte unzerlegbare Faktoren. Kievskogo un-ta. Izvestija **1912**, 1–6.

[643] SCHMIDT, O. JU.: Sur les produits directs. Bull. Soc. math. France **41** (1913), 161–164.

[644] SCHMIDT, O. JU. (Шмидт О. Ю.): Abstract Group Theory. Kiev 1916. WA in: Collected Papers, Moscow 1959.

[645] SCHMIDT, O. J.: Über unendliche Gruppen mit endlicher Kette. Math. Zeitschr. **29** (1929), 34–41.

[646] SCHMIDT, O. J.; TSCHERNIKOW, S. N.; NOWIKOW, P. S.: Endlichkeitsbedingungen in der Gruppentheorie. Ed. H. GRELL. Berlin 1963. (Translated from Russian.)

[647] SCHÖNEMANN, T.: De functionibus quibusdam, quae ad radices aequationum circuli sectionum, sive aequationis $x^p - 1 = 0$ pertinent, rationaliter determinandis. Journ. Crelle **17** (1837), 372–381.

[648] SCHÖNEMANN, T.: Über die Congruenz $x^2 + y^2 \equiv 1$ (mod. p). (Theorie der trigonometrischen Functionen in Bezug auf Congruenzen.) Journ. Crelle **19** (1839), 93–112.

[649] SCHÖNEMANN, T.: Theorie der symmetrischen Functionen der Wurzeln einer Gleichung. Allgemeine Sätze über Congruenzen, nebst einigen Anwendungen derselben. Journ. Crelle **19** (1839), 231–243, 289–308.

[650] SCHÖNEMANN, T.: Grundzüge einer allgemeinen Theorie der höheren Congruenzen, deren Modul eine reelle Primzahl ist. Journ. Crelle **31** (1846), 269–325.

[651] SCHÖNEMANN, T.: Über die Beziehungen, welche zwischen den Wurzeln irreductibeler Gleichungen stattfinden, insbesondere wenn der Grad derselben eine Primzahl ist. Denkschr. Akad. Wien (gelesen April 1852) **5** (1853), 143–156.

[652] SCHOENFLIES, A.: Klein und die nichteuklidische Geometrie. Die Naturwissenschaften. Jg. **7** (1919), 288–297.

[653] SCHREIER, O.: Über die Gruppen $A^a B^b = 1$. Abh. Math. Sem. Hamburg **3** (1924), 167–169.

[654] SCHREIER, O.: Abstrakte kontinuierliche Gruppen. Abh. Math. Sem. Hamburg **4** (1926), 15–32.

[655] SCHREIER, O.: Über die Erweiterung von Gruppen. I. Monatsh. Math. Phys. **34** (1926), 165–180.

[656] SCHREIER, O.: Über die Erweiterung von Gruppen. II. Abh. Math. Sem. Hamburg **4** (1926), 321–346.

[657] SCHREIER, O.: Die Untergruppen der freien Gruppen. Abh. Math. Sem. Hamburg **5** (1927), 161–183.

[658] SCHREIER, O.: Über den Jordan-Hölderschen Satz. Abh. Math. Sem. Hamburg **6** (1928), 300–302.

[659] SCHRÖDER, E.: Lehrbuch der Arithmetik und Algebra. Bd. 1. Die sieben algebraischen Operationen. Leipzig 1873.

[660] SCHRÖDER, E.: Über die formalen Elemente der absoluten Algebra. Stuttgart 1874.

[661] SCHRÖDER, E.: Operationskreis des Logikkalküls. Baden-Baden 1877.

[662] SCHRÖDER, E.: Ueber die Anzahl der Substitutionen, welche in eine gegebene Zahl von Cyklen zerfallen. Archiv für Mathematik und Physik, **68** (1882), 353–377.

[663] SCHRÖDER, E.: Vorlesungen über die Algebra der Logik. Vol. 3. Algebra und Logik der Relative. Leipzig 1895.

[664] SCHRÖDER, K.; NAAS, J.: Der Begriff des Raumes in der Geometrie. Berlin 1957.

[665] SCHUR, F.: Lehrbuch der analytischen Geometrie. Leipzig 1898.

[666] SCHUR, F.: Ueber die Grundlagen der Geometrie. Math. Ann. **55** (1902), 265–292.

[667] SCHUR, I.: Über die Darstellung der endlichen Gruppen durch gebrochene lineare Substitutionen. (I.) Journ. Crelle **127** (1904), 20–50.

[668] SCHUR, I.: Neue Begründung der Theorie der Gruppencharaktere. MB. Akad. Berlin **1905**, 406–432.

[669] SCHUR, I.: Arithmetische Untersuchungen über endliche Gruppen linearer Substitutionen. MB. Akad. Berlin **1906**, 164–184.

[670] SCHUR, I.; FROBENIUS, G.: Über die reellen Darstellungen der endlichen Gruppen. MB. Akad. Berlin **1906**, 186–208.

[671] SCHUR, I.; FROBENIUS, G.: Über die Äquivalenz der Gruppen linearer Substitutionen. MB. Akad. Berlin **1906**, 209–217.

[672] SCHUR, I.: Über die Darstellung der endlichen Gruppen durch gebrochene lineare Substitutionen. (II.) Journ. Crelle **132** (1907), 85–137.

[673] SCHUR, I.: Über die Darstellung der symmetrischen und der alternierenden Gruppe durch gebrochene lineare Substitutionen. Journ. Crelle **139** (1911), 155–250.

[674] SCHUR, I.: Über Gruppen linearer Substitutionen mit Koeffizienten aus einem algebraischen Zahlkörper. Math. Ann. **71** (1912), 355–367.

[675] SCHWARZ, H. A.: Ueber einige Abbildungsaufgaben. Journ. Crelle **70** (1869), 105–120.

[676] SCHWARZ, H. A.: Über diejenigen Fälle, in welchen die Gaussische hypergeometrische Reihe eine algebraische Function ihres vierten Elementes darstellt. Journ. Crelle **75** (1873), 292–335.

[677] SCHWEIKART, F. C.: Die Theorie der Parallellinien nebst dem Vorschlage ihrer Verbannung aus der Geometrie. Jena/Leipzig 1807.

[678] STÄCKEL, P.; ENGEL, F.: Die Theorie der Parallellinien von Euklid bis auf Gauß. Leipzig 1895.

[679] STÄCKEL, P.: Bericht über die Mechanik mehrfacher Mannigfaltigkeiten. JB. dt. Math. Ver. **12** (1903), 469–481.
[680] STAUDE, O.: Analytische Geometrie des Punktes, der geraden Linie und der Ebene. Ein Handbuch zu den Vorlesungen und Übungen über analytische Geometrie. Leipzig/Berlin 1905.
[681] STAUDT, CHR. v.: Geometrie der Lage. Nürnberg 1847.
[682] STAUDT, CHR. v.: Beiträge zur Geometrie der Lage. 3 Issues. Nürnberg 1856–1860.
[683] STEINER, J.: Gesammelte Werke. Ed. K. WEIERSTRASS. Vol. 1. Berlin 1881.
[684] STEINER, J.: Vorlesungen über synthetische Geometrie. Part 1: Die Theorie der Kegelschnitte in elementarer Darstellung. Auf Grund von Universitätsvorlesungen und mit Benutzung hinterlassener Manuscripte Jacob Steiner's. Ed. C. F. GEISER. 2nd ed., Leipzig 1875.
[685] STEINER, J.: Vorlesungen über synthetische Geometrie. Part 2: Die Theorie der Kegelschnitte, gestützt auf projectivische Eigenschaften. Auf Grund von Universitätsvorträgen und mit Benutzung hinterlassener Manuscripte Jacob Steiner's. Ed. H. SCHRÖTER. Leipzig 1876.
[686] STEINITZ, E.: Algebraische Theorie der Körper. Journ. Crelle **137** (1910), 167–309.
[687] STEINITZ, E.: Algebraische Theorie der Körper. Journ. Crelle **137** (1910). New ed. Ed. by H. HASSE and R. BAER. Berlin/Leipzig 1930.
[688] STEPHANOS, C.: Théorie des quaternions. Math. Ann. **22** (1883), 589–592.
[689] STICKELBERGER, L.; FROBENIUS G.: Ueber Gruppen von vertauschbaren Elementen. Journ. Crelle **86** (1879), 217–262.
[690] Strömungen, Die modernen, der wissenschaftlichen Forschung. Zusammengestellt von P.-V. AUGER, Sonderberater der UNESCO. Ed. in German: Abteilung Presse und Berichtswesen der Deutschen Akademie der Wissenschaften zu Berlin. o. J.
[691] TABER, H.: On Certain Sub-Groups of the General Projective Group. Bull. Amer. Math. Soc., **2** (1896), 221–233.
[692] TABER, H.: On Those Orthogonal Substitutions that Can Be Generated by the Repetition of an Infinitesimal Orthogonal Substitution. Proc. Math. Soc. London **26** (1895), 364–376.
[693] TATON, R.: Les relations d'Evariste Galois avec les mathématiciens de son temps. Revue d'histoire des sciences et leurs applications, **1** (1947), 114–130.
[694] TATON, R.: La géométrie projective en France de Desargues à Poncelet. Université de Paris, Palais de la Découverte, Série D, Nr. **4**, 1951.
[695] TATON, R.: L'Histoire de la géométrie descriptive. Université de Paris, Palais de la Découverte, Série D, Nr. 32, 1954.
[696] THOMAE, J.: Grundriß einer analytischen Geometrie der Ebene. Leipzig 1906.
[697] TOEPLITZ, O.: Die Entwicklung der Infinitesimalrechnung. Eine Einleitung in die Infinitesimalrechnung nach der genetischen Methode. Vol. 1. Aus dem Nachlaß ed. G. KÖTHE. Berlin/Göttingen/Heidelberg, 1949.
[698] TSCHERNIKOW, S. N.; SCHMIDT. O. J.; NOWIKOW, P. S.: Endlichkeitsbedingungen in der Gruppentheorie. Ed. H. GRELL. Berlin 1963. (Translated from Russian.)
[699] TSCHIRNHAUS, E. W. v.: Methodus auferendi omnes terminos intermedios ex data aequatione. Acta eruditorum **1** (1683), 204–207.
[700] VALSON, C. A.: La vie et les travaux du Baron Cauchy. 2 Vols. Paris 1868.
[701] VANDERMONDE, A.: Mémoire sur la résolution des équations. Hist. Acad. Sc. Paris, année 1771; Paris 1774, pp. 365–416.
[702] VANDERMONDE, A.: Abhandlungen aus der reinen Mathematik. German ed. C. ITZIGSOHN. Berlin 1888.
[703] VERONESE, G.: Grundzüge der Geometrie von mehreren Dimensionen. German translation by A. SCHEPP. Leipzig 1894.
[704] VERRIEST, G.: Evariste Galois et la théorie des équations algébriques. Paris 1934.

[705] VUILLEMIN, J.: La philosophie de l'algèbre de Lagrange (Réflexions sur le mémoire de 1770–1771). Université de France, Palais de la Découverte, Série D, Nr. 71, 1960.
[706] VOGT, H.: Leçons sur la résolution algébrique des équations. Paris 1895.
[707] VOSS, A.: Heinrich Weber. JB. dt. Math. Ver. **23** (1914), 431–444.
[708] VUCINICH, A.: Nikolai Ivanovich Lobachevskii: The Man Behind the First Non-Euclidean Geometry. Isis **53** (1962), 465–481.
[709] WAERDEN, B. L. VAN DER: Moderne Algebra. Part 1. 3rd ed., Berlin/Göttingen/Heidelberg 1950.
[710] WARING, E.: Miscellanea analytica de aequationibus algebraicis et curvarum proprietatibus. Cambridge 1762.
[711] WARING, E.: Meditationes algebraicae. Cambridge 1770.
[712] WEBER, H.; DEDEKIND, R.: Theorie der algebraischen Functionen einer Veränderlichen. Journ. Crelle **92** (1882), 181–290.
[713] WEBER, H.: Beweis des Satzes, dass jede eigentlich primitive quadratische Form unendlich viele Primzahlen darzustellen fähig ist. Math. Ann. **20** (1882), 301–329.
[714] WEBER, H.: Ueber die Galois'sche Gruppe der Gleichung 28ten Grades, von welcher die Doppeltangenten einer Curve vierter Ordnung abhängen. Math. Ann. **23** (1884), 489–503.
[715] WEBER, H.: Theorie der Abel'schen Zahlkörper. I. Abel'sche Körper und Kreiskörper. Acta math. **8** (1886), 193–221.
[716] WEBER, H.: Theorie der Abel'schen Zahlkörper. II. Über die Anzahl der Idealclassen und die Einheiten in den Kreiskörpern, deren Ordnung eine Potenz von 2 ist. Acta math. **8** (1886), 221–244.
[717] WEBER, H.: Theorie der Abel'schen Zahlkörper. III. Der Kronecker'sche Satz. Acta math. **8** (1886), 244–263.
[718] WEBER, H.: Theorie der Abel'schen Zahlkörper. IV. Über die Bildung Abel'scher Körper mit gegebener Gruppe. Acta math. **9** (1887), 105–130.
[719] WEBER, H.: Leopold Kronecker†. Math. Ann. **43** (1893), 1–25.
[720] WEBER, H.: Die allgemeinen Grundlagen der Galois'schen Gleichungstheorie. Math. Ann. **43** (1893), 521–549.
[721] WEBER, H.: Lehrbuch der Algebra. Vol. 1. Braunschweig 1895.
[722] WEBER, H.: Lehrbuch der Algebra. Vol. 2. Braunschweig 1896.
[723] WEBER, H.: Ueber Zahlengruppen in algebraischen Körpern (Erste Abhandlung). Math. Ann. **48** (1897), 433–473.
[724] WEBER, H.: Ueber Zahlengruppen in algebraischen Körpern (Zweite Abhandlung). Math. Ann. **49** (1897), 83–100.
[725] WEBER, H.: Ueber Zahlengruppen in algebraischen Körpern (Dritte Abhandlung). Math. Ann. **50** (1898), 1–26.
[726] WEBER, H.: Lehrbuch der Algebra. 2nd ed. Vol. 1. Braunschweig 1898.
[727] WEBER, H.: Lehrbuch der Algebra. 2nd ed. Vol. 2. Braunschweig 1899.
[728] WEBER, H.: Lehrbuch der Algebra. 2nd ed. Vol. 3. Elliptische Funktionen und algebraische Zahlen. Braunschweig 1908.
[729] WEBER, H.: Lehrbuch der Algebra. Kleine Ausgabe in einem Bande. Braunschweig 1912.
[730] WEIERSTRASS, K.: Theorie der Abel'schen Functionen. Journ. Crelle **52** (1856), 285–380.
[731] WEIERSTRASS, K.: Zur Theorie der bilinearen und quadratischen Formen. MB. Akad. Berlin, Mai **1868**, 310–338.
[732] WEIERSTRASS, K.: Zur Theorie der elliptischen Functionen. MB. Akad. Berlin **1882**, 443–451.
[733] WEIERSTRASS, K.: Briefe an P. du Bois-Reymond. Acta math. **39** (1923), 199–225.
[734] WEYL, H.: Gruppentheorie und Quantenmechanik. 2nd ed., Leipzig 1931.
[735] WEYL, H.: The Classical Groups. 2nd ed., Princeton 1946.

[736] WIMAN, A.: Ueber eine einfache Gruppe von 360 ebenen Collineationen. Math. Ann. **47** (1896), 531–556.

[737] WIMAN, A.: Zur Theorie der endlichen Gruppen von birationalen Transformationen in der Ebene. Math. Ann. **48** (1897), 195–240.

[738] WIMAN, A.: Endliche Gruppen linearer Substitutionen. In: Encyklopädie der mathematischen Wissenschaften mit Einschluß ihrer Anwendungen. Vol. **1**, Part 1, Leipzig 1898–1904, pp. 522–554.

[739] WINNECKE, F. A. T.: GAUSS – Ein Umriß seines Lebens und Wirkens. Braunschweig 1879.

[740] Wörterbuch, Mathematisches. Ed. J. NAAS and H. L. SCHMID. Vol. **1**. Leipzig/Berlin 1961.

[741] Wörterbuch, Mathematisches. Ed. J. NAAS and H. L. SCHMID. Vol. 2. Berlin/Leipzig 1961.

[742] WRONSKI, H.: Résolution générale des Equations de tous les degrés. Paris 1811.

[743] WUSSING, H.: Die Ecole Polytechnique – eine Errungenschaft der Französischen Revolution. Pädagogik, **13** (1958), 646–662.

[744] WUSSING, H.: Über den Einfluß der Zahlentheorie auf die Herausbildung der abstrakten Gruppentheorie. Beiheft Schriftenreihe NTM. Leipzig 1964, pp. 71–88.

[745] WUSSING, H.: August Ferdinand Möbius. In: Bedeutende Gelehrte in Leipzig, Vol. 2. Ed. G. HARIG. Leipzig 1965, pp. 1–12.

[746] WUSSING, H.: Zur Entstehungsgeschichte der abstrakten Gruppentheorie. NTM, **5** (1965), 1–16.

[747] WUSSING, H.: On the Genesis of the Abstract Group Concept. Istoriko-Matematičeskie Issledovanija **XVII** (1966), 11–30. (Russian.)

Name Index

Subject Index